Radio Frequency Integrated Circuits and Technologies

Frank Ellinger

Radio Frequency Integrated Circuits and Technologies

Second Edition

 Springer

Prof. Dr. Frank Ellinger
Dresden University of Technology
Helmholtzstrasse 18
01069 Dresden
Germany
Frank.Ellinger@tu-dresden.de

ISBN: 978-3-642-08885-8   e-ISBN: 978-3-540-69325-3

Cover design: Erich Kirchner, Heidelberg

Printed on acid-free paper

9 8 7 6 5 4 3 2 1

springer.com

To Karin

I feel the need, the need for speed.

Pete Mitchell,
in movie Top Gun, 1986

Preface

In the last decade wireless communications engineering has seen outstanding progress, making merged, enhanced and novel applications in the area of mobile phones, wireless networks, sensors and television feasible. Technologies have developed from hybrid systems to highly integrated solutions in silicon, SiGe, GaAs and InP. By aggressive scaling of device dimensions below 0.1 μm and employing advanced technologies such as SOI, strained silicon and low-k, circuits with operation frequencies and bandwidths up to approximately 100 GHz can now be fabricated. However, especially in silicon, the restrictions inherent in scaling make circuit engineering a demanding task. Examples of these drawbacks are the limited high frequency signal power, leakage effects and significant parasitics in passive devices. Enhanced circuit topologies and design techniques have to be applied to achieve maximum performance. In this context, designers must have profound skills in the following areas: circuit theory, IC technologies, communications standards, system design, measurement techniques, etc. The aim of this book is to address all these multidisciplinary issues in a compact and comprehensive form and in a single volume. Suitable for students, engineers and scientists, the manuscript provides the necessary theoretical background together with cookbook-like optimisation strategies and state-of-the-art design examples. Each chapter is accompanied by tutorial questions repeating the key issues of the treated subjects.

The manuscript is organised as follows: Chapter 1 preludes with an introduction concerned with the exciting history of integrated circuits, technologies and wireless communications. Moreover, an overview of the IC circuit design flow, tools, applications and markets is given. Chapter 2 reviews the key architectures of wireless systems. In Chap. 3 we study S-parameters and the Smith chart being instrumental for small signal circuit analyses and optimisations. Important RF basics including gain, stability, linearity and noise are treated in Chap. 4. Transistors and passive devices are discussed in Chaps. 5 and 6. Key circuit design techniques and components such as LNAs, PAs, VCOs, synthesisers, mixers, amplitude control elements and phase shifter are elaborated in Chaps. 7–14. Measurement methods and setups are outlined in Chap. 15.

Most of the subjects treated in this book are taught in lectures at the Dresden University of Technology (TUD) in Germany. Lecturers who might be interested in using the material of this manuscript for teaching purposes are encouraged to contact the author. An exchange of experiences is welcome.

This is the second edition of this book. However, the manuscript may still exhibit some unclear phrasings or errors awaiting to be discovered by careful readers like you! I would be very pleased to receive appropriate comments.

This book would not have been possible without the constructive impact of several great colleagues.

I have benefited very much from my teachers Prof. Dr. H. Jäckel, ETH Zürich, Prof. Dr. W. Bächtold, ETH Zürich, Dr. U. Lott, Founder of AnaPico Zürich, and Prof. Dr. H. Schumacher, University of Ulm. Moreover, I would like to thank Prof. Dr. G. Böck, University of Berlin, and Prof. Dr. Dr. habil. R. Weigel, Friedrich Alexander University of Erlangen-Nuremberg, for their support and nice collaboration in several projects. For his fruitful efforts concerning IBM/ETH Zürich CASE (Center For Advanced Silicon Electronics), I would like to acknowledge Dr. M. Schmatz at the IBM Zürich Research Laboratory. A big thank you goes to the members of his group, who are Dr. M. Kossel, Dr. T. Morf, Dr. T. Toifl, Dr. C. Menolfi, and Dr. J. Weiss for sharing their excellent expertise in high-speed circuit design. For the constructive teamwork in my former RFIC group at ETH Zürich, I would like to thank D. Barras, G. von Büren, J. Carls, Dr. C. Kromer, L. C. Rodoni, G. Sialm, and S. Wehrli. The joint afterwork parties were always very funny. I would like to express my gratitude to H. Benedickter and M. Lanz, both with ETH Zürich, for their very valuable help concerning challenging measurements and circuit assembling.

Dr. C. Baumann, Dr. D. Merkle and Ms. P. Jantzen, all with Springer, Heidelberg are acknowledged for their collaboration and support concerned with the publication of this book.

Prof. Dr. habil. U. Jörges, TUD, did an excellent job concerning the review of this manuscript making the content much more consistent and precise. Moreover, the fruitful and competent comments of Prof. Dr. habil. W. Schwarz, T. Ußmüller (he also kindly provided the IC cover photo), C. Keogh, M. Wickert, U. Mayer, D. Barras, G. von Büren, J. Carls, P. Haldi, Dr. C. Kromer, A. Lauterbach, Dr. D. Pasalic, Dr. S. Spiegel, G. Stark, and S. Wehrli regarding selected chapters of this manuscript were very helpful.

A large debt of gratitude is owed to my wonderful parents Margit and Wolfgang Ellinger. Finally, and most importantly, I want to express my deep appreciation to my inspiration and girl friend Karin Mächler for her unlimited patience and support. I'm very delighted to dedicate the book to her.

Most of this book has been written during my holidays in Brazil (Porto Galinhas and Fortaleza) and Spain (Canary Islands and Mallorca). These gorgeous locations gave me a fruitful balance and relaxation during the writing of the manuscript.

Since my student days I have found high-speed analogue and RF circuit design very interesting and enthralling. If this book succeeds in inspiring the same enthusiasm in others, then the efforts of its compilation have borne the desired fruit.

Frank Ellinger
Dresden, Germany
April 2008

Contents

1 Introduction

You see, wire telegraph is a kind of a very, very long cat. You pull the tail in New York and the head is meowing in Los Angeles. Do you understand this? And radio operates exactly the same way: you send signals here, they receive them there. The only difference is that there is no cat.

Albert Einstein

1.1 History

Today's wireless communications and IC design had to go a long way. The success was based on the enthusiastic efforts of brilliant scientists, creative engineers and clever businessmen. We learned an important lesson, namely that the belief in an idea and the consequent commitment can make the impossible possible. The following sections are devoted to a selection of key contributors and achievements in the area of wireless communications. Of course, these sections only present a small fraction of the inventions. A virtual acknowledgement is going to the pioneers not mentioned in these sections. Further information can be found in the literature [Huu03, Ber05, Léc05] and on web pages [Dcs05, Alb05, Tra05].

1.1.1 Timetable of Inventions

1745 **Georg von Kleist** found a method for **storing charge** in the so-called Leyden Jar.

1800 **Alessandro Volta** demonstrated the **existence of current** and invented the **battery**.

1820 **Hans Oersted** demonstrated that **current** and **electromagnetic field** are **related**.

1831 **Michael Faraday** explored electromagnetic induction.

1864 **James Maxwell** described electromagnetic waves using a consistent set of **equations**.

1887 **Heinrich Hertz** demonstrated **spark transmission** leading to the basic understanding of **dipoles** and **antennas**.

1894   **Olivier Lodge** invented the coherer, a **sensitive detector** of electro-magnetic waves applying **resonance** techniques.

1897   Successful wireless **data transmission** in England with a distance of around 5 km.

1906   **Lee De Forest** designed the **triode** in a **vacuum tube**, which was the first device that could **amplify** signals. However, disadvantages were large size, high power consumption and low reliability.

1914   **Edwin Howard Armstrong** developed the **regenerative receiver** offering high selectivity and sensitivity based on feedback.

1915   First **wireless transmission** between Europe and the United States.

1918   **Edwin Armstrong** and **Walter Schottky** invented the **super-heterodyne radio** featuring frequency conversion and bandwidth management.

1920   Experimental **broadcast services**, e.g. in Germany and England.

1924   The New York police successfully applied **commercial wireless communications systems**.

1926   **Patent** of an **FET-like semiconductor transistor** by **Julius Lilienfeld**. However, no working device was demonstrated.

1938   Investigations of **metal semiconductor contacts** by **Walter Schottky** leading to the fabrication of high-speed **Schottky diodes**.

1946   Introduction of **commercially available mobile phone services**.

1947   **John Bardeen** and **Walter Brattain**, staff members of the group of **Walter Shockley** at Bell Labs in California, experimentally demonstrated the **first working semiconductor transistor**, the **point contact transistor**, which is similar to a bipolar junction transistor. For details see Sect. 5.1.

1948   Independently from Bardeen and Brattain, the German scientists **Herbert Mataré** and **Heinrich Welker** invented the **transistron**, a working transistor, which was nearly identical to that produced by the scientists from Bell Labs, raising the question as to who was first.

1948   **Claude Shannon** published major mathematical communication theories determining data capacities and their impact factors.

1950   Realisation of the **field effect junction transistor** providing higher yield compared to that of Bardeen and Brattain. The associated concepts and theories were produced by Walter Shockley in 1948.

1953   **Mobile car phone** in Germany operating with a pre-version of the German A-Net demonstrated in a VW Beattle. The weight of such a system was 16 kg.

1956   John Bardeen, Walter Brattain and Walter Shockley receive the joint **Nobel Prize** in physics for the invention of the transistor.

1957   **A-Net with 850 subscribers**, 137 call cells and coverage range per call cell of 30 km. The price per system amounting to € 4000 was twice the price of a VW Beetle.

1958   The idea of the **integrated circuit** was developed by **Jack Kilby** and **Robert Noyce**. Multiple devices with different functionality in a single chip became feasible. At the same time costs, size and weight were significantly decreased. This was possible by implementing multiple layers with specific properties using **photo-lithography**.

1960   Bell scientist **John Atalla** developed the **MOSFET** based on Shockley's theories.

1965   **Wireless phone** transmission via **satellite**.

1971   **WLAN transmission** performed by University of Hawaii based on Alohanet.

1972   **B-Net** with up to **27 000 wireless mobile phone users**.

1981   **Nordic Mobile Telephone** cellular system at 450 MHz introducing handover.

1986   **C-Net** in Germany with 800 000 users operating at 450 MHz.

1989   Development of **GPS** (Global Positioning System) with positioning accuracy up to several meters and installation of required satellites.

1991   **GSM** (Global System for Mobile Communications) 900, a mainly digital mobile phone system using a spectrum around 900 MHz.

1997   **802** group defined several **WLAN standards**, e.g. 802.11 (a, b, etc.).

2003   **UMTS** (Universal Mobile Telecommunications System).

2007   **Local positioning** with centimetre accuracy on the basis of radar waves. The reader is referred to the EU project RESOLUTION.

1.1.2 Major Inventors

There's more than enough glory in this for everybody!
Walter Brattain

Ampere, Andre Marie (1775–1836) was born in Lyon, France. He repeated and continued Oersted's work verifying that the current flowing through a wire produces a magnetic field that can deflect a compass needle. This led to the development of the first instruments able to measure the flow of electricity in conductors.

Armstrong, Edwin Howard (1891–1954) born in the USA was instrumental in developing efficient detector architectures such as the regenerative and the superheterodyne receiver, which are still employed in modern wireless communications systems. Unfortunately, others such as L. de Forest copied and patented his ideas leading to several lawsuits. Totally discouraged and without getting profits out of his inventions, he committed suicide.

Bardeen, John (1906–1987) was an exceptionally gifted physicist and mathematician laying the theoretical background for the understanding of transistors. In

1956 and 1972, together with Walter Brattain and William Shockley, he received the Nobel Prize in physics for the invention of the transistor.

Barkhausen, Heinrich (1881–1956) studied at the Universities of Munich, Berlin and Göttingen. A photograph is shown in Fig. 1.1. At the young age of 29, he became a full professor at the Dresden University of Technology. Barkhausen's work in acoustics and magnetism led in 1919 to the discovery of the Barkhausen effect, providing evidence that magnetisation affects whole domains of ferromagnetic materials rather than individual atoms alone. Moreover, he developed groundbreaking theories in the area of communications and electronic devices, e.g. theories about oscillator conditions and the vacuum tube. Concerning industry funding he once commented:

Small devices you buy, large ones must be donated.

Fig. 1.1. Picture of Heinrich Barkhausen, courtesy of Dresden University of Technology.

Bell, Alexander Graham (1847–1922) is respected as one of the major fathers of the telephone. He demonstrated a phone communication over a 3 km wired distance between Boston and Cambridge in the USA. In addition to being a talented engineer, Bell was a clever businessman. He founded a worldwide trading company producing telephone apparatuses for the mass-market.

Brattain, Walter (1902–1987) was an excellent experimental physicist. Born in China he spent most of his life in California. Working with the theoretical ideas of Walter Shockley and John Bardeen, his hands built the first working transistor, the point contact device, which is similar to today's bipolar junction transistors. In 1956, he received the Nobel Prize together with Bardeen and Shockley.

Chappe, Claude (1763–1805) was born in Brulon, France, and was the inventor of optical telegraphy. One of his telegraphs consisted of towers with moving arms defining words and symbols to transmit data over sight-distance. Repeaters were used to achieve larger distances. On August 15, 1794, the system was able to transmit important strategic information between Paris and the 210 km distant Lille during the war.

Faraday, Michael (1791–1867), son of a blacksmith became one of the most brilliant scientists of the nineteenth century. Faraday mainly worked for the Royal Institute of Science in London He was an experimentalist rather than a mathematician. Faraday discovered that a voltage is induced when changing the strength of the magnetic field – a process well known today as magnetic induction. In honour of his findings, the Faraday cabinet was named accordingly.

Hertz, Heinrich Rudolf (1857–1894) saw the first light of day in Hamburg. He attended universities in Munich, Berlin and Karlsruhe, and was an assistant of Prof. Herman von Helmholtz. Using Maxwell's theory of electromagnetic forces, he proved the existence of electromagnetic waves by means of dipoles and reflectors, which are today known as Hertz's dipoles.

Hughes, David Edward (1831–1900) served as a professor of music. At the same time, he was an outstanding engineer. Hence, he can be described as a universal genius. His family emigrated from England to the USA. Examples of his inventions are the printing telegraph and the carbon microphone. He observed some unexpected results when using a telephone ear-piece as a detector on an inductive bridge. By coincidence and due to an open connection within an inductive circuit generating a spark, he heard a click every time the current was interrupted. He used this experimental setup for wireless transmissions over a remarkable distance of around 500 m.

Marconi, Guglielmo (1874–1937) son of an Irish mother was born in Italy. Based on the insights of Hertz, Marconi succeeded in transmitting radio signals over a few kilometres at Bologna in 1896. Like Bell, Marconi was able to exploit his ideas commercially, allowing for a prosperous life. His practical thinking, enthusiastic manner and ability to motivate people made him famous. Surprisingly, he did not manage to interest the Italian government in his work. Worldwide, the Marconi Company built many radio stations. When he died, all radio transmitters were shut down for minutes of silence.

Maxwell, James Clerk (1831–1879) was born in Edinburgh, Scotland. He served as a professor in London and Cambridge. At the age of 24, Maxwell translated and verified Faraday's theories in mathematical formulations known today as the Maxwell equations. He also showed that these equations implicitly require the existence of electromagnetic waves travelling at the speed of light. Maxwell demonstrated that light, as well as electric and magnetic phenomena are based on very similar relations.

De Forest, Lee (1873–1961) is recognised as one of the fathers of the electronic age since he developed the first amplifying device, the audion, or triode by inserting a metal grid in a vacuum tube. Applying voltage to the grid controls a current across the tube. De Forest was born in Alabama in the United States. He was involved in several court procedures since he copied ideas from others and patented them, e.g. the receiver principles of Armstrong.

Lodge, Oliver Joseph (1851–1940) served as a professor at the University of Liverpool. He was very interested in Maxwell's work and a friend of Hertz. Lodge established the principal of frequency dependant resonance tuning. He implemented a sensitive detector for electromagnetic waves, which was called the coherer. Due to enhanced sensitivity, signal transmissions became possible across larger distances. Marconi was forced to purchase one of Lodge's patents to continue his work.

Oersted, Hans Christian (1777–1851) served as a professor at the University of Copenhagen, Denmark. He discovered that a compass needle deflects when an electric current is switched in a nearby wire. This verified that electricity and magnetism are related phenomena – a finding that established the basics for the theory of electromagnetism and radio communications. In his honour, the unit of magnetic field strength was officially called the Oersted until 1978. Today, the use of the unit ampere/meter is more common.

Schottky, Walter (1886–1976) was born in Zürich, Switzerland, but he spent most of his life in Germany. His career consisted of movements between university and industrial research (e.g. University of Würzburg and Siemens in Berlin). The following unique achievements have established for him a place in history: investigations of charge transport mechanisms in vacuum and solid bodies, development and investigation of high-speed metal semiconductor diodes today referred to as Schottky diodes, invention of the ribbon microphone, and design of the superheterodyne receiver (independently from Edwin Armstrong). Moreover, he discovered the two fundamental sources of noise in semiconductor devices, which are thermal noise and shot noise.

Shockley, William Bradford (1910–1989) was born in London but spent the major part of his life in California. He was the leader of the famous transistor development group at Bell Laboratories. Together with John Bardeen and Walter Brattain he invented the transistor and won the Nobel Prize. After forming a symbiosis of brilliance the three guys went separate ways due to a clash of egos. Shortly after Bardeen and Brattain presented the first point-contact transistor, which is similar to the bipolar junction transistor, Shockley conceived the junction transistor, the first field effect transistor. Although Shockley was one of the most important originators of the Silicon Valley, he did not manage to exploit his brilliant ideas commercially. After leaving Bell Laboratories, he failed with own companies and projects.

Volta, Alessandro Giuseppe Antonio Anastasio (1745–n.a.) was born in Como and was appointed as a professor at the University of Padua in Italy. Volta demonstrated the existence of a flowing current. He used bowls of salt solutions connected by copper on one side and zinc on the other. Current was generated by the potential difference drawing current through a wire, being the load resistance. The battery was born! Mobile communications would not be possible without this invention. Throughout his life Volta was able to adapt to changing politics. After Napoleon was disempowered and Austria became dominant in Italy once more, Volta continued to excel and to obtain high honours. Volta received greatest appreciation from his fellow scientists. The unit of the potential difference that moves current is called "volt".

1.2 Generic Transceiver

A generic wireless transceiver consists of a transmitter, a receiver and the air-channel as illustrated in Fig. 1.2. The transmitter comprises the data source, a signal coder, the RF transmitter frontend and an antenna. In complementary order, the receiver involves an antenna, an RF receiver frontend, a signal decoder and finally the data interface. For further information, the reader is referred to the specific literature [Pro03].

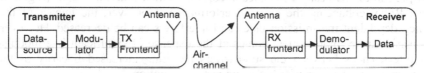

Fig. 1.2. Schematics of generic transceiver, TX: transmitter, RX: receiver

1.3 Modulation and Multiple Access

To establish a controlled transmission, most wireless communications systems modulate the signal on a carrier with exclusive frequency. Assuming that no other wireless system operates at the same frequency, the transmitter can emit high power in the air-channel ideally without disturbance of other systems. In other words, by proper modulation, each communications system has its own road.

According to Eq. (1.1), data can be coded by systematic variation of amplitude (A), angular frequency[1] (ω) or phase (φ) of the signal s. The corresponding modulation schemes are known as amplitude modulation (AM), frequency modulation (FM), and phase modulation (PM), respectively. The properties are represented in the following equation:

$$s = A \cdot \cos[\omega t + \varphi]. \tag{1.1}$$

Amplitude modulation is problematic for systems with varying channel losses. Frequency modulation has increased demands in terms of frequency bandwidth. Phase modulation is limited by the phase resolution of detectors. The different

---

[1] Throughout this manuscript, we may also loosely refer to as frequency instead of angular frequency.

techniques can be combined to relax the drawbacks of the individual approaches and to maximise the performance.

The schemes can be classified into constant and variable envelope modulations. Topologies with constant envelope amplitude allow amplification with moderate requirements in terms of component linearity. One example is the GMSK (Gaussian Minimum Shift Keying) modulation, which is based on frequency modulation. Typically constant envelope modulations are power efficient but not spectral efficient.

Variable envelope signals such as the QPSK (Quadrature Phase Shift Keying) exhibit both amplitude and phase variations. They can operate very spectral efficient. However, they need very linear components across a large amplitude range. The latter requirement is challenging for amplifiers and is traded off for the power efficiency as discussed in Sect. 9.5.

To decrease the risk of transmitting on a carrier frequency suffering from performance degradations, there is the trend to use multi carrier modulation. One example is OFDM (Orthogonal Frequency Division Multiplexing). The data rates of multiple sub-carriers can be matched according to their individual SNR (Signal to Noise Ratio) enhancing the overall performance. Moreover, the impact of frequency dependent transmission fades and interferers can be decreased. However, this performance improvement comes at the expense of significant intermodulation interferences between the sub-carriers requiring increased linearity of the power amplifier. This leads to higher power consumption.

Data transmission is also possible without carrier, e.g. by just simply transmitting pulses. However, those pulses cover a wide frequency band leading to undesired interferences with other communication systems. For those impulse radios, the transmit power must be very low to avoid corresponding problems.

Generally, modulation enables the transmission of different signals on the same channel. Further techniques are employed to handle the increasing number of users at given channel resources. To separate the different channels from each other, FDMA (Frequency Division Multiple Access), TDMA (Time Division Multiple Access), SDMA (Space Division Multiple Access) and CDMA (Code Division Multiple Access) techniques can be employed.

1.4 RF Frequency Bands

As listed in Table 1.1, the operation frequencies of RFICs are allocated within several frequency bands labelled by letters. The higher the frequencies, the higher the demands and costs for the employed technologies. Thus, most commercial circuits are operated at rather low frequencies. To date, the L, S and C bands have been intensively used for mobile and wireless communications.

Table 1.1. Microwave frequency allocations according to IEEE

Band	L	S	C	X	Ku	K	Ka	V	W
Frequency range	0.8–2 GHz	2–4 GHz	4–8 GHz	8–12 GHz	12–18 GHz	18–27 GHz	27–40 GHz	40–75 GHz	75–110 GHz

1.5 Channel Capacity

Shannon has demonstrated that the maximum channel capacity C of an ideal communication system is given by

$$C = BW \cdot \left[\log_2 \left(1 + SNR \right) \right]. \tag{1.2}$$

The signal to noise ratio

$$SNR = \frac{P_r}{BW \cdot N_0} \tag{1.3}$$

is determined by the signal bandwidth BW, the received RF power P_r and the power spectral density of the noise N_0. In Eq. (1.3) we assume that the noise has constant characteristics versus frequency. This type of noise is referred to as white noise. From the last two equations we can derive two important insights. The data rate can be improved by increasing the SNR, which can, e.g. be done by raising the transmit power P_t determining P_r. However, since the dependency in C is logarithmic, a strong power boost is required to show a relevant impact. More efficiently, the data capacity can be improved by making BW large. Because of interferences with other systems, the later approach is frequently limited within predefined frequency allocations.

1.6 Air Propagation

In the previous section, we have observed that an increased SNR improves the maximum possible data capacity in an ideal channel. Let us now calculate a first order approximation for the free space transmission properties to conclude on the impact of key parameter such as the associated transmit power P_t and the link distance r. The power density after the transmitter amounts to

$$S = \frac{P_t}{4\pi r^2} \cdot G_t \tag{1.4}$$

and considers the antenna gain of the transmitter G_t, which by definition is unity for an ideal isotropic radiator and above unity for directional antennas. An isotropic antenna is an ideal antenna that radiates power with unit gain uniformly in all directions. Also, for the received power, we have to take into account the gain of the receive antenna. The power available after the receive antenna is given by

$$P_r = S \cdot A_r \cdot G_r, \qquad (1.5)$$

with G_r as the antenna gain in the receiver. The effective antenna area A_r depends on the wavelength of the signal and can be approximated by

$$A_r = \frac{\lambda^2}{4\pi}. \qquad (1.6)$$

From Eqs. (1.4) and (1.5), we can now write

$$P_r = \frac{1}{16\pi^2} \cdot P_t \cdot G_t \cdot G_r \cdot \left(\frac{\lambda}{r}\right)^2. \qquad (1.7)$$

The received power and, consequently, the SNR scale with the square of the factor $\frac{\lambda}{r}$. This means that the maximal possible data capacity degrades significantly with increased link distance and signal frequency. Doubling of the link distance and frequency cuts by a factor of 16 the received power. In other words, to obtain the same SNR we would require 16 times higher transmit power. In our calculations we have neglected second order effects like the frequency dependent behaviour of the atmosphere. The approximated free-space attenuation based on measurements is illustrated in Fig. 1.3. We can observe attenuation peaks because of water and oxygen resonances. Equation (1.7) has been derived for ideal free-space conditions and does not include effects such as indoor fading, interferers and multipath propagation arising in real indoor environments. Hence, in reality, the attenuations are much higher, leading to $\frac{\lambda}{r}$ exponents of around 4 or even higher.

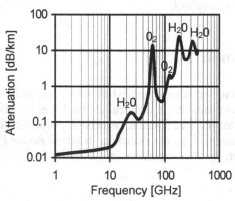

Fig. 1.3. Approximated free-air attenuation at sea level, T=20 °C, H_2O: 7.5 g/m^3 [Fcc97]

1.7 Overview of RF Applications and Markets

It is much easier to make science out of money
than to make money by means of science.
Prof. R. Hütter, ETH-Zürich

According to Fig. 1.4, the main markets for RFICs involve wireless phones, data networks, positioning, RFIDs (Radio Frequency Identification) and sensors. There is the trend to merge several applications into one device by reusing potential synergies between the different systems. This demands reconfigurable circuits. Different types of reconfigurability can be exploited including: 1. Re-use of circuits, 2. Multifunctional circuits, and 3. Switching of components [Böc03]. Advantages of reconfigurability are lower overall costs, smaller system size and higher market potential. Nevertheless, the overall performance of multifunctional devices will always be below those optimised for a single specific application.

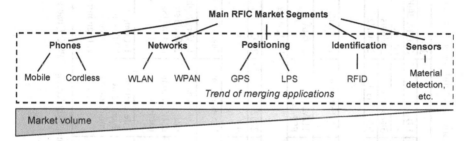

Fig. 1.4. Overview of the most important wireless systems and corresponding markets

1.8 Mobile Phones

Life is not like a shop.
It does not withdraw anything.
Annett Louisan, German singer

The most prominent mobile standards are listed and compared in Table 1.2. Key parameters such as the possible data rates, market shares, access modes, regions of utilisation, life cycles, strengths and weaknesses are analysed. In accordance with their technological status, they are divided into 1 G (Generation), 2 G, 2.5 G and 3 G standards.

Table 1.2. Key characteristics of mobile phone standards

Gen.	2G			2.5G		3G	
Name	GSM	PDC	IS95	GPRS	EDGE	CDMA 2000	UMTS
Mod.	GMSK	QPSK	QPSK	GMSK	8PSK	QPSK	QPSK
Max. DR	14.4 kbit/s	6.7 kbit/s	1.2–9.4 kbit/s	115 kbit/s	384 kbit/s	2 Mbit/s	2 Mbit/s
Frequency	~0.5W/1..100W 0.9/1.8 GHz	0.9 GHz	1.9 GHz	0.9/1.8 GHz	0.9/1.8 GHz	2.4 GHz	1.9-2 GHz and 2.1-2.2 GHz
RF power mobile/BS	~0.5W/1..100W						
Realistic data rate	9.6–14.4 kbit/s	6.7 kbit/s	1.2–9.4 kbit/s	20–50 kbit/s	60–150 kbit/s	144 kbit/s	144 kbit/s
Access mode	TDMA/FDMA	TDMA/FDMA	CDMA	TDMA/FDMA	TDMA/FDMA	CDMA	WCDMA
Main regions	Europe, worldwide	USA	USA, Canada, Asia	Europe, worldwide		USA, Asia	Europe, worldwide
Launch	1989	1990	1992	2001	2002	2002	2003
Market share 2007	~50% together with GPRS	~8%	~15%	50% with GSM	n.a.	~3%	~6%
Life-cycle	Decrease	Saturation	Growth/ saturation	Saturation	Development	Onset/growth	Onset
Strengths	1. Market size 2. Cash cow	-	1. Good channel separation 2. Good position in USA and Asia	1. GSM compatibility 2. Data rate	1. High data rate 2. GSM compatibility	1. Very high data rate 2. Good starting position in USA and China	1. Top standard in Europe, very high data rate 2. Good starting position worldwide
Weaknesses	1. Relative low data rate 2. Competition with 2.5G and 3G	1. Low data rate 2. End of life-cycle	1. Relative low data rate 2. End of life-cycle	Life-cycle influenced by launch of 3G	Life-cycle influenced by launch of 3G	1. Competition with UMTS, no chances in Europe 2. High infrastructure costs	1. Extremely high licence costs in Europe 2. Competition with CDMA2000 in USA and China 3. High infrastructure costs

Mod.: modulation, DR: Data Rate, BS: Basestation, GMSK: Gaussian Minimum Shift Keying, QPSK: Quadrature Phase Shift Keying, 8PSK: Eight Times Phase Shift Keying

1.8.1 First Generation

The 1 G standards, such as, e.g. AMPS (Enhanced Mobile Phone Service) only provide very low data rates and are mainly based on analogue techniques. Since they are at the end of their life cycle, they are not discussed further.

1.8.2 Second Generation

The 2 G standards, dominated by GSM (Global System for Mobile Communication) can be seen as cash cows, which are at the maximum of their life cycle. To increase performance, more and more functions are relocated into the digital domain. The main region of utilisation for GSM is Europe. IS95 and PDC (Pacific Digital Cellular) are further 2 G standards mainly employed in the USA.

1.8.3 Second and a Half Generation

Increasingly the 2 G standards are being substituted by 2.5 G standards such as GPRS (General Packet Radio Service) and EDGE (Enhanced Data Rates for Global Evolution), allowing higher data rates. A significant advantage of the GPRS standard is the compatibility with the existing GSM networks. Only minor modifications have to be performed for its implementation, therefore keeping additional expenses low.

1.8.4 Third Generation

The life-cycles of the 2.5 G standards are limited by the 3 G standards, such as CDMA 2000 (Code Division Multiple Access 2000) and UMTS (Universal Mobile Telecommunications System) featuring further data-speed improvements. For CDMA 2000, the major market regions will be USA and Asia. UMTS plays a dominant role within Europe and several other regions in the world. An excellent book about UMTS has been published [Spr04].

In the year 2000, UMTS service providers spent extremely high amounts on regional licences [Eur01] (e.g. € 50 billion in Germany). Furthermore, for both UMTS and CDMA 2000, high costs have to be considered for the base-stations and mobile phones, which have to be completely replaced. New designs are required due to the totally different access modes based on CDMA techniques. Thus, high cost and time pressures are expected for the service providers to accumulate their expenses for the 3 G standards. In addition, it has to be considered that the data rates of the 2.5 G standards may be sufficient for most applications, thereby delaying the success of the 3 G standards. In any case, we have learned a lot from the economic mismanagement associated with UMTS.

1.8.5 Fourth Generation

Fourth generation standards are capable of providing further data rate improvements together with an increased level of reconfigurability. Multimode operation may enable mobile phone, WLAN, and positioning functionality, etc. in one single device. In this context, the communication with different standards, operation at different frequencies, global roaming, handover and interface design are challenging issues. Synergies inherent in different systems can be exploited to minimise the system complexity, costs, size and power consumption. One fruitful scenario could be as follows. Within a large coverage range, a moderate data transfer can be performed by a mobile phone standard. Given the availability of a local WLAN connection, the data transfer can be handed over to the local network providing enhanced performance. For further reading refer to the literature [Hui03].

1.8.6 Market Forecast and Major Vendors

A mobile phone market forecast per unit shipments is plotted in Fig. 1.5a. Worldwide, approximately 1.1 billion mobile phones have been sold in 2007. Due to the saturation of the market, the growth is only moderate. The introduction of new functions such as Internet, GPS (Global Positioning System) video and audio streaming with high data rates make increased selling prices and revenues feasible. Today, the total IC costs for one mobile phone are approximately € 15, out of which around 20 % are for the analogue part. The market shares of the mobile top vendors are compared in Fig. 1.5b.

a)

b)

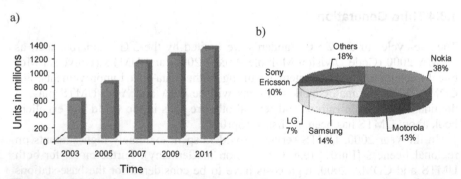

Fig. 1.5a,b. Market data, source e.g. from Imran's Everything Cellular [Imr08]: **a** forecast of worldwide cellular handset shipments; **b** unit sale shares of top mobile phone vendors in 2007

1.9 Wireless Local Area Networks

While mobile phone data transfer is designated for global communications with large coverage range, higher performance can be obtained in local environments equipped with WLANs (Wireless Local Area Networks). In addition to data communication via base stations, WLAN devices can also operated peer-to-peer. WLAN systems with very short coverage below about 10 m are segmented into WPANs (Wireless Personal Area Networks) or WBANs (Wireless Body Area Networks). Among the most important WLANs are the 802.11 standards. Bluetooth and the UWB (Ultra-Wideband) standards are candidates for WPANs. Potential applications are high-speed short-range communication, internet access, video streaming, traffic control, medical imaging, security systems, sensors, etc. There will be the trend to integrate the WLAN functionalities into mobile phones and PDAs (Personal Digital Assistants). Some frequently employed WLAN systems are outlined in the following subsections and in Table 1.3.

1.9.1 WiFi (IEEE 802.11)

The 802.11 standards are employed since nearly a decade for indoor coverage ranges up to 100 m. For marketing reasons and to avoid the long name, they are also labelled as WiFi (Wireless Fidelity). Up to now we can find three sub-standards on the market, namely 802.11a operating at frequencies allocated around 5 GHz, and 802.11 b and g transmitting at 2.4 GHz. The b standard achieves a data throughput of around 11 Mb/s. The more advanced a and g standards are based on OFDM (Orthogonal Frequency Division Multiplexing) and offer higher data rates of around 54 Mb/s. Moreover, a future n standard has been announced featuring wideband OFDM, and eventually adaptive antenna combining, which is also known as MIMO (Multiple In Multiple Out). Further data rate enhancements of above 100 Mb/s are expected.

1.9.2 Bluetooth

Bluetooth has been developed as a low cost and power consuming wireless replacement of short cable connections, e.g. headphone connections of mobile phones. Coverage range and data rate are below 10 m and 1 Mb/s. The allocated frequencies are around 2.4 GHz.

1.9.3 Ultra-Wideband

According to the theorem of Shannon treated in Sect. 1.5, the data capacity can be significantly enhanced by increasing of the bandwidth motivating for ultra-wideband systems [Por03, Uwb06]. Unfortunately, parts of the available bandwidth are already occupied by several standards as illustrated in Fig. 1.6. To avoid undesired interferences, the transmit power of wideband system has to be very low. According to the rules of the FCC (Federal Communications Commission) in the USA [Fcc02], the maximum emitted power spectral density of ultra-wideband systems is limited to –41.3 dBm/MHz within the assigned maximum bandwidth ranging from 3.1 GHz to 10.6 GHz. The maximum corresponding power is below 1 mW compared to a permitted power of up to 1 W for the 802.11 standards. Per definition, ultra-wideband devices have a bandwidth of at least 500 MHz or a fractional bandwidth

$$FBW = \frac{2(f_H - f_L)}{(f_H + f_L)} \qquad (1.8)$$

higher than 0.2. The upper and lower frequencies are denoted by f_H and f_L, respectively. Recently, the 802.15.3a standardisation group considers two major proposals, namely the impulse radio and the OFDM multiband approach.

The impulse radio is a traditional principle employing pulses with very short duration [Paq04]. To get a small pulse duration, a large bandwidth is required, e.g. several GHz. No carrier and frequency mixers are required leading to reduced transceiver complexity, power consumption and costs.

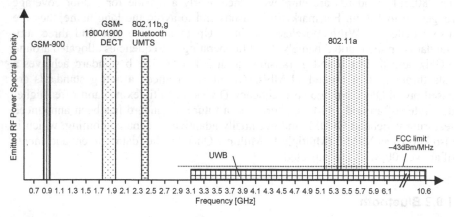

Fig. 1.6. Illustration of spectrum and power spectral density of ultra-wideband and narrowband systems

Several smaller bands are used for the OFDM approach [Mul06], which is similar to the one used for the 802.11a and n standards. Since a carrier is needed, this approach is more complex, costly and power consuming. However, the supporters expect better performance as a result of the enhanced rejection of frequency dependent interferers. Due to the much lower emitted power compared to the 802.11 standards, the full bandwidth between 3.1 GHz and 10.6 GHz can be exploited, allowing higher data rate. This comes at the expense of a lower coverage range. For further reading about UWB refer to [Opp05, Ree05, Man03, Por03].

1.9.4 Millimetre Wave WLAN

The demand for higher data rates and frequency bandwidths forces the market to move to higher frequencies located in the millimetre wave band [Har02]. Some of these high speed radio LAN standards belong to the group of MBS (Mobile Broadband Services) providing a large bandwidth from 27 GHz to 60 GHz. Due to their wide bandwidth very high data rates well above 100 Mb/s should be feasible. According to Sect. 1.6, absorption and air transmission losses increase with frequency. Thus, millimetre wave systems are mainly applied for short-range communications. However, this has a useful side effect, since the interference associated with neighbouring systems are reduced.

The LMDS (Local Multipoint Distribution Service) with an allocated frequency band at 27.5–29.5 GHz and 38–39 GHz [Lmd04], and the MVDS (Multipoint Video Distribution System) operating between 40.5 GHz and 43.5 GHz [Mvd04] belong to the group of MBS standards. The 802.15.3 working group for WPAN aims to establish a standard with at least 3 GHz bandwidth around 60 GHz allowing for very high data rates above 1 Gb/s. First CMOS circuits operating up to 60 GHz have been demonstrated [Doa04, Raz06, Ell04].

1.9.5 Market Forecast

Within the next few years the WLAN market will be dominated by the 802.11 (WiFi) standards. The counted and expected shipments are shown in Fig. 1.7. In total, 170 and 450 million chipsets are counted and forecasted for 2006 and 2009, respectively. The corresponding growth is significant. More than a billion shipments are expected in 2012 [Abi08]. In comparison, the growth in the mobile phone market is saturated and therefore much lower.

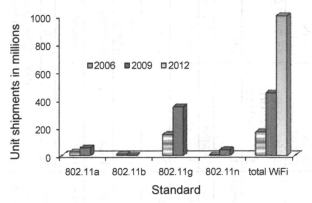

Fig. 1.7. Worldwide WiFi chipset shipments [Hor05] and [Abi08]

Table 1.3. Comparison of major wireless network standards

Type	WLAN			WPAN	
Group	802.11 (WiFi)			Bluetooth	UWB
Sub-group	a	b	g		Multicarrier	Impulse
Modulation/ spreading	BPSK/ OFDM	QPSK/ CCK	OFDM/ CCK	GFSK/ DSFH	PSK/ OFDM	PPM/ n.a.
Frequency band	5.15–5.35 GHz 5.47-5.725 GHz 5.725–5.825 GHz	2.4–2.483 GHz	2.4–2.483 GHz	2.4– 2.483 GHz	3.1–10.6 GHz Multiband	-
Maximum emitted power	1 W	1 W	100 mW/1 W	1 mW/ 100 mW	0.5 mW	
Max. data speed	54 Mb/s	11 Mb/s	54 Mb/s	0.7 Mb/s	110–200 Mb/s	
Max. coverage range	50–100 m	100 m	100 m	10 m/100 m	10 m	
Market launch	2001	1999	2005	2003	2002 (USA) 2005 (Europe)	
Recent life cycle status	Growth	Growth/ saturation	Growth	Growth	Development	
Complexity and costs	High	Moderate	High	Low	High	Very low
Power consumption	High	Moderate	High	Low	Low	Very low
Standardisation	IEEE	IEEE	IEEE	1.2	Draft IEEE 802.15.3a	

PSK: Phase Shift Keying, BPSK: Binary PSK, QPSK: Quadrature PSK, GFSK: Gaussian Frequency Shift Keying, CCK: Complementary Code Keying, OFDM: Orthogonal Frequency Division Multiplexing, DSFH: Direct Sequence Frequency-Hopping

1.10 Positioning

> *At the beginning of a lucrative innovation there is never a business plan*
> *but the spirit of curiosity and discovery.*
> Peter Derleder

During the Second World War, there was the aim to develop systems capable of locating planes, ships and submarines. This led to the invention of the radar (Radio Detection and Ranging) based on the transmission and detection of RF waves [Jam89]. Today, positioning services have been expanded and enhanced for guiding of cars and pedestrians, interactive maps, automated factories, robotics, augmented reality, etc.

Generally, the distance of an object can be measured by means of reference transmitters with known location. As illustrated in Fig. 1.8, three different principles based on the propagation delay, received signal strength and phase interpolations can be applied to extract the distance [Vos03]. To determine the three-dimensional (3-D) position defined by three coordinates, reference measurements associated with at least three transmitters placed at different locations are performed.

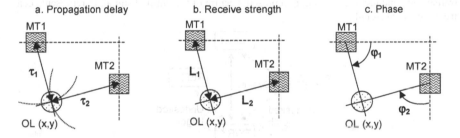

Fig. 1.8. Measurement principles to extract the position of an object, MT: measurement transmitter, OL: object to be tracked, τ_n: roundtrip time of flight, L_n: propagation loss, φ_n: angle. At least three MTs are required for three-dimensional tracking

1.10.1 Propagation Delay

We can determine the distance d between an object and a reference transmitter on the basis of the propagation time of arrival (TOA). For example, this principle is used for the pulse positioning radar [Sko90] illustrated in Fig. 1.9. With given signal velocity c, d is measured by means of the time difference Δt between transmitted and reflected signal:

$$d = \frac{1}{2} \cdot c \cdot \Delta t . \tag{1.9}$$

The extraction of d is performed in the time domain. According to Fig. 1.10, we can identify three types of reflections. First, the desired one carrying the distance information. Second, unwanted multipath reflections. If the delay of these multipath reflections is small compared to the direct path, it is not possible to distinguish between target and multipath response, thereby significantly worsening the localisation accuracy. Third, we have to consider reflections at undesired objects not hitting the target object. It is difficult to distinguish between different objects. Similar problems exist, e.g. for the received strength and angle of arrival approaches.

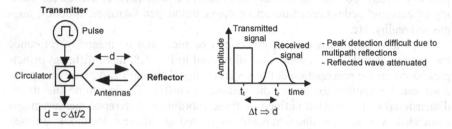

Fig. 1.9. Pulse positioning radar, d: distance, c: signal velocity, Δt: time delay between transmitted and received signal, t_t: represents centre of transmitted signal, t_r: represents centre of received signal

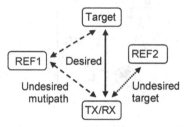

Fig. 1.10. Types of reflections

1.10.2 Received Signal Strength

Given that we know the attenuation in a specific communication channel, we can approximate the distance by means of the propagation loss. According to Eq. (1.7), the free space transmission is proportional to $\frac{1}{d^n}$ with n=2. For realistic scenarios, $2 < n < 5$. Most modern radio modules already have a received strength indicator (RSSI). Thus, this approach may be the simplest and consequently the cheapest. However, especially in indoor environments with strong multipath fading and shadowing, the applicability and accuracy of this method is limited.

1.10.3 Angle of Arrival

A further possibility for positioning is the calculation by means of goniometry considering the angles of the signals. With directional antennas, the angles of arrival (AOA) relative to points located at known positions are measured. The intersection of the pointers yields the position. High requirements have to be met regarding the directivity of the measurement antennas. Antenna steering and combining can be applied to enhance the directivity.

1.10.4 FMCW Radar

An advanced but more complex positioning approach is the FMCW (Frequency Modulated Continuous Wave) radar [Lam99]. The functionality is illustrated in Fig. 1.11.

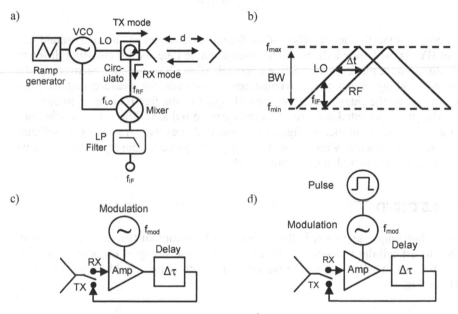

Fig. 1.11a–d. FMCW radar with three different types of reflectors, d: distance, c: signal velocity, Δf: frequency offset, BW: bandwidth, f_{mod}: modulation frequency, Δτ: time delay.

An oscillator is modulated by a ramp generator yielding a reference signal, which is transmitted and reflected back (see Fig. 1.11a). The transmitted and received signals denoted by LO and RF, respectively are depicted in Fig. 1.11b. Due to the time delay Δt, the two signals have a frequency offset f_{IF}, which can be extracted by mixing. Suppose that the mixer acts as frequency subtractor yielding

$$f_{IF} = f_{LO} - f_{RF} .$$

(1.10)

Attributed to the linear dependency between and d, Δt and f_{IF}, the distance can be determined by

$$d \sim f_{IF}. \tag{1.11}$$

Multipath components can be suppressed by a lowpass filter, since the delayed multipath components have a higher IF frequency than the target object. Recall the problem of identifying different objects. Employing an active modulated reflector as sketched in Fig. 1.11c can solve this problem. The feedback amplifier acts as an oscillator excited with the frequency of the input signal. After filtering, the distance can be extracted on the basis of the spacing between the two remaining frequency components located around the modulation frequency f_{mod}. The corresponding relation yields [Wie03]

$$d = \frac{\Delta f \cdot c}{f_{mod} \cdot 8 \cdot BW}, \tag{1.12}$$

where Δf is the frequency offset of the frequency components located around f_{mod} and BW is the bandwidth. Multiple objects can be detected by choosing different modulation frequencies. The active approach allows an amplitude recovery of the signal resulting in enhanced coverage range. However, the measurement accuracy is limited by the jitter inherent in the LO signal of the free running oscillator. By pulsing the modulated reflector, this jitter can be reduced. The corresponding circuit schematic is outlined in Fig. 1.11d. Note that at every switch-on, the oscillator frequency is coherent with respect to the input frequency. Consequently, the signal frequency is recovered at every pulse cycle.

1.10.5 Cell-ID

A further simple approach is the positioning by means of the known location of the closest cellular base-stations together with signal strengths measurements. Depending on the density of base stations, this method may yield accuracies up to 10–500 m.

1.10.6 Global Positioning System

GPS (Global Positioning System) is based on satellites acting as reference transmitters and trilateralisation principles. Maximum free-space accuracy is in the region of 1 m. However, due to the large distance between the satellites and the object, the received power is too small to allow further losses arising in indoor environments with strong fading and multipath. Thus, the functionality of GPS is limited in buildings, or not possible at all, opening the market for local positioning systems.

1.10.7 Local Positioning

Within a coverage range limited to a few hundred meters, enhanced performance can be achieved by local positioning systems. Since the object to be localised and the reference antennas are much closer than for GPS, enhanced performance can be achieved in indoor environments. Moreover, due to the small propagation times, real-time applications are possible. The base-stations of WLAN systems can be used for the implementation of an FMCW radar [Wie03]. It is the aim of the EU project RESOLUTION to investigate the FMCW radar with active reflectors for local positioning with highest accuracy [Res06]. A localisation accuracy below 10 cm should be feasible, making novel and enhanced applications such as smart factories, interactive and cultural guiding possible. By means of different modulation frequencies, various objects can be identified.

1.10.8 Combination of Positioning and Communications

As pointed out in Sect. 1.7, there is a trend to merge different applications into reconfigurable devices. A typical example is the combination of data communication and positioning allowing a variety of novel and useful applications. According to Fig. 1.12, the tradeoff between positioning accuracy and coverage range has to be considered for optimum system design.

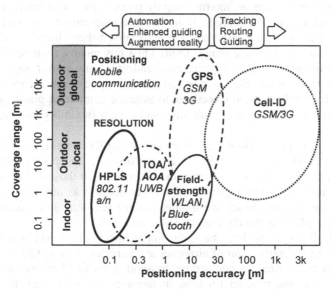

Fig. 1.12. Positioning accuracy and coverage range of systems merging mobile communications and positioning, TOA: Time of Arrival, AOA: Angle of Arrival, HPLS: High Precision Localisation System based on FMCW

1.11 Radio Frequency Identification

How would you like it if, for instance, one day you realized your underwear was reporting on your whereabouts?

California State Senator Debra Bowen, 2003

Radio-frequency identification (RFID) is a low cost identification system based on the wireless transmission of radio waves. The purpose of an RFID system is to enable data to be transmitted by a compact mobile device, called a tag, which is read by an RFID reader and processed according to the needs of particular applications.

The tag consists of the RF unit and a digital memory that is given a unique electronic identification code typically having a length around 12 bits. When an RFID tag passes the reader, it detects an activation signal and transmits the data to the reader. A host computer processes the data. A highly integrated circuit chip and a compact antenna are required for the tag.

Table 1.4. Frequencies, where RFID technology typically can be used without licence

LF (low frequency)	**HF** (high frequency)	**UHF** (ultra high frequency)
125-134.5 kHz	13.56 MHz	868 MHz-
140-148.5 kHz		928 MHz

Active approaches require an internal supply power. They can provide reliable performance up to several hundreds of meters. Wake-up circuits are implemented to reduce the power consumption and the associated battery lifetime. At the expense of much lower coverage ranges limited to a few meters, their passive counterparts have the advantages of being independent of any power supply, lower costs, smaller form factor and lower weight. Even chip-less approaches are feasible allowing tags to be printed directly onto assets, e.g. made of plastic, resulting in lower costs but also lower performance and design flexibility than traditional tags.

Three major frequency bands are employed for RFID systems. They are listed in Table 1.4 and do not require any licence. However, the specific frequencies used for RFID in the USA are currently incompatible with those of Europe or Japan. Furthermore, no emerging standard has yet become as universal as for the simpler barcode systems. Since the frequencies are relatively low, cheap CMOS technology is applied for the chip implementations.

First RFID-like systems have been developed during the World War II for espionage tools and identification of airplanes and vehicles. Today, there is a growing market in areas such as contactless passport identification, payment cards, transport toll systems, product tracking in factories and supermarkets, baggage tracking at airports, sport items tracking (e.g. soccer and golf balls), drug identification, collision avoidance, and animal identification. Moreover, RFID tags serve as high-tech barcode replacement. These applications can be very fruitful and may increase the live quality of citizens.

On the other hand, RFID technology can also be used for identification of humans by implants under the skin, and military purposes, e.g. for friend and foe detection and deadly interaction. Thus, these applications raise questions concerning ethical issues, privacy limitations and undesired surveillance. Hence, we can find oppositions and campaigns against RFID, e.g. in US (refer to the saying at beginning of this section) and Germany.

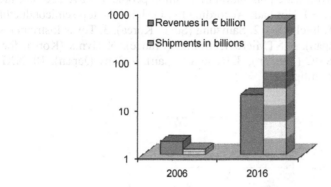

Fig. 1.13. RFID market, source IDTechEx [Idt08]

As illustrated in Fig. 1.13, a very strong growth is forecasted for RFID systems. Revenues of around € 18 billion are expected in 2016 for RFID chips and systems. Additional revenues in the same order are forecasted for corresponding services. The number of tags delivered in 2016 is expected to be over 450 times higher than the number in 2006. Depending of the performance, the current costs per RFID tag currently ranges from 0.05 - 20 €. According to Fig. 1.13, the costs significantly decrease versus time.

For further information concerning RFID systems, the reader is referred to the specific literature [Leh08, Dob07, Fin03].

1.12 IC Foundries and Markets

I check every offer, it could be the deal of my life.
Henry Ford

The total IC sales in 2007 have amounted to approximately € 200 billion [Eet08]. Considering a total worldwide population of 7 billion people, the average revenue per person is around € 30. Not bad! According to Fig. 1.14, the top semiconductor players in 2007 are 1. Intel (US), 2. Samsung (South Korea), 3. Texas Instruments (US), 4. Toshiba (Japan), 5. STMicroelectronics (France), 6. Hynix (Korea, former Hyundai), 7. TSMC (Taiwan), 8. Renesa (Japan), 9. Sony (Japan), 10. NXP (Netherlands, former Philips).

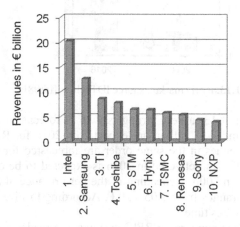

Fig. 1.14. Ranking of top ten IC foundries regarding total IC market in 2007, data from IC Insights [Eet08]

In 2007, the analogue market segment including RFIC circuits exhibited a revenue of approximately € 25 billion. The global players are Texas Instruments, STMicroelectronics, Infineon (Germany), Analog Devices (US), NXP, National (US), Maxim (US), Matsushita (Japan), Renesas, Freescale (US, former Motorola), Rohm (Japan), and Toshiba. In Fig. 1.15, the market volumes and expectations are plotted. The major business of around 60 % is generated by means of ASICs (Application Specific Circuits) optimised for telecommunication, automotive, consumer, computer and industrial applications. The remaining 40 % are due to standard ICs including voltage regulators and references, data converters, amplifiers and interface circuits.

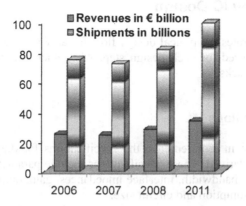

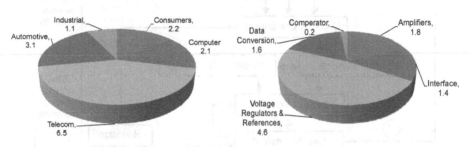

Fig. 1.15. Analogue IC market, data from IC Insights [Ici08]

1.13 Overview IC Design

Figure 1.16 illustrates the typical design flow for an RF integrated circuit including the time required for each design step. Altogether, the overall design cycle may take 12-35 weeks.

1.13.1 Specifications

According to system requirements, the specifications for the individual components have to be derived. These specifications include parameters such as the operation frequency, bandwidth, interface impedances, gain, output power, linearity, noise, power consumption and circuit size.

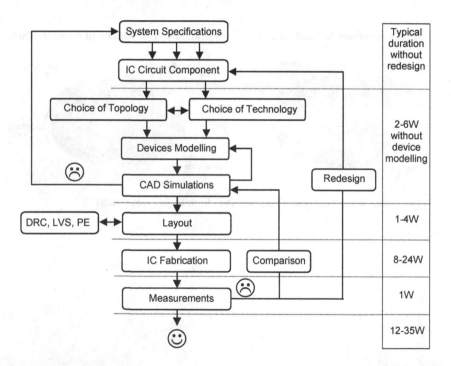

Fig. 1.16. IC design flow, W: weeks, DRC: Design Rule Check, LVS: Layout Versus Schematics check, PE: Parasitic Extraction

1.13.2 Choice of Technology and Topology

Based on experiences, literature studies or simplified hand calculations using key data of the technology, a suitable circuit topology is chosen with respect to system specifications. E.g. by knowing the transit frequency (f_t), the maximum supply voltage and the noise figure (NF) of the technology, we can estimate the gain, the maximum output power and the noise performance of an amplifier at a certain frequency. Subsequently, we can estimate if, e.g. a common source, cascode, or cascaded configuration is superior.

1.13.3 CAD Simulations

Based on the circuit schematics and device models, CAD simulations are performed using software tools such as Agilent ADS (Advanced Design Simulator) or Cadence Virtuoso Spectre. All relevant impact and device parameters can be swept, tuned and optimised, e.g. the bias, element values, frequencies, stimuli powers and temperature. Optimisation routines can be fruitful to get the maximum out of the circuit. Gradient optimisation is well suited if the characteristics are monotone. Unfortunately, several local maxima may exist making gradient optimisation difficult. In this case, we may employ a random optimisation routine for coarse-tuning followed by a gradient optimisation for fine-tuning. Experience has shown that it is mandatory to verify the computer-based optimisations by manual optimisations or first order hand calculations. Despite all the computer power available, the designer should have a profound theoretical background in circuit theory. Often the designer has to find a tradeoff between many parameters such as gain, noise, power consumption, etc., making the definition of optimisation goals and corresponding priorities challenging. Thus, in many cases, it may be reasonable to optimise the circuit manually. Simulations should take into account process variations and the impact of temperature changes. To break down complexity, hierarchical design can be employed. Sub-circuits with symbolic references are defined and implemented in the complex design. Any changes of the sub-blocks are automatically considered in the overall design.

1.13.3.1 Simulator Types

Referring to Table 1.5, different types of simulators are available to solve individual tasks [Rob01, Ven05, Maa03]. They can be segmented into small signal and large signal simulations, which can be accomplished in the time or frequency domain. In the small signal case, a linear approximation is made. The corresponding simulations are mainly performed in the frequency domain. By means of an FFT (Fast Fourier Transformation), the result obtained in the time domain can be transformed in the frequency domain and vice versa. Prerequisites for accurate transformations are the simulation of a high number of periods and high frequency resolutions. It is clear that these prerequisites lead to long simulation times. To minimise simulation times, investigations in individual domains can be advantageous.

DC simulations are conducted to evaluate and set the correct bias point of transistors and circuits. To be demonstrated later the applied bias has a very significant impact on the properties of active devices. Thus, DC simulations are frequently performed as a first step. All device models can be traced back to their DC equivalents. Capacitors become open and inductors can be approximated by short circuits with a certain DC resistance. With these simple models, DC simulations are very fast. In most cases they are calculated in the frequency domain and extracted at a frequency of zero. After having determined a reasonable bias, further simulations can be performed.

Table 1.5. Simulator types, FFT: Fast Fourier Transformation

Type	Domain	Stimuli	Calculation	Applications
Small signal				
DC	DC	DC	I/V look up table, interpolation	Bias points, supply power of linear circuits, e.g. of LNAs
Linear	Frequency	One small signal sinus function	Linear algebraic equations	S-, Z-, Y-parameters, stability
Noise	Frequency	Noise parameter	Cascade of noise matrices	Small signal noise, e.g. of LNAs
Large signal				
DC	Frequency or time	Multiple DC	Harmonic balance or transient + FFT DC at f=0	Bias points, supply power of large signal circuits such as oscillators, mixers, power amplifiers
Transient	Time, frequency data can be converted via FFT	Multiple time varying signals	Nonlinear equations	Large signal circuits
Harmonic balance	Frequency and time	Multiple large signal sinus functions	Linear and nonlinear equations, Fourier transformation	Large signal circuits, harmonics, intermodulation, power amplifiers, mixers
Envelope	Frequency and time	Multiple large signal sinus functions	Harmonic balance at multiple times	Intermodulation, nonlinear responses
Noise	Frequency or time	Large signal functions	Noise power	Oscillators, mixers
Mixed mode	Time	Large signal and digital inputs	Logic and transient simulator	System on a chip combining analogue and digital

a. Small Signal Simulations

Small signal simulations are suited to analyse the steady-state characteristics of a circuit fed by an input stimuli with very low power. The circuit is assumed to be fully linear meaning that nonlinearities are neglected. A single stimulus frequency input is used. Small signal simulations are carried out in the frequency domain. Due to the linear calculations and the single tone stimulus, the simulations are very fast. They are frequently employed for low power circuits such as LNAs. Since the received power in LNAs is weak, small signal simulations are sufficient to calculate the gain or the noise. Passive devices such as inductor, metal resistors and plate capacitors are very linear elements. Hence, in most cases, small signal considerations are sufficient for the latter devices. In Fig. 1.17, the difference between large signal and small signal characteristics is illustrated in the IV curves of a transistor.

b. Large Signal Simulations

Large signal simulations are required for circuits operating under input stimuli with high power. In this case, the variation of the input amplitude versus time is large enough to change the characteristics of active devices and circuits. These variations are, e.g. very strong when the input power is high enough to drive the devices into the IV (Current Voltage) boundaries of transistors. One of the consequences is that the g_m of transistors exhibits a strong nonlinear variation versus time. Corresponding applications involve power amplifiers, drivers, mixers and oscillators. Large signal analysis allows multiple stimuli frequencies required to calculate the intermodulation characteristics of circuits and the conversion gain of mixers and frequency multipliers. As, e.g. discussed in the chapter concerned with power amplifiers, the supply current drawn in circuits can depend on the input power and consequently on the signal conditions (small signal or large signal).

The transient and the harmonic balance engines belong to the most efficient large signal simulators. As its name implies, the transient engine is a pure time domain simulator. Both frequency and time calculations are performed for the harmonic balance engine. The harmonic balance simulation is based on a calculation of the DC start conditions followed by linear simulations to achieve coarse values. Both simulations are performed in the frequency domain. Subsequently, the exact values are calculated in the time domain. Fourier transformation yields the results in the frequency domain. Harmonic balance simulations can be faster than pure transient simulations. However, due to the multiple Fourier transformations and the associated approximations, harmonic simulations tend to introduce more significant simulation errors than transient simulations considering a high number of periods.

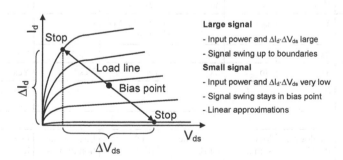

Fig. 1.17. Difference between large signal and small signal characteristics

For large signal circuits, impedance matching is often performed by small signal simulations in a bias point representing average characteristics. The reasons are as follows: small signal simulations are much faster, the impedance matching is still reasonably accurate and the results can be conveniently visualised in the Smith chart.

1.13.3.2 Commercial CAD Software

In Table 1.6, the most frequently employed commercial software tools are outlined. Well suited for analogue RF simulations are Agilent ADS (Advanced Design System) and AWR (Applied Wave Research) MW (Microwave) Office. In addition to common linear and transient simulators, they provide a powerful harmonic balance engine.

Table 1.6. List of frequently used CAD software

Simulations	Circuit design				EM Simulation	Device Modelling
Tool	ADS	Virtuoso Spectre	MW office	Genesys	HFSS	IC CAP
Supplier	Agilent	Cadence	AWR	Eagleware	Ansoft	Agilent
DC	Yes	Yes	Yes	Yes	No	Extraction of linear models
Small signal	Yes	Yes	Yes	Yes	Yes	
Linear noise	Yes	Yes	Yes	Yes	No	
Transient	Yes	Yes	Yes	Yes	No	Extraction of nonlinear models based on linear data arrays
Harmonic balance	Yes	Possible via ADS	Yes	Yes	No	
Nonlinear noise	Yes	Yes	Yes	Yes	No	
Envelope	Yes	Yes, by using harmonic balance				-
Mixed mode	Yes	Yes	No	Yes	No	-
EM	Yes, 2.5D	No	No	Yes, 3D	Yes, 3D	-
DRC	Yes	Yes	Yes	Yes	-	-
LVS	Yes	Yes	Yes	e.g. via MW office	-	-
Unix	Yes	Yes	No	No	Yes	Yes
PC	Yes	No	Yes	Yes	Yes	Yes
Parameter extraction	No					Yes
Layout editor	Yes					
Major application	Analogue, RFIC	Mixed signal, digital	Analogue, RFIC	Analogue, RFIC	EM simulation of passive devices	Device modelling

MW: Microwave, EM: Electromagnetic

a)

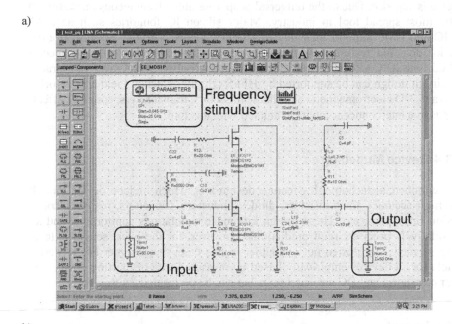

b)

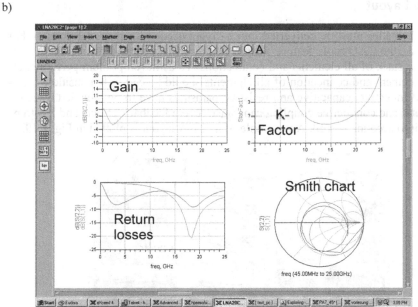

Fig. 1.18a,b. Agilent ADS user interface: **a** schematic input including linear frequency sweep stimulus; **b** results

In Fig. 1.18, the ADS user interface for the schematic input and the result are depicted. For the development of complex mixed mode systems, Cadence Virtuoso

Spectre is superior. Due to the universal properties and advantageous marketing, it is the most spread tool in industry. Major silicon IC foundries such as IBM, TSMC and UMC support their design kits in Cadence Virtuoso Spectre. If a design kit does not feature appropriate device models, modelling may be accomplished with IC CAP (Integrated Circuit Characterisation and Analysis Program). The circuit design can be supported by HFFS (High Frequency Field Simulator), a full-wave 3-D electromagnetic simulator. Inductors, transformers, coupling effects and passive interconnects can be evaluated with this tool.

1.13.4 Device Modelling

Most commercial technologies feature design kits, which include device models with high accuracy for transistors, diodes, inductors, capacitors, resistors, transmission lines, pads, etc. These design kits make simulations comfortable and allow relatively precise circuit predictions. However, for the latest scientific technologies, models are often not available. In this case, the designer has to derive models for all devices empirically based on measured data of existing devices. S-parameter, noise and DC measurements have to be performed for this task.

1.13.5 Layout

The schematics must be converted into a layout needed for mask fabrication. For this task, the CAD tools feature a layout editor. These three-dimensional layouts define a stack of all elements and connections. To achieve optimum performance, the placement and connection of all elements should be carefully considered. Undesired inductive, capacitive and resistive parasitics have to be included. Electromagnetic and conductive coupling can cause severe feedback effects and has to be taken into account as well. An example of an LNA layout is plotted in Fig. 1.19.

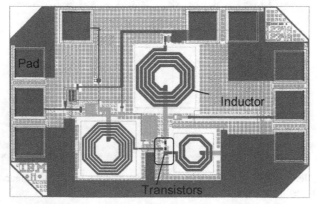

Fig. 1.19. Layout of LNA in SiGe BiCMOS technology, kindly provided by David Barras

In advanced design kits, the libraries feature so-called P-cells providing prearranged device layouts. Convenient setting of the device dimensions such as transistor width and length is possible. The range of material stacks, dimensions and distances of all material layers is carefully specified by the design rules reflecting the limitations in terms of processing. In any case, those design rules have to be met. For example the distance between two metals must be exactly within a certain range. Otherwise, an undesired short-circuit could result. The minimum dimension of a gate is determined by the lithography. Smaller gate length would introduce significant process variations and subsequently yield problems, which would not be acceptable. Foundries only accept layouts fulfilling the complete set of design rules. A DRC (Design Rule Check) routine is included in most design kits, allowing the automatic localisation and manual revisions of the errors. To meet all the design rules may be the most time consuming job regarding the layout.

Moreover, there is a helpful tool called LVS (Layout versus Schematics) which is able to compare the schematics with the drawn layout. Most deviations are identified and can be corrected.

The connections between the devices and the ports are not modelled yet. Especially at high frequencies, these connections can have a significant impact and must be taken into account. We can do that by means of simple LRC models. Long connections are mainly inductive, whereas wide and short connections are rather capacitive. A further helpful tool, namely the parasitic extraction tool can be applied for assistance. Some extractor tools consider only RC parasitics calculated on the basis of approximated equations for series resistances and plate capacitors. More advanced parasitic extractors also include the inductive impact of the connections, which may be the most important one. However, these inductive contributions are difficult to compute due to their complex electromagnetic nature.

If the design is DRC and LVS error free, and the effect of all parasitics has been taken into account properly, the designer may book the next holidays.

1.13.6 Tape-Out and Fabrication

After having a DRC and LVS clean layout, we are ready to submit the design file to the foundry fabricating the hardware. This procedure is called the tape-out. Depending on the IC foundry, these tape-outs are scheduled in periods of two to six months. To speed up the overall design cycle, we have to consider these dates in the planning. The duration required for specific design tasks have been indicated in Fig. 1.16. Dependent on the circuit complexity and the experience of the designer, we have to plan three to ten weeks for the circuit simulations and the layout. The duration of fabrication is in the range of 8-24 weeks. Commercial hardware requires much less fabrication time than research oriented technologies aiming to exploit the limits of processing.

1.13.7 Measurements

When the hardware returns, an exciting task is performed, namely the measurement of the circuit. These DC, small signal and large signal measurements may take approximately a week. Now, the real performance of the circuit can be determined. Hopefully, everything is fine and the circuit works as simulated. Usually, due to process variations and the limited accuracy of the device models, we observe some deviations between the measurements and the simulations. As long as the deviations are not out of the design specifications, everything is fine and you may go for a beer or ask your boss to raise your salary. Details about measurement techniques and the required equipment will be discussed in Chap. 15.

1.13.8 Redesign

If the measurements show significant degradations with respect to simulations and specifications, we have to find the reasons for the deviation and revise the circuit. Such a redesign leads to a significant delay in system development and subsequently to high costs. Therefore redesigns have to be avoided as much as possible by careful design, considering layout errors, connection parasitics and coupling, as well as by verifying the device models.

1.14 Tutorials

1. Recapitulate the history of wireless communications. Which were the major achievements? Who were the main inventors?
2. Discuss the major steps towards the invention of the transistors and the integrated circuit. List the main inventors.
3. How is a generic transceiver system assembled?
4. Comment on three ways to modulate a signal. Why do we have to modulate a signal? Why do we perform frequency conversion?
5. How can we perform multiple user access?
6. Outline the major frequency bands used today for data transmission.
7. What determines the basic limit of the data capacity in a channel? How can we increase the data rate of wireless systems?
8. Outline the main applications of RFICs in the area of wireless communications.
9. Describe key characteristics (e.g. data rate, applications, market share, life cycle) of the major mobile phone standards.
10. What is or was the economic problem of UMTS?
11. Briefly discuss the most important WLAN and WPAN standards and consider their advantages and disadvantages.
12. What are the advantages and disadvantages of UWB systems? Explain the two different types of UWB systems considered today. What is the advantage of the impulse radio based approach in terms of power consumption?
13. How can we perform global positioning? Explain the functionality of GPS.
14. How can we perform local positioning? What are the pros and cons compared to global positioning?
15. Explain the basic time domain positioning principle.
16. How does an FMCW radar operate?
17. What does reconfigurable mean? How could we realise a reconfigurable amplifier? Propose scenarios and ideas. Consider the operation frequencies and power constraints of different standards.
18. Outline the IC design process and subsequently explain all required steps. Comment on the times required for each step.
19. What is a DRC? Why is it important?
20. What is an LVS?
21. What are the most important simulator types? What is the difference between small signal and large signal? How do they correlate?
22. Which commercial CAD software tools are available? If you design a power amplifier and a complex digital signal processor, which kind of commercial software would you choose?
23. Who are the major worldwide semiconductor players?
24. Give an idea of worldwide IC sales in 2007.

References

[Abi08] www.abiresearch.com

[Alb05] www.albury-field.demon.co.uk/radio.htm

[Ber05] L. Berlin, The Man Behind the Microchip, Oxford University Press, New York, 2005.

[Böc03] G. Böck, D. Pienkowski, R. Circa, M. Otte, B. Heyne, P. Rykaczewski, R. Wittmann, R. Kakerow, "RF front-end technology for reconfigurable mobile systems", IEEE International Microwave and Optoelectronic Conference, pp. 863-868, Sept. 2003.

[Dcs05] www-groups.dcs.st-and.ac.uk/~history/Indexes/A.html

[Doa04] C. H. Doan, S. Emami, D. A. Sobel, A. M. Niknejad, R. W. Broderson, "Design considerations for 60 GHz CMOS radios", IEEE Communications Magazine, pp. 132–140, Dec. 2004.

[Dob07] D. M. Dobkin, The RF in Rfid: Passive UHF Rfid in Practice: Passive UHF RFID in Practice, Newnes, 2007.

[Eet08] www.eetimes.eu/uk/

[Ell04] F. Ellinger, Analog SOI CMOS Integrated Circuits at Millimeter Wave Frequencies, ETH Zürich, Hartung-Gorre Verlag, Konstanz, ISBN 386628-007-6, 2005.

[Eur01] "European 3G auction update", Microwave Engineering, pp. 14–16, June 2001.

[Fcc97] Federal Communications Commission, "Millimeter Wave Propagation: Spectrum Management Implications", Bulletin, No. 70, July 1997.

[Fcc02] Federal Communications Commission, 47 CFR Part 15, Sec. 503, Federal Register, Vol. 67, no. 95, May 2002.

[Fin03] K. Finkenzeller, Fundamentals and Applications in Contactless Smart Cards and Identification, Wiley & Sons, 2002.

[Har02] D. Harame, A. Joseph, D. Coolbaugh et al., "The emerging role of SiGe BiCMOS technology in wired and wireless communications", IEEE International Caracas Conference on Devices, Circuits and Systems, Aruba, CDROM, April 2002.

[Hor05] C. L. Horney, W. Strauss, "Global cellular handset & chip markets", Forward Concepts, April 2005.

[Hui03] S. Y. Hui, "Challenges in the migration to 4G mobile phones", IEEE Communications Magazine, Dec. 2003.

[Huu03] A. A. Huurdeman, The Worldwide History of Telecommunications, Wiley-Interscience, 2003.

[Ici05] IC Insights, www.icinsights.com/news/

[Idt08] www.idtechex.com/

[Imr08] www.mobileisgood.com/

[Jam89] R. J. James, "A history of radar", IEE Review, Vol. 35, No. 9, pp. 343–349, 1989.

[Lam99] J. R. Lamberg, M. J. Gawronski, J.J. Geddes, W. R. Carlyon, R. A. Hart, G. S. Dow, E. W. Holmes, M. Y. Huang, "A compact high performance

W-band FMCW radar front-end based on MMIC technology", IEEE International Microwave Symposium, Vol. 4, pp. 1797–1800, June 1999.

[Léc05] C. Lécuyer, Making Silicon Valley, MIT Press, Cambridge, 2005.

[Leh08] H. Lehpamer, RFID Design Principles, Artech House, 2008.

[Lmd04] www.lmdswireless.com/

[Maa03] S. A. Maas, Nonlinear Microwave and RF circuits, Artech House, 2003.

[Man03] K. Mandke, H. Man, L. Yerramneni, C. Zuniga and T. Rappaport, "The evolution of ultra wide band radio for wireless personal area networks", High Frequency Electronics, Sept. 2003.

[Mvd04] www.elva-1.spb.ru/

[Mul06] www.multibandofdm.org

[Opp05] I. Oppermann, M. Hämäläinen, J. Iinatti, UWB Theory and Applications, Wiley, 2005.

[Paq04] S. Paquelet, L. M. Aubert and B. Uguen, "An impulse radio asynchronous transceiver for high data rates", Joint Conference on Ultrawideband Systems and Technologies, May 2004.

[Por03] D. Porcino, W. Hirt, "Ultra-wideband radio technology: potential and challenger ahead", IEEE Communications Magazine, July 2003.

[Pro03] J. G. Proakis, M. Salehi, Communication Systems Engineering, Prentice Hall, 2003.

[Raz06] R. Razavi, "A 60 GHz CMOS frontend", IEEE Journal on Solid-State-Circuits, Vol. 41, No. 1, pp. 17–21, Jan. 2006.

[Ree05] J. H. Reed, An Introduction to Ultra-wideband Communication Systems, Prentice Hall, 2005.

[Res06] www.ife.ee.ethz.ch/RESOLUTION/

[Rob01] I. D. Robertson, S. Lucyszyn, RFIC and MMIC Design and Technology, IEE Circuits, Devices and Systems Series 13, 2001.

[Sko90] M. Skolnik, Radar Handbook, McGraw-Hill, 1990.

[Spr04] A. Springer and R. Weigel, UMTS- The Universal Mobile Telecommunication System, Springer, 2004.

[Tra05] Transistorized, www.pbs.org/transistor/album1/

[Uwb06] www.uwbforum.org

[Ven05] G. D: Vendelin, A. M. Pavio, U. L. Rohde, Microwave Circuit Design using Linear and Nonlinear Techniques, Wiley, 2005.

[Vos03] M. Vossiek, L. Wiebking, P. Gulden, J. Wieghardt, C. Hoffmann, P. Heide, "Wireless local positioning", IEEE Microwave Magazine, Vol. 4, No. 4, pp. 77–86, Dec. 2003.

[Wie01] L. Wiebking, L. Vossiek, M. Nalezinski and P. Heide, "Transpondersystem und Verfahren zur Entfernungsmessung", Patent 101 55 251.3, 2001.

[Wie03] L. Wiebking, Entwicklung eines zentimetergenauen mehrdimensionalen Nahbereichs-Navigations-Systems, VDI Verlag, 2003.

2 Transceiver Architectures

There is no ingenuity without passion.
Theodor Mommsen, University of Leipzig

Consumer markets demand miniaturised and low-cost transceivers with low power consumption and weight. These goals mandate the consequent integration of all functions and devices on a minimum number of ICs, ideally on one single chip without requiring any external components. However, this is not as trivial as replacing the external elements by on-chip components. Due to significant performance differences between the on-chip and off-chip components, complete overhauls of the transceiver architecture may be necessary. Over the last few years there has been a trend to relocate functions and modulation schemes from the analogue to the digital domain. Reasons are higher flexibility, simpler portability regarding new technologies and standards, and robustness against interferers and noise. A transceiver consists of a receiver and a transmitter. The architecture and key characteristics of different types of receivers and transmitters are covered in this chapter, helping the designer to choose the optimum for a specific application. For detailed information, the reader is encouraged to study the specific literature [Raz03, Meh01, Raz98, Raz96, Spr02].

2.1 Receiver

The main function of a receiver is the demodulation of a wanted signal in the presence of undesired interferers and noise. Due to the strong attenuation during air transmission, the RF signal has to be amplified and recovered. Taking into account scenarios with varying attenuation, a wide dynamic range is required for the detection of signals with high data-rates. To be treated in Sect. 4.5, the dynamic range is determined by noise as the lower bound, and nonlinearities caused by saturation effects as the upper limit.

Due to strongly increasing data traffic, the associated frequency bandwidths are limited. To support a high number of users, these frequency bands are divided into narrow channels typically having a bandwidth in the range of 100 kHz to 100 MHz. Filters with high out-of-band attenuation are required to select those narrow channels. Unfortunately, this high off-band attenuation increases the complexity of filters requiring an increased number of elements. The quality factor

Q of these components must be large to minimise the attenuation of the desired signal.

Power consumption is an important issue for receivers. Even if there is no active communication, receivers can't be switched off completely. They have to detect when a transmitter requests a data transmission and subsequently must switch on the receiver chain by means of a wake-up circuit. Consequently, if not active, a receiver has to be operated in a standby mode, where the DC power is reduced. Nevertheless, accumulation of the drawn DC power over a long stand-by-time can result in significant power consumption. Thus, mobile receivers must have a low power consumption in the stand-by mode.

In the following, we discuss the most important concepts used for wireless communication.

2.1.1 Regenerative Receiver

Regenerative receivers have been a great milestone in radio history since they provided sensitivity and selectivity far beyond that available by the former crystal radio. In 1914, at the age of 21, while he was a student in college, Edwin Armstrong invented the first regenerative radio [Arm14]. The idea behind the enhanced performance was the careful control of the positive feedback between the antenna input and the triode output as depicted in Fig. 2.1. Due to the constructive combining of signal power at the input and the subsequent amplification relatively high gain and output power has been achieved. The feedback allows the generation of negative resistance within the devices leading to controlled instability and oscillation with maximum amplitude at a specific frequency.

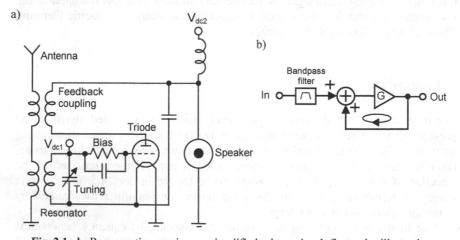

Fig. 2.1a,b. Regenerative receiver: **a** simplified schematics; **b** first order illustration

A patent was filed in 1922 by Armstrong for the principle called the super-regenerative receiver. To prevent the devices from getting saturated and stuck in a previous signal period, the circuit is periodically shut down by an additional quench oscillator. This circuit opens and closes the connection between the resonator and the active device. Due to the high performance, only a few components are required making the super-regenerative receiver a low cost solution, which is still employed today for low data rate applications such as walkie-talkies.

2.1.2 Super-heterodyne Receiver

In 1917, Armstrong invented a further receiver principle, which is still used for a majority of wireless systems. It is the super-heterodyne topology as illustrated in Fig. 2.2. In the literature we frequently find the shortcut simply named heterodyne. At the same time and independently from Armstrong, a similar architecture was proposed by Walter Schottky.

The signal is received by the antenna, coarse filtered by a bandpass filter, amplified by an LNA and converted down to an intermediate frequency (IF) by means of a mixer fed by a local oscillator (LO) signal. The demanding channel filter is employed at IF frequency, followed by an analogue to digital converter and a digital signal processor performing the demodulation and the data decoding. As discussed in Sect. 10.1 and verified by trigonometry, the mixer acts as a signal multiplier yielding

$$\omega_{IF} = \omega_{RF} - \omega_{LO} \qquad (2.1)$$

after proper filtering where we assume that $\omega_{RF} > \omega_{LO}$. The demodulation, channel filtering and a part of the amplification can now be performed at the low IF frequency. This relaxes the demands for the components, which typically exhibit raised performance at lowered frequencies. The LO frequency is tuned to fix the IF frequency at varying RF frequency. Consequently, the filter frequency remains constant simplifying the filter complexity.

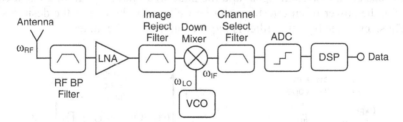

Fig. 2.2. Simplified architecture of the super-heterodyne receiver with single down-conversion, BP: Bandpass, LNA: Low Noise Amplifier, VCO: Voltage Controlled Oscillator, ADC: Analogue Digital Converter, DSP: Digital Signal Processor

On one hand, the in-band loss of filters has to be minimised demanding for low order filters with weak resonances. On the other hand, high selectivity with strong attenuation towards interferers requires high filter orders and high out-of-band attenuation slopes. The requirements for the later parameter depend on the distance between the desired and the unwanted signal frequency related to the desired frequency component. Before and after down-conversion the relation yields $\frac{\omega_{RF} - \omega_{RF*}}{\omega_{RF}}$ and $\frac{\omega_{IF} - \omega_{IF*}}{\omega_{IF}}$, respectively, with potential interferes labelled by *. Since frequency distances are preserved during frequency conversion, we get $\omega_{RF} - \omega_{RF*} = \omega_{IF} - \omega_{IF*}$. By recalling that $\omega_{IF} \ll \omega_{RF}$, we can conclude that down conversion significantly relaxes the demands for the filter with respect to the out-of-band attenuation. Let us review an example with a desired signal at ω_{RF}=15.0 GHz and an undesired interferer at ω_{RF*}=14.4 GHz. The relative frequency difference $\omega_{RF} - \omega_{RF*}$ is only 4 % of ω_{RF}. Therefore, filtering of the interferer is challenging. After down-conversion with ω_{LO}=13.5 GHz, we get ω_{IF}=1.5 GHz and ω_{IF*}=0.9 GHz, respectively. Now, the frequency difference with respect to ω_{IF} is 40 %, which is by a factor of 10 beyond that without frequency conversion. This is illustrated in Fig. 2.3a for filters with the same selectivity at the IF and the RF frequency. By depicting the frequency axis in logarithmic scale, the attenuation slope of the filters appears equal. At IF, the undesired interferer is completely filtered, whereas it can't be entirely filtered at RF.

A severe problem may arise at the undesired image frequency of the RF signal denoted by RFi. Suppose frequencies symmetrically located above and below the LO frequency as illustrated in Fig. 2.3b. In this case, the RF and RFi frequencies are converted to exactly the same IF frequency given by

$$\omega_{IF} = \omega_{RF} - \omega_{LO} \qquad (2.2)$$

as wanted, and the component

$$\omega_{IF} = \omega_{LO} - \omega_{RFi} \qquad (2.3)$$

incorporating undesired content associated with ω_{RFi}. After mixing, there is no way to separate the original signal and the undesired image signal. In a worst-case scenario, the power of an interferer can be well above the one of the desired signal. Thus, an image rejection filter is required in front of the mixer.

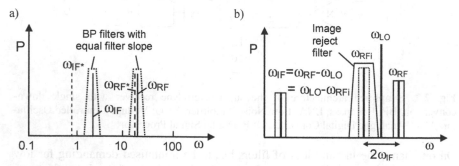

Fig. 2.3a,b. Filtering: **a** of undesired interferer denoted by * at IF and RF using same filter selectivity, BP: Bandpass; **b** of undesired image frequency ω_{RFi}, ω_{RFi} and desired RF frequency ω_{RF} are converted to the same intermediate frequency ω_{IF}

The frequency difference between the desired RF signal and the undesired image signal RFi is given by $2\omega_{IF}$. From this point of view, a large IF frequency is favourable to relax the requirements for the image rejection filter. Unfortunately, this is in contradiction with the requirements for the channel selection filter. Therefore, a reasonable trade-off concerning the IF frequency has to be found for the super-heterodyne receiver with single down conversion.

This tradeoff can be mitigated by using the dual super-heterodyne topology as illustrated in Fig. 2.4. Two different IF frequencies are used. Image rejection is carried out at high IF, whereas channel selection is accomplished at low IF, thereby relaxing the requirements for both filters simultaneously. A constant frequency can be used for the demanding first VCO. Frequency tuning can be performed by the second VCO operating at lower frequency. Unfortunately, two mixers and two oscillators are required increasing the circuit complexity, power consumption and costs.

Nevertheless, systems with narrow channel distances and low IF frequencies typically need image reject filters comprising elements with Q factors well above 100. Such a performance is not feasible with on-chip elements. In this case, external SAW (Surface Acoustic Wave) filters are used. The parasitic off-chip connection, e.g. by bonding wires is not severe given that the IF frequency is below several GHz. Due to filter and connection constraints, the interface impedance between the output of the filters and the input of the active circuits must be around 50 Ω limiting the exploitation of the active device gain, which typically increases with raised load impedance.

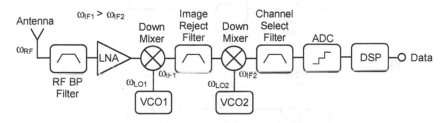

Fig. 2.4. Simplified architecture of super-heterodyne receiver with double down-conversion

The RF band filter preceding the LNA operates as coarse filter and rejects strong interferers, which may saturate the receiver. A certain level of filtering can be achieved by narrowband design of the LNA. Due to their high Q-factors, off-chip filters allow enhanced performance compared to on-chip realisations. However, they consume much more circuit size. Moreover, low loss connections are challenging at high frequencies. For these RF filters, low resistive losses are very important since they directly add to the system noise figure. According to the equation of Friis, see Eq. (4.29), the noise contribution of the filters located after the LNA do not have a significant noise contribution as long as the LNA gain is high.

2.1.3 Image Rejection

The problems associated with the image rejection have motivated designers to invent smart techniques for the rejection of the image frequency without requiring sophisticated filters. Such techniques are especially useful for applications where the desired RF and the undesired image signal are so close in frequency that conventional filtering is not possible. Among the most often used approaches are the Hartley and Weaver image reject techniques developed in 1928 and 1956, respectively.

These techniques are based on the idea of producing two paths having the same polarity for the desired signal, and the opposite polarity for the undesired image signal. Subsequent combining of the paths recovers the desired signal and cancels the image signal.

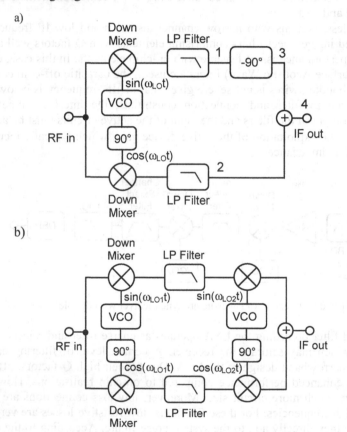

Fig. 2.5a,b. Image reject techniques, LP: Lowpass: **a** Hartley; **b** Weaver

In Fig. 2.5a, the Hartley architecture is illustrated. Let us assume an input signal of $v_{in}(t) = V_{RF} \cos \omega_{RF} t + V_{RFi} \cos \omega_{RFi} t$ with V_{RFi} and ω_{RFi} denoting the voltage amplitude and frequency of the image component. We assume that $\omega_{RF} > \omega_{LO}$.

After multiplication with $\sin \omega_{LO} t$ and $\cos \omega_{LO} t$ in the upper and lower path, respectively, and low pass filtering of harmonics and undesired intermodulation products, we get the following intermediate results at points 1 and 2:

$$v_1(t) = -\frac{V_{RF}}{2} \cdot \sin(\omega_{RF} - \omega_{LO})t + \frac{V_{RFi}}{2} \cdot \sin(\omega_{LO} - \omega_{RFi})t \qquad (2.4)$$

$$v_2(t) = \frac{V_{RF}}{2} \cdot \cos(\omega_{RF} - \omega_{LO})t + \frac{V_{RFi}}{2} \cdot \cos(\omega_{LO} - \omega_{RFi})t . \qquad (2.5)$$

The 90° phase shift in the VCO can be generated by a quadrature VCO as presented in Sect. 11.6.2. Of course, optionally, a common VCO can be used together with a 90° phase shifter. Considering a further phase-shift of −90° in $v_1(t)$ we obtain

$$v_3(t) = \frac{V_{RF}}{2} \cdot \cos(\omega_{RF} - \omega_{LO})t - \frac{V_{RFi}}{2} \cdot \cos(\omega_{LO} - \omega_{RFi})t . \qquad (2.6)$$

Adding $v_2(t)$ and $v_3(t)$ yields $v_4(t) = V_{RF} \cdot \cos(\omega_{RF} - \omega_{LO})t$, verifying that the desired signal is recovered and the image is rejected. However, full image rejection mandates ideal matching of the phases and amplitudes in the paths. A gain mismatch would be caused by the 90° phase shifter. To reduce corresponding amplitude mismatches it is advantageous to split the total phase shift of 90° into 45° in the upper path and −45° in the lower path. For these phase shifters, RC or LC filters can be used. RC filters exhibit wider bandwidths, whereas the LC counterparts yield lower losses.

An optional approach is offered by the Weaver architecture depicted in Fig. 2.5b. The bandwidth limiting and process variation dependent phase shifters are replaced by a second pair of quadrature mixers fed by an additional VCO typically having lower frequency than the preceding one. Larger bandwidth and better image rejection can be achieved at the expense of higher power consumption and circuit complexity. Optionally, the second IF can be mixed down to DC leading to the direct conversion approach, which is subject to discussions in the next section.

As for other I/Q based architectures, phase and amplitude mismatches in the quadrature mixers can significantly degrade the performance. Given fully symmetrical designs, the mismatches are mainly determined by process variations. Those process variations are relatively small for fully integrated solutions. Practical implementations typically exhibit image rejection of more than 30 dB, which is sufficient for many applications.

2.1.4 Direct Conversion Receiver

The motivation of increased integration has led to the direct conversion receiver, which is also referred to as homodyne or zero-IF approach [Raz97, Zha03]. The idea is to translate the RF signal directly to zero-IF frequency thereby exhibiting the following advantages. First, the channel filtering can be performed by a lowpass filter. Recall that a more complex bandpass filter is necessary for the superheterodyne receiver. Second, the IF frequency of zero eliminates the image problem. Hence, no external high-Q image reject filter is required making fully integrated solutions feasible.

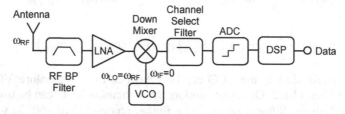

Fig. 2.6. Illustration of zero-IF approach

In Fig. 2.6, a simple direct conversion architecture is illustrated, which can be used for processing of amplitude modulated signals featuring the same information at the two sidebands allocated around the carrier frequency. For more sophisticated frequency and phase modulations schemes, the information within the two sidebands can be different. However, after conversion around DC, these sidebands can't be separated leading to a loss of information. This can be prevented by using quadrature mixing with in-phase (I) and quadrature (Q) signals as illustrated in Fig. 2.7. Consequently, the information of both sidebands can be preserved allowing efficient modulation schemes.

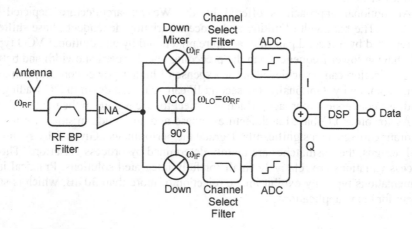

Fig. 2.7. Illustration of the zero IF approach with I/Q quadrature mixing

The zero-IF approach seems to be superior compared with other architectures. However, the following issues impede its widespread use in today's radios. The RF carrier and the local oscillator are at the same frequency. Thus, LO leakage to the mixer input can lead to self mixing resulting in a time-varying DC offset at the output of the mixer. This DC offset may corrupt the signal and can lead to a saturation of the following stages thereby significantly degrading the upper boundary of the dynamic range. Consequently, sophisticated offset cancellation techniques are required in practical implementations. We will learn in Sect. 4.3.3 that the flicker noise of the active devices becomes significant at low frequencies. Thus, low noise amplification and active filtering is difficult for zero-IF topologies degrading the lower limit of the dynamic range.

2.1.5 Low-IF Receiver

One integrated receiver solution, which mitigates some problems associated with the direct conversion receiver is the low-IF receiver [Cro98]. Similar to the direct conversion receiver, a quadrature mixer is used to translate the desired channels to a low IF frequency. Typically, an IF frequency in the order of one up to two channel bandwidths corresponding to 50 kHz to 10 MHz are used as IF frequency. The image rejection can be performed by mixer topologies similar to the Hartley or Weaver architecture. Due to the low IF frequency, channel filtering is relatively simple. Common switched capacitor filters can be applied. Consequently, all relevant filters can be implemented on-chip.

Unlike the zero-IF architecture, the low-IF receiver is not sensitive to the parasitic DC offset, LO leakage and flicker noise. We can conclude that the low IF topology is an excellent compromise between the zero-IF and the super-heterodyne architecture. Thus, the low-IF approach is quite popular in today's receivers. However, as for the zero-IF topology, process variations can introduce I/Q imbalances, which degrade the performance. Corresponding compensation techniques can be applied [Win04].

2.1.6 Digital-IF Receiver

To make systems more flexible, as much signal processing as possible is transferred into the digital domain. Figure 2.8 depicts a realisation of a digital-IF receiver. The idea is to perform the demanding channel filtering completely in the digital domain. Thus, the requirements for the RF filters are relaxed. Simple RF filters may be employed for coarse band selection. The major advantage is the flexibility of the architecture. The receiver can be reconfigured for a variety of systems with different modulation types, channel frequencies and bandwidths meeting the demands of different standards. Moreover, the digital approach avoids the phase and amplitude mismatch problems of analogue I/Q signals. Generally, the impact of process tolerances is less significant.

However, digital-IF receivers are still in their infancy. One of the main critical issues is the extremely high dynamic range required for the ADC, or the AGC (Automatic Gain Control) circuit located in front of the ADC. A wanted channel, which may be significantly attenuated during air propagation, mandates a very high sensitivity with respect to the inherent noise properties, whereas an unfiltered non-desired channel with high power can saturate the ADC.

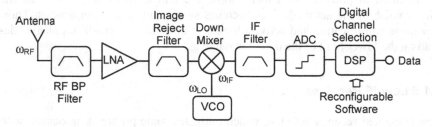

Fig. 2.8. Simplified digital-IF receiver

Typical IF values are around 100 MHz, demanding sampling rates of 200 MHz, which are easily achievable with plain vanilla technologies. Speed and resolution of an ADC have to be traded off. Considering typical dynamic range requirements, resolutions in the range of 12–16 bits are necessary, which is very difficult to obtain together with the high speed. The subsequent baseband filtering requires enormous processing power. With present technology, the excessive power consumption limits the use of digital IF receivers for mobile applications.

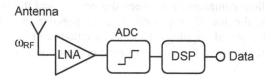

Fig. 2.9. Vision of digitalised receiver with RF analogue to digital conversion

A further architecture to be envisioned is illustrated in Fig. 2.9 [Hen99]. Given that the challenges associated with the figure of merit described by

$$\text{FOM} = \frac{\text{dynamic range} \cdot \text{resolution}}{\text{power consumption}} \tag{2.7}$$

can be solved, the analogue to digital conversion may be accomplished at RF without requiring any frequency conversion in the analogue domain. Considering the current state-of-the-art this seems to be impossible. However, in the area of microelectronics, we have witnessed that the impossible has been made possible for many times by employing enhanced technologies and techniques. Let us meet the challenge.

2.1.7 Impulse Radio Receiver

In recent years, impulse based radios receive a revival due to its promising properties for short range, low power and high speed applications [Por03, Uwb06, Weis04, Paq04, Opp04, Zas03, Sto04, Bar06, Ba206]. In the USA, corresponding UWB (Ultra-Wideband) standards have already been published by the FCC (Federal Communications Commission) [Fcc02], whereas in Europe the community is still waiting for adequate standards. A complete UWB chip set is already available [Fre06].

Let us recall that the first efficient radio transmissions have been performed on the basis of impulse transmission. Today, the UWB standard employs impulse transmission within a frequency band between 3.1 GHz and 10.6 GHz. Pulse position modulation (PPM) can be employed for data transmission. The major benefit of the impulse radio is the low complexity. Compared to conventional receivers, no power consuming PLL synthesisers are required. Due to the large bandwidth available, the demands for the frequency accuracy are much more relaxed. Frequency down-conversion is not necessary. However, mixers may be applied for correlation purposes.

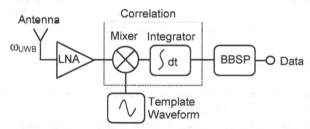

Fig. 2.10. Example of top-level schematics of impulse radio receiver, BBSP: Baseband Signal Processing

An example of an architecture is outlined in Fig. 2.10 [Opp05]. It consists of an antenna, an LNA, a correlation circuitry and the baseband processing. After amplification, the received signal is correlated with a template waveform. A mixer and a template waveform generator can be applied for this task. The output signal of the mixer is integrated to maximise the received signal level with respect to the inherent noise power. To encode the data, the output of the correlation circuit is processed in the baseband. Challenges in terms of circuit design are the speed and the bandwidth of the RF components. Due to the large bandwidth, wideband receivers may be susceptible to interferers.

2.1.8 Receiver Comparison

In Table 2.1, the advantages and disadvantages of the different receiver architectures are summarised. The final choice strongly depends on the individual specifications and the available technology.

Table 2.1. Comparison of receiver architectures

Architecture	Complexity	Full integration	Power cons.	Comments	
Super-regenerative	Low	Possible	Very low	No carrier, impulse with wide bandwidth can cause interferences with other systems if emitted power not limited	High sensitivity because of resonant feedback
Impulse radio					High bandwidth of 3.11–10.6 GHz allows high data rates
Super-heterodyne	Moderate	Off-chip image reject and channel select filter required	Moderate	IF has to be traded off for image rejection and channel selection	
Dual super-heterodyne	High		High	Good image rejection and channel selection possible	
Direct conversion	Low	Possible	Low	DC offsets can significantly degrade the performance, sensitive to flicker noise	
Low-IF	Low/moderate	Possible	Low/moderate	Good overall performance	
Digital-IF	RF: very low Baseband: very high	Possible	Very high	Very flexible architecture, can handle different standards, modulation and frequencies, ADC limits dynamic range which is a major drawback	

2.2 Transmitter

The three primary functions of common transmitters are modulation, frequency conversion and power amplification. Consequently, the key performance parameters are modulation accuracy, spectral purity and RF output power. Since a strong signal is locally available, band selection and noise are not as critical as in receivers. Moreover, the variation of the signal level is small relaxing the requirements in terms of the dynamic range. Thus, transmitters are less complex and are found in a smaller variety of approaches than receivers. The generation of high output power leads to a high DC power consumption. Thus, in active operation, the power consumption of transceivers is determined by the transmitter rather than by the receiver. However, a transmitter can be completely shut down after signal transmission to save power.

To transmit data, modulation modes with both constant and variable signal amplitude can be employed. The first scheme is more power efficient, whereas the latter one is more spectral efficient at the expense of challenging requirements in terms of linearity.

2.2.1 Direct Conversion Transmitter

Figure 2.11 illustrates the principle of a direct conversion architecture. The baseband signal is up-converted to RF, bandpass filtered, amplified and lowpass filtered before the signal is emitted by the antenna. The direct up-conversion architecture suffers from the so called injection pulling, where a part of the strong power amplifier signal is coupled back to the oscillator operating at the same RF frequency. Thus, undesired DC components are generated. Reasons for the coupling are the non-ideal substrate isolation and reflections at the component interfaces.

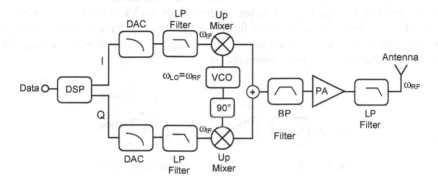

Fig. 2.11. Architecture of direct conversion transmitter, PA: Power Amplifier, DAC: Digital Analogue Converter

Injection pulling may lead to spectral interferences, additional noise and frequency drifts. Sophisticated shielding methods may be used to alleviate this problem. One solution is the separation of the VCO and the PA on different chips. However, the aim for single chip solutions makes this idea unattractive.

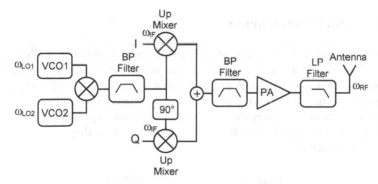

Fig. 2.12. Illustration of offset direct conversion transmitter

The injection pulling can be avoided if the frequencies of the power amplifier and the oscillator are different. To this end, the up-conversion can be performed in two steps, where the two VCOs run at the different frequencies. Alternatively, as illustrated in Fig. 2.12, the frequencies $\omega_{LO1} \neq \omega_{LO2}$ of two VCOs can be added or subtracted to obtain the desired oscillation frequency.

2.2.2 Direct Modulation Transmitter

A typical architecture of a direct modulation transmitter is illustrated in Fig. 2.13. The baseband signal is modulated and up-converted in one single step. By means of the frequency control voltage, the VCO is modulated by the applied data. Subsequently, the signal is amplified, low pass filtered and emitted via antenna. Amplitude modulated signals can't be transmitted since the VCO is always in saturation. The architecture is well suited for frequency and phase modulations. Among the advantages of this approach are the low complexity, the increased ability for integration and the low power consumption.

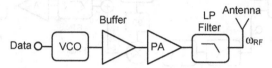

Fig. 2.13. Architecture of direct modulation transmitter

One critical issue associated with the direct modulation is the frequency stability of the VCO during the transmission. The following disturbances may change the VCO frequency and corrupt the modulation: injection pulling, supply voltage variations, and impedance variations between the VCO and the PA. The latter effect can be mitigated by implementation of a buffer with high isolation. A PLL (Phase Locked Loop) is often added to improve the frequency stability and to reduce the content of harmonics and noise.

2.2.3 Impulse Radio Transmitter

In Fig. 2.14, the simple architecture of an impulse radio transmitter is illustrated consisting of a pulse generator, a timing circuit and a clock oscillator [Opp05]. PPM is used for data modulation. A programmable delay circuit can be employed to determine the timing. The desired waveform is produced by the pulse generator, while the clock oscillator defines the pulse repetition frequency. Step, Gaussian or monocycle pulses are suited for UWB communication since they have a broadband frequency spectrum.

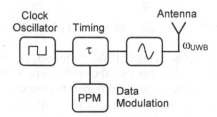

Fig. 2.14. Example of top-level schematics of impulse radio transmitter

2.2.4 Transmitter Comparison

In Table 2.2, the advantages and disadvantages of the different transmitter approaches are summarised. As for the receiver, the final choice strongly depends on the individual specifications, applications and the employed technology.

Table 2.2. Comparison of transmitter architectures

Architecture	Complexity	Full integration	Power consumption	Comments
Direct conversion	Low	Possible since no sophisticated filters are required. However technology must be capable of providing enough output power	Similar since major power drawn by PA, offset approaches slightly higher	Sensitive to injection pulling
Offset direct conversion	Moderate			Injection pulling alleviated
Direct modulation	Very low			Modulation can be corrupted by frequency variations
Impulse radio	Very low		Very low since output power restricted due to potential for interferences with other standards	High bandwidth allows high data range at low coverage range

2.3 Transceiver Example

Obviously, transceivers consist of both a receiver and a transmitter. Figure 2.15 depicts a simple super-heterodyne transceiver. In many cases, a transceiver needs only one multifunctional VCO since it may be used for both the receiver and transmitter. This holds also for the antenna. SPDT (Single Pole Double Throw) switches can be employed to change between the receive- and transmit- modes. Drawback of these switches is the additional losses of around 0.5–2 dB, which directly add to the overall noise figure in the receiver and reduce the effective PA power. However, the benefit regarding the saved space and costs with respect to a second antenna may be considerable. For detailed information concerning switches, the reader is referred to Sect. 13.1.

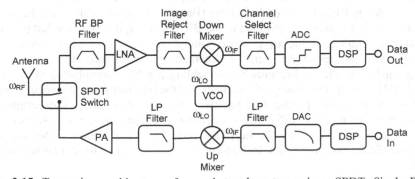

Fig. 2.15. Transceiver architecture of super-heterodyne transceiver, SPDT: Single Pole Double Throw

2.4 Smart Antenna Transceivers

Recall that maximum antenna gain can be achieved with antennas providing a high directivity. On the other hand, the probability of spatially dependent fading increases with raised directivity. The probability of fading can be decreased by employing multiple antennas, which are properly spaced apart from each other. To make sure that one antenna receives a non-faded signal in case that one antenna is in a fading whole, antenna distances in the order of fractions of a wavelength are reasonable. Based on the signal detected by a RSSI (Received Strengths Indicator), the system may choose the optimum antenna. Passive or active switches can be applied for this task. The drawback of this switched antenna approach is the increased system size. We have to keep in mind that antennas consume significant space. An advantage of this approach is the low control complexity. However, only one antenna is active at the same time. Thus, the potential of the other antennas is not exploited.

To achieve diversity gain and to combine the signals available at all antennas in the most efficient way, the complex weighting vector w_i of the signal path must be adjusted [Lib99, Witt00]. The weighting vector w_i of each active antenna path is a function of phase and amplitude. All vectors are optimised to maximise the quality of the available signal, which can be specified by the received power or more meaningful by the bit error rate available in the baseband. This approach is widely known as adaptive antenna combining or MIMO (Multiple In Multiple Out) approach and is a promising technique to reach wireless data rates of beyond 50 Mb/s in realistic environments. Since the impact of inter-symbol interferences can be mitigated, the coverage range and indoor penetration is enhanced. Through range extension, initial costs for system installations can be reduced. Moreover, the number of simultaneous subscribers supported in each cell can also be increased.

The weighting factor can be adjusted in both the transmitter and the receiver. Typically, in active operation, receiver paths consume less power than transmitter branches. Thus, for mobile applications, it can be advantageous to perform the weighting in the receiver only, a concept referred to as MISO (Multiple In Single Out). As illustrated in Fig. 2.16 for a receiver, w_i can either be set and combined in the analogue RF, LO, IF, or digital BB (Baseband) section. Due to the flexibility, this task is frequently performed in the BB. However, it is obvious that the latter solution requires the highest power consumption since all components from RF to BB have to be operated in parallel. Alternatively, to minimise the power consumption for mobile applications, the weighting can be accomplished in the RF part [Ell21, Ell29]. Promising enhancements of the bit error rates in environments with strong multipath propagation has been demonstrated [Ell2]. The design of high performance phase shifter ICs is challenging. Referring to Sect. 14, they introduce amplitude variations vs the phase control making precise vector adjustments difficult. Moreover, low loss phase shifter ICs tend to have a small bandwidth.

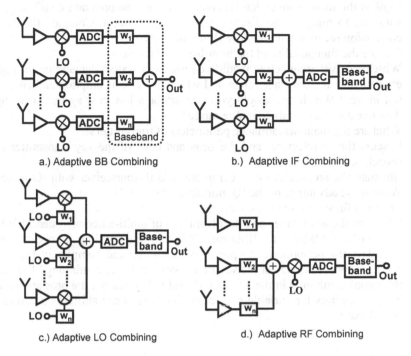

Fig. 2.16a–d. Adaptive antenna receivers, weighting factor w_i is a function of amplitude and phase with i {1..n} as number of antenna paths, combining can be performed in: **a** BB; **b** IF; **c** LO; **d** RF section

2.5 Tutorials

1. What are the general tasks of wireless transceivers? Why do we need transceivers at high and allocated frequencies? What are the main performance parameters?
2. Explain the single super-heterodyne receiver architecture.
3. What is the problem if an interferer is located close to the desired RF frequency? Consider the filtering. How can we mitigate this problem? How do we have to choose the IF frequency?
4. How is the image signal at IF generated? What is the problem of this image signal? How do we have to choose the IF frequency to simplify the suppression of the image frequency?
5. Explain further concepts allowing image rejection.
6. For the single heterodyne receiver, what are the design tradeoffs in terms of the IF frequency?
7. Explain the double heterodyne receiver. What is the advantage in terms of channel selectivity and image rejection? What is the economic disadvantage?

8.   Explain the direct conversion receiver. What are the pros and cons? Suggest solutions to mitigate the disadvantages of the concept. Compare the direct conversion receiver with the low-IF architecture.
9.   Outline the functionality of the impulse based receiver.
10. Which receiver architecture would you use for a system demanding for highest data rate and coverage range and where power consumption and costs do not matter? Which one would you choose for a low cost system with high data rate and very small coverage range?
11. What are the main performance parameters of transmitters?
12. Discuss the architecture, and the pros and cons of the key transmitter approaches.
13. Illustrate the architecture of a complete zero-IF transceiver with IQ mixers. What is the advantage of the IQ modulation?
14. Envision future transceiver concepts.
15. What are the advantages and disadvantages of multi-antenna systems? What does MIMO, SIMO and MISO mean? Which one provides a good tradeoff regarding the performance to power consumption and complexity figure of merit? How can we adjust the vector for smart antenna combining? Illustrate the signal combining in the RF, LO, IF and BB. What are the pros and cons? Suggest circuits for variable gain amplifiers and phase shifters operating in the RF path.

References

[Arm14] E. H. Armstrong, "Some recent developments in the audion receiver", Proc. IRE, V. 3, pp. 215–247, 1915.
[Bar06] D. Barras, F. Ellinger, H. Jäckel, W. Hirt, "A robust front-end architecture for low-power UWB radio transceivers", IEEE Transactions on Microwave Theory and Techniques, pp. 1713–1723, April 2006.
[Ba206] D. Barras, F. Ellinger, H. Jäckel, W. Hirt, "Low-power ultra-wideband wavelets generator with fast start-up circuit", IEEE Transactions on Microwave Theory and Techniques, pp. 2138–2145, April 2006.
[Cro98] J. Crols, M. S. Steyaert, "Low-IF topologies for high-performance analog front ends of fully integrated receivers", IEEE Transactions on Circuits and Systems II: Analog and Digital Signal Processing, Vol. 45, No. 3, pp. 269-282, March 1998.
[Ell2] F. Ellinger, Monolithic Integrated Circuits for Smart Antenna Receivers, Diss. ETH 14063, Jan. 2001.
[Ell21] F. Ellinger, W. Bächtold, "Adaptive antenna receiver module for WLAN at C-Band with low power consumption", IEEE Microwave and Wireless Components Letters, Vol. 12, No. 9, pp. 348–350, Sept. 2002.

[Ell29] F. Ellinger, R. Vogt and W. Bächtold, "Calibratable adaptive antenna combiner at 5.2 GHz with high yield for PCMCIA card integration", IEEE Transactions on Microwave Theory and Techniques, Special Issue, Vol. 48, No. 12, pp. 2714–2720, Dec. 2000.

[Fcc02] Federal Communications Commission, 47 CFR Part 15, Sec. 503, Federal Register, Vol. 67, no. 95, May 2002.

[Fre06] Freescale, "XS110 UWB solution for media-rich wireless applications," www.freescale.

[Hen99] T. Hentschel, M. Henker, G. Fettweis, "The digital front-end of software radio terminal", IEEE Personal Communications, Vol. 6, No. 4, pp. 40–46, Aug. 1999.

[Lib99] J. L. Liberti, T. S. Rappaport, "Smart Antennas for Wireless Communication", Prentice Hall, 1999.

[Meh01] J. L. Mehta, "Transceiver architectures for wireless Ics", RF Mixed Signal Magazine, pp. 76–96, Feb. 2001.

[Opp04] I. Oppermann, L. Stoica, A. Rabbachin, Z. Shelby and J. Haapola, "UWB wireless sensor networks: UWEN – a practical example," IEEE Communications Magazine, Vol. 42 , Iss.12, pp. 27–32, Dec. 2004.

[Opp05] I. Oppermann, M. Hämäläinen, J. Iinatti, "UWB Theory and Applications", Wiley, 2005.

[Paq04] S. Paquelet, L. M. Aubert and B. Uguen, "An impulse radio asynchronous transceiver for high data rates," Joint Conference on Ultrawideband Systems and Technologies, May 2004.

[Por03] D. Porcino and W. Hirt, "Ultra-wideband radio technology: potential and challenges ahead", IEEE Communication Magazine, Vol. 4, No. 7, pp. 66–74, July 2003.

[Raz96] B. Razavi, "Challenges in portable RF transceiver design", IEEE Circuits & Devices, pp. 12–25, Sept. 1996.

[Raz97] B. Razavi, "Design considerations for direct-conversion receivers", IEEE Transactions on Circuits and Systems-II: Analog and Digital Signal Processing, Vol. 44, No. 6, June 1997.

[Raz98] B. Razavi, RF microelectronics, Prentice Hall, New York, 1998.

[Raz03] B. Razavi, "RF CMOS transceivers for cellular telephony", IEEE Communications Magazine, pp. 144-149, Aug. 2003.

[Spr02] A. Springer, L. Maurer, R. Weigel, "RF system concepts for highly integrated RFICs for W-CDMA mobile radio terminals", IEEE Transactions on Microwave Theory and Techniques, Vol. 50, No. 1, Part 2, pp. 254–267, Jan. 2002.

[Sto04] L. Stoica, S. Tiuraniemi, A. Rabbachin and I. Oppermann, "An ultra-wideband tag circuit transceiver architecture", Joint Conference on Ultra-wideband Systems and Technologies, May 2004.

[Uwb06] www.uwbforum.org

[Weis04] M. Weisenhorn and W. Hirt, "Robust noncoherent receiver exploiting UWB channel properties", Joint Conference on Ultra-wideband Systems and Technologies, May, 2004.

[Win04]   M. Windisch, G. Fettweis, "Adaptive I/Q imbalance compensation in low-IF transmitter architectures", IEEE Vehicular Technology Conference, Vol. 3, pp. 2096–2100, Sept. 2004.

[Witt00] A. Wittneben, "Smart antennas for low cost wireless communications", Frequenz, vol. 54, no. 1–2; p. 58-64, Jan.-Feb. 2000.

[Zas03]   T. Zasowski, F. Althaus, M. Stäger, A. Wittneben and G. Tröster, "UWB for noninvasive wireless body area networks: channel measurements and results", IEEE Conference on Ultra-Wideband Systems and Technologies, Nov. 2003.

[Zha03]   P. Zhang, T. Nguyen, C. Lam et al., "A 5-GHz direct-conversion CMOS transceiver", IEEE Journal of Solid-State circuits, Vol. 38, No. 12, Dec. 2003.

3 S-Parameters and Impedance Transformation

In case that you are bitten by a snake you should always carry a small bottle of
Whiskey – moreover, you should always carry a snake.
W.C. Fields

With regard to the characterisation of devices and circuits, the definition of the terminal impedances is mandatory. In the low-frequency domain, input and output ports are frequently described by means of the well-known Z or Y parameters. However, for high frequencies, the application of theses parameter is limited since short and open circuits are required for the measurements. Unfortunately, ideal shorts with $\underline{Z}$ =0 and opens with $\underline{Z}$ =∞ cannot be realised at RF frequencies. Due to the significant reactive series impedance $\underline{Z}$ =jωL of interconnects it is impossible to implement a real short. Capacitive coupling approximated by the admittance $\underline{Y}$ =jωC does not allow the definition of a real open. Moreover, we have to consider antenna effects, which may pick up undesired signals.

Consequently, to obtain accurate measurements up to high frequencies, another approach must be considered. In this context, the S-parameter (Scattering Parameter) approach is applied, which is based on the measurement of forward and reflected voltage and current waves at 50-Ω terminations. Impedances of 50 Ω can be realised much more precisely than shorts and opens, or in other word, the 50-Ω impedance represents a good compromise between both extremes keeping the impact of non-ideal grounds and capacitive opens to a minimum. The reference impedance of 50 Ω has also a historical background. According to Sect. 15.2, coaxial cables have a loss minimum around 50 Ω. Thus, measurement equipment and system interfaces have been standardised for this impedance.

With state-of-the-art NWAs (Network Analysers), measurements of the scattering parameters are possible up to approximately 300 GHz. Such NWAs are based on the small signal approach, which describes the characteristics of a device around an operation point with small input power. This means that the signal swing around the operating point does not change the device properties. For many circuits such as low noise amplifiers, the input signal is small enough to make small signal considerations feasible. At high signal swings, the characteristics and associated parameters are a function of time. This is referred to as large signal operation. Examples of circuits operating under large signal conditions are oscillators, power amplifiers and mixers. Large signal models are much more complicated since they cannot be described by fixed values as is possible for the small signal approach. However, large signal operation can be attributed to an array of small signal models representing the characteristics at each individual time fraction and operating point. Even under large signal conditions, for linear circuit analyses, reasonable accuracy can be obtained by using an average small signal operation point.

Prior to the definition of the S-parameter, we introduce the reflection coefficient and the Smith chart. All of the following discussions are based on the small signal approach.

3.1 Reflection Coefficient

The derivation of the reflection coefficient can be performed by using a loaded transmission line as shown in Fig. 3.1. The resistive reference impedance is denoted by Z_0. At every location x on the line we find the following voltage:

$$\underline{V}(x) = \underline{V}_t + \underline{V}_r \tag{3.1}$$

and current:

$$\underline{I}(x) = \underline{I}_t - \underline{I}_r . \tag{3.2}$$

By definition, $\underline{V}_t$ and $\underline{V}_r$ have the same direction, whereas $\underline{I}_t$ and $\underline{I}_r$ have opposite directions. At the end of the line where x=l, the following equation has to be satisfied:

$$\underline{Z} = \frac{\underline{U}(l)}{\underline{I}(l)} = \frac{\underline{V}_t + \underline{V}_r}{\underline{I}_t - \underline{I}_r} = \frac{\underline{V}_t + \underline{V}_r}{(\underline{V}_t - \underline{V}_r)/Z_0} = Z_0 \cdot \frac{1 + \underline{V}_r/\underline{V}_t}{1 - \underline{V}_r/\underline{V}_t} = Z_0 \cdot \frac{1 + \underline{\Gamma}}{1 - \underline{\Gamma}}, \tag{3.3}$$

where the reflection coefficient is given by

$$\underline{\Gamma} = \frac{\underline{V}_r}{\underline{V}_t} = \frac{\underline{I}_r}{\underline{I}_t} . \tag{3.4}$$

Usually we want to maximise the power transferred to the load. Thus, we have to minimise the reflected power and consequently the reflection coefficient. Expressed in terms of impedances we get

$$\underline{\Gamma} = \frac{\underline{Z} - Z_0}{\underline{Z} + Z_0} . \tag{3.5}$$

If $\underline{Z} = Z_0$, $\underline{\Gamma}$ equals 0, and in the case of $\underline{Z} = 0$ or $\underline{Z} = \infty$, $\underline{\Gamma}$ equals -1 and 1, respectively.

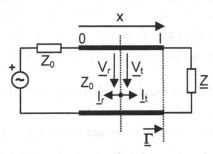

Fig. 3.1. Transmitted and reflected currents and voltages

3.2 Smith Chart

Fifty years ago, CAD tools were not available in the present form. Thus, graphical tools were frequently applied for circuit calculations. The Smith chart [Smi44] has been invented by P. H. Smith and simplifies the calculation of impedance transformations. Even today, in the age of high performance computing, the Smith chart is still a very useful tool for the understanding of passive networks and transmission lines. With the Smith chart, we can graphically determine the reflection coefficient between impedance transitions. The normalised impedance plane

$$\underline{z} = \frac{\underline{Z}}{Z_0} \tag{3.6}$$

is transformed into the reflection coefficient plane

$$\underline{\Gamma} = \frac{\underline{Z} - Z_0}{\underline{Z} + Z_0} = \frac{\underline{z} - 1}{\underline{z} + 1} \tag{3.7}$$

as shown in Fig. 3.2.

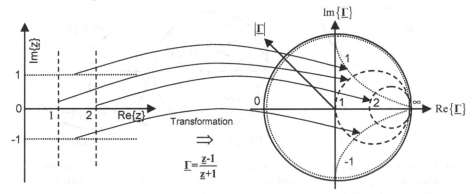

Fig. 3.2. Transformation from impedance plane $\underline{z}$ to reflection coefficient plane $\underline{\Gamma}$

Mathematically, this transformation complies with the so-called Möbius transformation. This operation comes along with the following key characteristics:

- The transformation can be performed in the $\underline{z}$- or $\underline{y}$-plane. Z-parameters can easily be converted in Y-parameters by 180° rotation around the origin $\underline{Z} = 1/\underline{Y} = 1$.

- Constant Re{$\underline{z}$} values are located on circles from $0 \le$ Re{$\underline{z}$} $\le \infty$; the same applies for Re{$\underline{y}$}.

- Constant Im{$\underline{z}$} values are located on arcs from $0 \le$ Im{$\underline{z}$} $\le \pm$ j∞; the same applies for Im{$\underline{y}$}.

- Circles stay circles.

- Angles are preserved.

The Smith chart has the advantages that we can concisely illustrate all complex values from 0 to ∞. In this context, we may point out similarities with a logarithmic scale. In Fig. 3.3, the circles of constant real parts of $\underline{z}$ and the arcs of constant imaginary parts of $\underline{z}$ are illustrated. The centre of the Smith chart equals the reference impedance. Detailed Smith charts can be downloaded from the web [Sss06; Rfc06]. On these homepages, we can find further information, useful tutorials and historical background concerning the Smith chart.

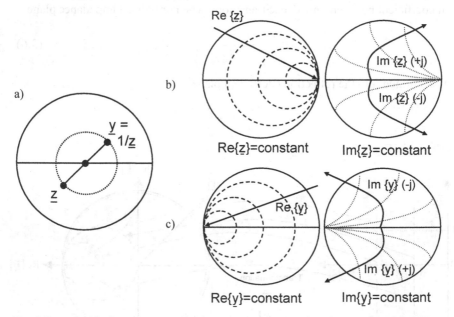

Fig. 3.3a–c. Smith chart: **a** conversion from z to y parameter by mirroring at the origin; **b** $\underline{z}$-plane showing circles with constant real part and arcs with constant imaginary part; **c** $\underline{y}$-plane showing circles with constant real part and arcs with constant imaginary part

The use of the Smith chart depicted in Fig. 3.4 is now explained by a typical application example shown in Fig. 3.5. We want to transfer a signal from the resistive source impedance Z_0 to the load with complex impedance $\underline{Z}_L$. The load impedance is not matched to the source impedance. We want to know how much of the signal is reflected and consequently not transferred to the load.

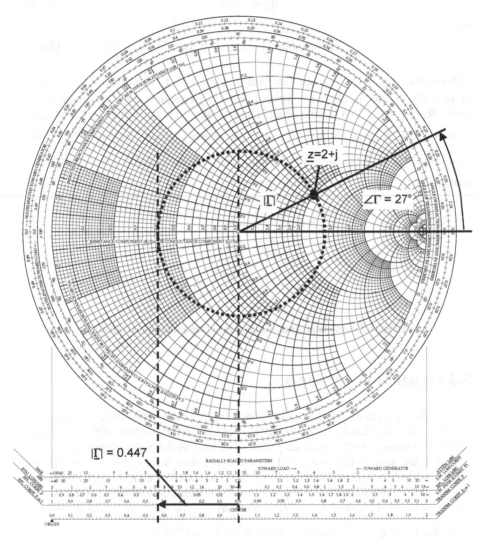

Fig. 3.4. Smith chart, courtesy of Analog Instruments Company of New Providence

For this simple example we can calculate

$$\underline{\Gamma} = \frac{\underline{Z} - Z_0}{\underline{Z} + Z_0} = \frac{\underline{z} - 1}{\underline{z} + 1} = \frac{2 + j - 1}{2 + j + 1} = \frac{1 + j}{3 + j}. \tag{3.8}$$

Separated into magnitude and phase, $\underline{\Gamma}$ is given by

$$|\underline{\Gamma}| = 0.447 \tag{3.9}$$

and

$$\angle\Gamma = \arctan(1) - \arctan\left(\frac{1}{3}\right) = 27°. \tag{3.10}$$

Referring to Fig. 3.4, the magnitude and the phase of $\underline{\Gamma}$ can also be determined graphically in the Smith chart without requiring complex calculations or CAD tools. Recall that the reflection coefficient is based on the relation of voltages. Consequently, the magnitude of the reflected power is proportional to $|\underline{\Gamma}^2|$. In our example, 20% of the power is reflected at the impedance transition, whereas the remaining 80% ($1 - |\underline{\Gamma}^2|$) is fed into the load.

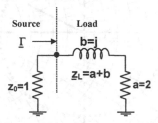

Fig. 3.5. Impedance transition from source to unmatched load

3.3 Signal Flow Analysis

Signal flow diagrams are frequently used for the universal characterisation of transfer and reflection functions [Bäc94]. Both the determination of the reflection factor as discussed in the preceding section as well as the S-parameters to be presented in the next section are based on the signal flow theory. The systematic approach of the signal flow analysis allows the calculation of complex networks with many nodes. As illustrated in Fig. 3.6a, the transfer and reflection functions of networks can be described by the relations between the incoming wave $\underline{a}_j$ and the outgoing wave $\underline{b}_k$. With direction from $\underline{a}_j$ to $\underline{b}_k$ we define

$$\underline{S}_{kj} = \frac{\underline{b}_k}{\underline{a}_j}. \tag{3.11}$$

For complex networks as shown in Fig. 3.6b, the Mason's rule [Mas56] can be applied for simple determination of $\underline{S}_{kj}$. First let us make some definitions. A path $\underline{P}_v$ extends from node j with signal wave $\underline{a}_j$ to the node k with the signal wave $\underline{b}_k$. The number of paths are denoted by v [1..n].

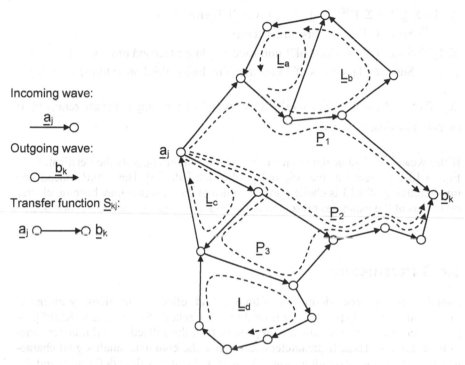

Incoming wave:

Outgoing wave:

Transfer function $\underline{S}_{kj}$:

Fig. 3.6a,b. Signal flow diagram of generic n-port

If there are sub-paths, the transfer function of a path can be calculated by the product of the transfer functions $\underline{S}_{xy}$ of each sub-path:

$$\underline{P}_v = \Pi \, \underline{S}_{xy} \, . \tag{3.12}$$

Within a path, all directions of the sub-paths must be directed towards $\underline{b}_k$. Otherwise, the path is not active and does not have to be considered. A loop $\underline{L}_v$ is a path, which is closed. Based on these definitions, we can calculate any transfer function via

$$\underline{S}_{kj} = \frac{\underline{b}_k}{\underline{a}_j} = \frac{\sum\limits_{v=1}^{n} \underline{P}_v \underline{\Delta}_v}{\underline{\Delta}} \tag{3.13}$$

where

$\underline{S}_{kj}$: Transfer function directed from $\underline{a}_j$ to $\underline{b}_k$

v : Integer representing the number of each forward path

$\underline{P}_v$: Path gain of forward path v from $\underline{a}_j$ to $\underline{b}_k$

$\underline{\Delta}$: $1 - \Sigma \underline{L}^{(1)} + \Sigma \underline{L}^{(2)} - \Sigma \underline{L}^{(3)}$ ... (overall determinant)

$\Sigma \underline{L}^{(1)}$: Sum of all loops (first order term)

$\Sigma \underline{L}^{(2)}$: Sum of all products of 2 non-touching loops (second order term)

$\Sigma \underline{L}^{(3)}$: Sum of all products of 3 non-touching loops (third order term)

....

$\underline{\Delta}_v$: Part of $\underline{\Delta}$, which remains if path v and all its touching paths are removed. If no path remains, we get $\underline{\Delta}_v = 1$

If the weather is bad or the reader is bored, the author suggests the verification of Eq. (3.13) by means of the original publication [Mas56]. Have fun! Indeed, the understanding of 3.13 is challenging. The use of the equation may become clearer by means of the examples presented at the end of Section 3.4 and in Section 4.2.1.

3.4 S-Parameters

Based on the universal definition of transfer and reflection functions by means of the signal flow analysis, we will now define the related S-parameters [Kur65]. S-parameters are very well suited to represent both the reflection and transfer function of devices. Thus, S-parameters can define the complete small signal characteristics of a device excluding noise. Similar to the signal flow definitions and the reflection coefficient, the S-parameters are based on the relationship between the incoming and outgoing waves.

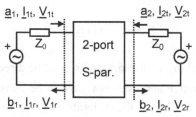

Fig. 3.7. Source and load transmission and reflection power wave parameters of a 2-port

As shown in Fig. 3.7 for a typical 2-port, S-parameters are based on the power wave parameters $\underline{a}_j$ and $\underline{b}_k$, which determine the transmitted and reflected powers, respectively, of the signal at or between a node. In our 2-port example, the possible values for j and k are 1 or 2. The power wave parameters are a function of the transmitted and reflected voltages $\underline{V}_t$ and $\underline{V}_r$, and are defined by

$$\underline{a}_j = \frac{\underline{V}_{jt}}{\sqrt{Z_0}},$$

(3.14)

and

$$\underline{b}_k = \frac{\underline{V}_{kr}}{\sqrt{Z_0}}.$$

(3.15)

Analogously, the last two equations could also be determined by the transmitted and reflected currents. The products $\underline{a}_j\underline{a}_j^*$ and $\underline{b}_k\underline{b}_k^*$ are proportional to the corresponding real powers. Per definition, only one reference source is active for the calculation of a specific S-parameter, while all others are set to zero. Note that the signal of the transfer function is directed from node j to node k. The full S-parameter set of a network consists of all possible relations between the power wave parameters excluding the relationship between the same parameters. Thus, the number of power wave parameters and S-parameters amounts to 2^n where n denotes the number of ports.

Table 3.1. Definition of 2-port S-parameters

Input reflection coefficient	Output reflection coefficient
$\underline{S}_{11} = \frac{\underline{b}_1}{\underline{a}_1} \quad \left\|\underline{a}_2 = 0\right.$	$\underline{S}_{22} = \frac{\underline{b}_2}{\underline{a}_2} \quad \left\|\underline{a}_1 = 0\right.$
Ratio of reflected power to incoming power at port 1	Ratio of reflected power to incoming power at port 2
No power fed from port 2	No power fed from port 1
Port 2 terminated with Z_0	Port 1 terminated with Z_0
Forward transmission coefficient	**Reverse transmission coefficient**
$\underline{S}_{21} = \frac{\underline{b}_2}{\underline{a}_1} \quad \left\|\underline{a}_2 = 0\right.$	$\underline{S}_{12} = \frac{\underline{b}_1}{\underline{a}_2} \quad \left\|\underline{a}_1 = 0\right.$
Ratio of outgoing power at port 2 to incoming power at port 1	Ratio of outgoing power at port 1 to incoming power at port 2
No power fed from port 2	No power fed from port 1
Port 2 terminated with Z_0	Port 1 terminated with Z_0

In many cases, 2-ports are applied, e.g. also for transistors, where the third port serves as the common ground. A 2-port can be fully described by a set of 4 S-parameters, namely the input reflection, output reflection, forward transmission and reverse transmission coefficient. The corresponding S-parameter definitions are summarised in Table 3.1. These parameters can be found in typical transistor data sheets and depend on the frequency and the applied DC bias.

Attentive readers may have noticed that the reflection coefficients $\underline{\Gamma}_{11}$ and $\underline{S}_{11}$ are very similar. The only difference is that as per definition $\underline{S}_{11}$ is terminated with Z_0 at port 2, whereas $\underline{\Gamma}_{11}$ considers any termination at port 2. Given that $\underline{Z}_2 = Z_0$, which is the case in typical measurement environments we obtain $\underline{\Gamma}_{11} = \underline{S}_{11}$.

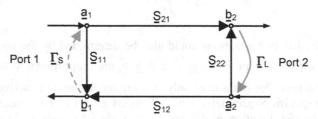

Fig. 3.8. Calculation of $\underline{\Gamma}_{11} = \dfrac{\underline{b}_1}{\underline{a}_1}$ of a device characterised by its S-parameters and terminated by an output load with $\underline{\Gamma}_L$ using Mason's rule

Let us prove the latter claim by applying the Mason's rule. As an exercise, we calculate $\underline{\Gamma}_{11}$. Referring to Fig. 3.8, the input characteristics of a device are determined by its S-parameter and the load impedance represented by $\underline{\Gamma}_L$. Please note that $\underline{\Gamma}_S$ is not considered since there is no source termination when evaluating $\underline{\Gamma}_{11}$. Directed from $\underline{a}_1$ to $\underline{b}_1$, there are the forward paths $\underline{P}_1 = \underline{S}_{11}$ and $\underline{P}_2 = \underline{S}_{21}\underline{\Gamma}_L\underline{S}_{12}$. We can identify $\underline{S}_{22}\underline{\Gamma}_L$ as the only loop yielding $\underline{\Delta} = 1 - \underline{S}_{22}\underline{\Gamma}_L$. $\underline{S}_{22}\underline{\Gamma}_L$ touches $\underline{P}_2$ but not $\underline{P}_1$. Hence, we get $\underline{\Delta}_2 = 1$ and $\underline{\Delta}_1 = 1 - \underline{S}_{22}\underline{\Gamma}_L$. Now we can calculate the input reflection coefficient by means of Eq. (3.13):

$$\underline{\Gamma}_{11} = \frac{\underline{b}_1}{\underline{a}_1} = \frac{\underline{S}_{11}\left(1 - \underline{S}_{22}\underline{\Gamma}_L\right) + \underline{S}_{21}\underline{\Gamma}_L\underline{S}_{12}}{1 - \underline{S}_{22}\underline{\Gamma}_L} = \underline{S}_{11} + \frac{\underline{S}_{12}\underline{S}_{21}\underline{\Gamma}_L}{1 - \underline{S}_{22}\underline{\Gamma}_L}. \tag{3.16}$$

On the condition that $\underline{Z}_2 = Z_0$ we obtain $\underline{\Gamma}_L = 0$, and consequently $\underline{\Gamma}_{11} = \underline{S}_{11}$ as we have set out to prove. Analogously, we can determine the output characteristics with respect to the input termination $\underline{\Gamma}_S$. For $\underline{\Gamma}_{22}$ we can deduce

$$\underline{\Gamma}_{22} = \frac{\underline{b}_2}{\underline{a}_2} = \underline{S}_{22} + \frac{\underline{S}_{21}\underline{S}_{12}\underline{\Gamma}_S}{1 - \underline{S}_{22}\underline{\Gamma}_S}. \tag{3.17}$$

Given that $\underline{Z}_S = Z_0$, we obtain $\underline{\Gamma}_{22} = \underline{S}_{22}$ as expected.

3.5 Parameter Conversion

Sometimes, the representation of device properties by means of impedance- (Z), admittance- (Y) or chain- (ABCD) parameters can be useful for circuit analyses. However, recall that at high frequencies, direct measurement of these parameters is challenging. As show in Table 3.2, these parameters can be converted from S-parameters, which can be accurately measured up to very high frequencies.

Table 3.2. Relationships between S, Z, Y and ABCD parameters

	S	Z	Y	ABCD
S_{11}	S_{11}	$\dfrac{(Z_{11}-Z_0)(Z_{22}+Z_0)-Z_{12}Z_{21}}{\Delta Z}$	$\dfrac{(Y_0-Y_{11})(Y_0+Y_{22})+Y_{12}Y_{21}}{\Delta Y}$	$\dfrac{A+B/Z_0-CZ_0-D}{A+B/Z_0+CZ_0+D}$
S_{12}	S_{12}	$\dfrac{2Z_{12}Z_0}{\Delta Z}$	$\dfrac{-2Y_{12}Y_0}{\Delta Y}$	$\dfrac{2(AD-BC)}{A+B/Z_0+CZ_0+D}$
S_{21}	S_{21}	$\dfrac{2Z_{21}Z_0}{\Delta Z}$	$\dfrac{-2Y_{21}Y_0}{\Delta Y}$	$\dfrac{2}{A+B/Z_0+CZ_0+D}$
S_{22}	S_{22}	$\dfrac{(Z_{11}+Z_0)(Z_{22}-Z_0)-Z_{12}Z_{21}}{\Delta Z}$	$\dfrac{(Y_0+Y_{11})(Y_0-Y_{22})+Y_{12}Y_{21}}{\Delta Y}$	$\dfrac{-A+B/Z_0-CZ_0+D}{A+B/Z_0+CZ_0+D}$
Z_{11}	$Z_0\dfrac{(1+S_{11})(1-S_{22})+S_{12}S_{21}}{(1-S_{11})(1-S_{22})-S_{12}S_{21}}$	Z_{11}	$\dfrac{Y_{22}}{\det Y}$	$\dfrac{A}{C}$
Z_{12}	$Z_0\dfrac{2S_{12}}{(1-S_{11})(1-S_{22})-S_{12}S_{21}}$	Z_{12}	$\dfrac{-Y_{12}}{\det Y}$	$\dfrac{AD-BC}{C}$
Z_{21}	$Z_0\dfrac{2S_{21}}{(1-S_{11})(1-S_{22})-S_{12}S_{21}}$	Z_{21}	$\dfrac{-Y_{21}}{\det Y}$	$\dfrac{1}{C}$
Z_{22}	$Z_0\dfrac{(1-S_{11})(1+S_{22})+S_{12}S_{21}}{(1-S_{11})(1-S_{22})-S_{12}S_{21}}$	Z_{22}	$\dfrac{Y_{11}}{\det Y}$	$\dfrac{D}{C}$

Table 3.2. Continued

	(S)	(Z)	(Y)	(ABCD)
$\underline{Y}_{11}$	$Y_0 \dfrac{(1-\underline{S}_{11})(1+\underline{S}_{22})+\underline{S}_{12}\underline{S}_{21}}{(1+\underline{S}_{11})(1+\underline{S}_{22})-\underline{S}_{12}\underline{S}_{21}}$	$\dfrac{\underline{Z}_{22}}{\det \underline{Z}}$	$\underline{Y}_{11}$	$\dfrac{\underline{D}}{\underline{B}}$
$\underline{Y}_{12}$	$Y_0 \dfrac{-2\underline{S}_{12}}{(1+\underline{S}_{11})(1+\underline{S}_{22})-\underline{S}_{12}\underline{S}_{21}}$	$\dfrac{-\underline{Z}_{12}}{\det \underline{Z}}$	$\underline{Y}_{12}$	$\dfrac{\underline{BC}-\underline{AD}}{\underline{B}}$
$\underline{Y}_{21}$	$Y_0 \dfrac{-2\underline{S}_{21}}{(1+\underline{S}_{11})(1+\underline{S}_{22})-\underline{S}_{12}\underline{S}_{21}}$	$\dfrac{-\underline{Z}_{21}}{\det \underline{Z}}$	$\underline{Y}_{21}$	$\dfrac{-1}{\underline{B}}$
$\underline{Y}_{22}$	$Y_0 \dfrac{(1+\underline{S}_{11})(1-\underline{S}_{22})+\underline{S}_{12}\underline{S}_{21}}{(1+\underline{S}_{11})(1+\underline{S}_{22})-\underline{S}_{12}\underline{S}_{21}}$	$\dfrac{\underline{Z}_{11}}{\det \underline{Z}}$	$\underline{Y}_{22}$	$\dfrac{\underline{A}}{\underline{B}}$
$\underline{A}$	$\dfrac{(1+\underline{S}_{11})(1-\underline{S}_{22})+\underline{S}_{12}\underline{S}_{21}}{2\underline{S}_{21}}$	$\dfrac{\underline{Z}_{11}}{\underline{Z}_{21}}$	$\dfrac{-\underline{Y}_{22}}{\underline{Y}_{21}}$	$\underline{A}$
$\underline{B}$	$Z_0 \dfrac{(1+\underline{S}_{11})(1+\underline{S}_{22})-\underline{S}_{12}\underline{S}_{21}}{2\underline{S}_{21}}$	$\dfrac{\det \underline{Z}}{\underline{Z}_{21}}$	$\dfrac{-1}{\underline{Y}_{21}}$	$\underline{B}$
$\underline{C}$	$Y_0 \dfrac{(1-\underline{S}_{11})(1-\underline{S}_{22})-\underline{S}_{12}\underline{S}_{21}}{2\underline{S}_{21}}$	$\dfrac{1}{\underline{Z}_{21}}$	$\dfrac{-\det \underline{Y}}{\underline{Y}_{21}}$	$\underline{C}$
$\underline{D}$	$\dfrac{(1-\underline{S}_{11})(1+\underline{S}_{22})+\underline{S}_{12}\underline{S}_{21}}{2\underline{S}_{21}}$	$\dfrac{\underline{Z}_{22}}{\underline{Z}_{21}}$	$\dfrac{-\underline{Y}_{11}}{\underline{Y}_{21}}$	$\underline{D}$

$\det \underline{Z} = \underline{Z}_{11}\underline{Z}_{22} - \underline{Z}_{12}\underline{Z}_{21}$, $\det \underline{Y} = \underline{Y}_{11}\underline{Y}_{22} - \underline{Y}_{12}\underline{Y}_{21}$, $\Delta \underline{Y} = (\underline{Y}_{11}+Y_0)(\underline{Y}_{22}+Y_0) - \underline{Y}_{12}\underline{Y}_{21}$, $\Delta \underline{Z} = (\underline{Z}_{11}+Z_0)(\underline{Z}_{22}+Z_0) - \underline{Z}_{12}\underline{Z}_{21}$,

$Y_0 = \dfrac{1}{Z_0}$.

3.6 Impedance Transformation

Impedance matching is required for many purposes. In the previous sections we have learned that impedance mismatches generate signal reflections at the junction between devices. In the interest of transferring the maximum power to a circuit, undesired signal losses arising from these reflections have to be avoided as much as possible. We have to keep in mind that instability may occur due to uncontrolled reflections. Impedance matching is important for circuits requiring high gain. LNAs belong to this group of circuits.

Moreover, the large signal current and voltage characteristics and the corresponding maximum output power of circuits are strongly dependent on the load impedance seen by the circuit. Since the output termination impedance is fixed in most cases, the termination impedance has to be transferred to an optimum load impedance. Such a transformation is important for power amplifiers.

We can conclude that high performance circuits mandate impedance matching. Within a narrow frequency band and assuming ideal elements, loss-less impedance matching is possible by using reactive elements such as inductors, capacitors or transmission lines.

Transmission lines, which have distributed properties are used for operating frequencies above 10 GHz and into the 100-GHz range [Bur01]. Below 10 GHz, their size and losses are too large for ICs. In this case, lumped elements such as inductors and capacitors are frequently employed. The use of lumped elements is mainly limited by the inductors. Below 1 GHz they are too large, whereas above 50 GHz the self-resonance frequencies, layout constraints and process yields are determining factors. For frequencies lower than 1 GHz, purely active circuit topologies are preferred, since the area consumed for lumped elements or transmission lines is too large and thus too costly. Moreover, due to the high performance conditions of the transistors at low frequencies, reactive elements are not necessarily required. The frequency ranges of the individual approaches are illustrated in Fig. 3.9.

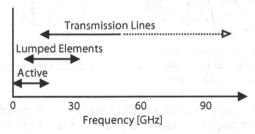

Fig. 3.9. Typical frequency range of transmission lines, lumped elements and purely active elements

Impedance matching and reduction of reflections can also easily be achieved by adding resistors in series or in parallel. However, in this case significant resistive losses would result, which can be unacceptable for low power applications, where the signal power has to be preserved as much as possible. In addition, matching resistors generate considerable noise.

3.6.1 Lumped LC Elements

Theoretically, every complex impedance can be matched to every complex impedance by means of at least one of the LC network configurations illustrated in Fig. 3.10. Since the reactive matching is based on a resonance, matching can only be performed within a limited bandwidth. Reactive series elements move the imaginary part of the impedance on circles with constant real part in the $\underline{z}$-plane, whereas reactive shunt elements move the imaginary part of the impedance on circles with constant real part in the $\underline{y}$-plane. Consequently, it is advantageous to apply the $\underline{z}$-plane for series networks and the $\underline{y}$-plane for parallel networks. Recall that we can change between the $\underline{z}$- and $\underline{y}$-planes by mirroring with respect to the origin.

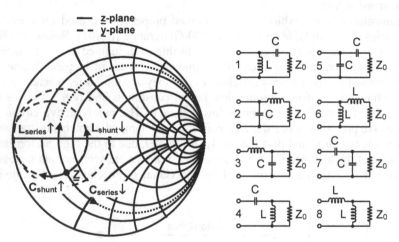

Fig. 3.10. Reactive impedance matching using lumped elements. At one frequency, every impedance can be matched to Z_0 using one lumped element network with 2 elements. Every network has a certain matching range

In practice, the benefit of impedance matching has to be traded off with the associated losses of the employed passive elements. This is especially the case for circuits in silicon technologies. Due to their low substrate resistivities, they introduce relatively high losses for the passive devices and interconnects. In this context, proper choice of the matching network is important. Given the flexibility of design parameters such as the gate width of the transistors and by allowing a reasonable return loss, different networks can be used for matching. In any case, the number of lossy inductors has to be minimised. Consequently, networks such as configurations 6 and 8, where two inductors are required should be avoided if possible.

Furthermore, the matching network with the lowest inductance values and the highest Q-factor at the target frequency is preferred. However, for applications with high bandwidth, the introduction of resistive elements and losses can be advantageous, since it broadens the bandwidth of the impedance matching resonance.

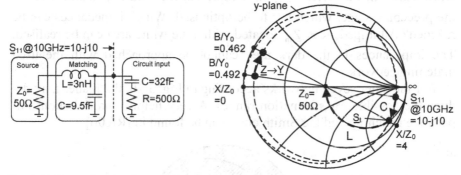

Fig. 3.11. Example of an impedance transformation of a complex device or circuit impedance represented by $\underline{S}_{11}$ to the reference impedance Z_0. Values and characteristics are in accordance with the detailed Smith chart shown in Fig. 3.4

The graphical impedance matching by means of the Smith chart can be very convenient. Generally, transformation of an arbitrary device impedance represented by $\underline{S}_{11}$ to the reference impedance $Z_0 = 50\ \Omega$ can be performed in two steps:

1. Transformation of $\underline{S}_{11}$ to either the $\underline{z}$ - or $\underline{y}$ -unity circle by means of a series or shunt reactive element, respectively. Note that on these unity-circles, the real part of the impedance stays constant and equals Z_0.
2. Compensation of the imaginary part along the $\underline{z}$ - or $\underline{y}$ -lines with either a series or shunt reactive element.

We treat a quantitative example. According to Fig. 3.11, at 10 GHz, the input of a FET transistor represented by a series connection of a capacitance and a resistance should be matched to Z_0. We can use a shunt capacitor C to transform $\underline{S}_{11}$ onto the z-unity circle. Since this shunt capacitance is in parallel to the impedance represented by $\underline{S}_{11}$, we have to perform the transformation in the y-plane. Mirroring of $\underline{S}_{11}$ and the transit point $\underline{S}_t$ with respect to Z_0 leads to

$$j(0.492 - 0.462)\cdot Y_0 = j\omega C, \qquad (3.18)$$

yielding C=9.5 fF. The transfer from $\underline{S}_t$ to Z_0 is accomplished in the $\underline{z}$ -plane with a series inductance. With

$$j(4-0)\cdot Z_0 = j\omega L, \qquad (3.19)$$

we obtain a value of L=3 nH. With these two elements the transistor input and the source are matched. No signal reflection will occur at the input.

LC networks are only capable of transforming an impedance to a limited impedance range. For illustration, we want to match now the reference impedance Z_0 to an arbitrary input impedance $\underline{Z}_{in}$ by means of the network shown in Fig. 3.12a. $\underline{Z}_{in}$ may represent the load impedance seen by the preceding stage, which has to be optimised. Which impedances can be reached? All impedances $\underline{Z}_{in}$ located within the white area can be realised. The impedances in the dark area cannot be approached with the used matching network.

For the other seven networks according to Fig. 3.10, the reader may deduce the possible transformation ranges. A very helpful tutorial about impedance matching and the Smith chart can be found in [Rfc06].

a)   b)

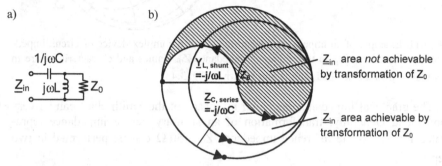

Fig. 3.12a,b. Example of the possible transformation from Z_0 to $\underline{Z}_{in}$: **a** LC matching network; **b** transformation range in the Smith chart

3.6.2 LC Equivalent Transmission Lines

For frequencies above 20 GHz, lumped element matching may become difficult due to the limited self-resonance frequencies of the inductors.

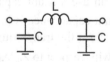

Fig. 3.13. Idealised lumped element equivalent circuit for short transmission lines with $1 < \dfrac{\lambda}{4}$, resistive losses are neglected

The lumped elements can be replaced by capacitive and inductive transmission lines. For short lengths ($1 < \frac{\lambda}{4}$) and by neglecting the resistive losses, the transmission lines can be modelled using the simple equivalent circuit as shown in Fig. 3.13 with

$$L = \frac{Z_w}{2\pi f} \sin\left(\frac{2\pi l}{\lambda}\right) \tag{3.20}$$

and

$$C = \frac{1}{2\pi f Z_w} \tan\left(\frac{\pi l}{\lambda}\right). \tag{3.21}$$

The characteristic impedance of the transmission line Z_w has a strong impact on the reactive characteristics. Referring to Sect. 6.1, inductive transmission lines require small widths (corresponding to high Z_w), whereas capacitive transmission lines have large widths (corresponding to low Z_w).

3.6.3 Reference Line

As illustrated in Fig. 3.14, an impedance or admittance can be rotated by a line with reference impedance. The reflection coefficient at position x=0 can be calculated by

$$\underline{\Gamma}(0) = \underline{\Gamma}(1) \cdot e^{-2j\beta l} \cdot e^{-2\alpha l}, \tag{3.22}$$

where $\underline{r}(1)$ is the reflection coefficient at x=l, β represents the phase factor and α is the attenuation factor. A line with a length of $\lambda/4$ and $\lambda/2$ transforms the impedances/admittances by an angle of 180° and 360°, respectively. Given that the line is lossless, the magnitude of the reflection coefficient stays the same. If the line exhibits losses, the magnitude of the reflection coefficient decreases since a part of the reflected energy is absorbed in the resistive parasitics.

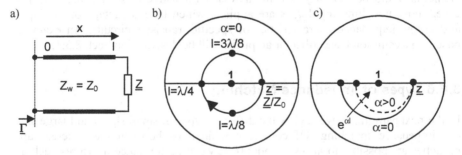

Fig. 3.14a–c. Transformation by transmission line: **a** line with length l; **b** lossless transformation; **c** consideration of loss with attention factor α

3.6.4 Quarter Wave Transmission Line

Every resistive impedance Z_{R1} can be transformed to every resistive impedance Z_{R2} by using a series transmission line with length of $\lambda/4$ and characteristic impedance of [Bäc94]

$$Z_T = \sqrt{Z_{R1} \cdot Z_{R2}} \ . \tag{3.23}$$

An example is illustrated in Fig. 3.15. In practice, the impedances are not purely resistive. Therefore, the imaginary part has to be resonated by a transmission-line or a lumped element.

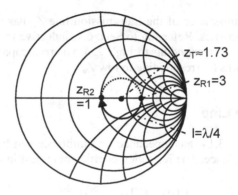

Fig. 3.15. Transformation of impedance Z_{R1}=150 Ω to Z_{R2}=Z_0=50 Ω by mean of $\lambda/4$ line with impedance of 86.6 Ω

3.6.5 Active Transformation

For frequencies below approximately 0.5 GHz, transmission lines and lumped elements would be too large and the associated resistive losses too high. In this case, active matching topologies are applied, which take advantage of the input and output impedance characteristics of specific transistor circuits. An example based on a common base/collector amplifier will be described in Sect. 8.3.1.

3.6.6 Types of Impedance Matching

In the next paragraphs we review the different types of small signal and large signal impedance matching. Of course, combinations between the different approaches are possible to allow a tradeoff between performance measures such as gain, noise, output power and linearity.

3.6.6.1 Small Signal Gain and Return Loss Matching

To minimise the power reflected at an impedance interface, the reflection factor must be made as low as possible. This kind of matching is frequently applied to maximise the transferred small signal power and correspondingly, the small signal gain. Moreover, undesired reflections are avoided, which may cause oscillation if fed back and amplified.

One typical example of such a matching has been given in Sect. 3.6.1 and Fig. 3.11. By means of an LC network the input impedance of a transistor has been matched to the resistive reference impedance Z_0. In this case, the impedances at the interfaces between the source and the matching network (refer to point S in Fig. 3.16) must be equal. Expressed by means of their equivalent reflection coefficients we can write

$$\underline{\Gamma}_s = \underline{S}_{11m} .\tag{3.24}$$

where $\underline{\Gamma}_s$ and $\underline{S}_{11m}$ denote the reflection coefficient of the source and the transformed transistor impedance, respectively. The representation in terms of reflection coefficients is advantageous for the use in the Smith chart.

However, which conditions have to hold if we consider the interface between the complex device input and the matching network (refer to point D in Fig. 3.16)? In order to compensate the reactive power and to maximise the real part of the power, the impedance must be conjugate complex leading to

$$\underline{\Gamma}_{sm} = \underline{S}_{11}^* .\tag{3.25}$$

The last equation is a more general expression for the impedance matching, which holds also for complex impedances. Equation (3.24) can be deduced from Eq. (3.25). In the literature it is also referred to as small signal power matching.

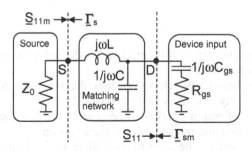

Fig. 3.16. Impedance transformation evaluated at interface between source and matching network, and device input and matching network

3.6.6.2 Noise Matching

Conjugate matching between the device and the transformed source impedance as required for maximum small signal gain does not necessarily result in minimum noise figure. As treated in Sect. 4.3.6 and shown in Fig. 8.4b, the associated optimum source impedances are not equal. This is attributed to the fact that the impedance for minimum noise depends on the impedances allowing maximum compensation of the correlated noise sources, whereas gain matching is achieved at conjugate impedance transitions. The latter demand does not depend on any noise sources. To be derived in Sect. 4.3.6, the noise figure depends only on the source impedance and not on the device impedance. Consequently, the termination is transformed to the desired impedance finally seen by the device. Fortunately, in practice, we observe that a good compromise between noise matching and gain matching can be reached without significant degradation of the other measure.

3.6.6.3 Large Signal Power Matching

The properties under small signal and large signal conditions are different. In the latter case, the choice of the load resistance is limited by the large signal boundaries in the IV curves and the nonlinearities of the used transistors. Corresponding details concerned with power amplifiers can be found in Sect. 9.2.8. The characteristics of the load at a broad frequency range have to be considered since in addition to the fundamental frequency, we have to take into account harmonic frequencies. Consequently, large signal power matching and small signal gain matching are not necessarily equal.

A typical design procedure is as follows: In a first step, the circuit is matched considering the simple small signal case. This can either be done by analytical calculations using a Smith chart or by CAD simulations. In a second step, the matching is improved by means of large signal CAD simulations.

3.7 Tutorials

1. What are the practical problems concerned with the RF measurement of the Z, Y or ABCD parameters?
2. What is the solution? Which value does the reference impedance for S-parameters typically has? Why is it reasonable?
3. What does the reflection coefficient represent? What is the relationship between the reflection and transmission of signal power?
4. Derive the equation for the reflection coefficient with respect to the port impedances.
5. Draw a simplified version of a Smith chart and identify circles/arcs/lines with constant impedance values. What is the benefit of the Smith chart in terms of the possible range of values?

6. The Smith chart is based on the Möbius transformation. What are the main characteristics of this transformation? How can we convert data from the Z- to Y-parameters?

7. Which representation do we use in the Smith chart for shunt elements and which for series elements?

8. An impedance is transformed by a lossless line with length of $\lambda/4$ and $\lambda/8$. What is the corresponding phase rotation in the Smith chart? What happens if the line has significant loss? Why?

9. Match a load impedance $\underline{Z}_L = 200\,\Omega - j100\,\Omega$ to a resistive source impedance $Z_S = 50\,\Omega$ by using one inductor and one capacitor in the Smith chart.

10. Match a load impedance $\underline{Z}_L = 200\,\Omega + j25\,\Omega$ to a source impedance $Z_S = 50\,\Omega$. The matching network should work for high frequencies above 20 GHz. What limits the maximum frequency of operation? Hint: use a $\lambda/4$ transformer and a lumped element with high SRF.

11. Describe the set of S-parameters for a 3-port.

12. What are the similarities and differences between the reflection coefficient and the S-parameters?

13. Why are we not allowed to include $\underline{S}_{21}$ and $\underline{S}_{12}$ in a Smith chart? What about $\underline{S}_{11}$ and $\underline{S}_{22}$?

14. What does it mean if $\underline{S}_{11}$ or $\underline{S}_{22} > 1$?

15. Illustrate the frequency dependent input and output S-parameters of a typical FET and bipolar transistor in the Smith chart. Comment on the differences.

16. At f=5.25 GHz, the S-parameters of a MOSFET transistor are given by $\underline{S}_{11} = 0.933\angle -66°$, $\underline{S}_{12} = 0.085\angle 49°$, $\underline{S}_{21} = 2.39\angle 131°$ and $\underline{S}_{22} = 0.539\angle -38.3°$. Perform return loss matching at the input and the output of the device to 50 Ω according to Fig. 3.17 using the Smith chart.
    a) Discuss all possible matching networks using two lumped elements (C, L). How many networks are possible?
    b) Which one is the superior matching network with lowest complexity? Try to reuse the bias elements and assume that they include shunt inductors.
    c) Calculate the values for all required elements. Draw a conclusion concerning possible limitations in terms of the SRF assuming that the total parasitic capacitance of an inductor is around 30 fF/nH.

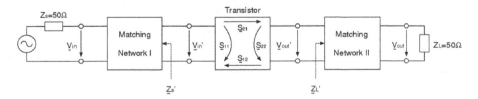

Fig. 3.17. Matching of a transistor represented by S-parameters

References

[Bäc94]   W. Bächtold, Lineare Elemente der Höchstfrequenztechnik, VDF Verlag, 1994.

[Bur01]   J. N. Burghartz, K. T. Ng, N. P. Pham, B. Rejaei, P. Sarro, "Integrated RF passive components -discrete vs. distributed", IEEE Device Research Conference, 2001, pp. 113–114.

[Kur65]   K. Kurokawa, "Power waves and the scattering matrix", IEEE Transactions on Microwave Theory and Techniques, Vol. 13, pp. 194–202, March 1965.

[Mas56]   S. J. Mason, "Feedback theory: further properties of signal flow graphs", Proceedings of the I.R.E., Vol. 44, No. 7, pp. 920–926, July, 1956.

[Rfc06]   www.rfcafe.com/references/electrical/smith.htm

[Smi44]   P.H. Smith, "An improved transmission line calculator", Electronics, Vol. 17, p. 130, Jan. 1944.

[Sss06]   www.sss-mag.com/pdf/smithchart.pdf

4 RF Basics

4.1 Stability

Stability is a very important issue for the circuit design. If a circuit is unstable, it may become useless since major properties including bandwidth, gain, noise, linearity, output power, DC power consumption and impedance matching can be significantly degraded. Moreover, proper control is not possible any more.

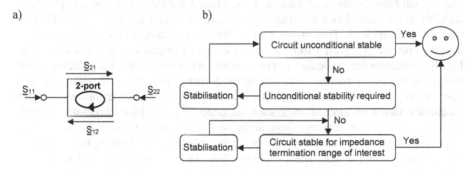

Fig. 4.1a,b. Stability: **a** signal flow; **b** design strategy

Stability analyses have to be performed to investigate if and under which condition the circuit is unstable. If the circuit is unstable, we have to apply stabilisation methods to guarantee stability within the impedance termination range of interest.

4.1.1 Analysis

Stability can be investigated by reviewing the S-parameters. For a 2-port, the following parameters are relevant:

1. $\underline{S}_{21}$: Gain of device directed from input to output.

    Matched active devices typically provide $|\underline{S}_{21}| > 1$, whereas generally $|\underline{S}_{21}| < 1$ for passive devices.

2. $\underline{S}_{12}$: How much energy is transferred from device output back to input.

    Usually for both active and passive devices we observe that $|\underline{S}_{12}| < 1$.

3. $\underline{S}_{11}$: How much energy is reflected at the input.

4. $\underline{S}_{22}$: How much energy is reflected at the output.

Let us review the signal flow on basis of Fig. 4.1a. A part of the signal is amplified by the active device, is reflected at the output, fed back, reflected at the input, amplified again and fed back, etc. Depending on the amplitude and phase of the S-parameters, the energy may increase for each signal cycle. In this case, the circuit becomes instable. It is obvious that the tendency for instability can be decreased by lowering the magnitude of the S-parameters, corresponding to good impedance matching, low gain and reverse transmission. However, high gain is required for many circuits. Thus, high gain and stability are conflicting goals. The reverse transmission is determined by parasitic feedback capacitance of the applied device.

The stability is strongly influenced by the impedance terminations. We can distinguish between unconditional and conditional stability. The first one provides stability at any impedance terminations, whereas the latter one depends on the impedance terminations. That means that in the event of conditional stability, stability can only be achieved within a certain range of impedance terminations. It is highly recommended to design circuits, which are stable at any frequencies. Even when the frequency range where instability occurs is not allocated within the frequency range of interest, the instability can degrade the circuit performance in the frequency range of interest by means of undesired DC bias variations, energy losses, and nonlinear mixing. Depending on the application, conditional stability with respect to the terminations may be acceptable, e.g. for challenging and scientific oriented applications, where highest performance is important. For most consumer products, unconditional stability is required since the customer may apply any terminations. A strategy for stability considerations is illustrated in Fig. 4.1b. Concerning the implementation of stabilisation techniques and networks into circuits, the reader is referred to Section 7.2.

By employing the signal flow theory and the S-parameters, a helpful and frequently employed criterion can be derived, which shows if a circuit is unconditional stable or not. It is the Rollet's factor given by [Rol62, Gon97]

$$K = \frac{1 - |\underline{S}_{11}|^2 - |\underline{S}_{22}|^2 + |\det \underline{S}|^2}{2|\underline{S}_{12}\underline{S}_{21}|} \qquad (4.1)$$

with

$$\det \underline{S} = \underline{S}_{11}\underline{S}_{22} - \underline{S}_{12}\underline{S}_{21}. \qquad (4.2)$$

Since the S-parameters are a function of frequency, the K-factor is frequency dependent. This is illustrated in Fig. 4.2 for a typical non-resonated transistor. Due to the low-pass effect, the losses rise towards high frequencies thereby decreasing $|\underline{S}_{21}|$. Thus, the tendency for instability decreases towards high frequencies. The transistor is unconditionally stable if $K \geq 1$, $|\underline{S}_{11}| < 1$ and $|\underline{S}_{22}| < 1$. That means that there is no input or output termination possible, which could make the transistor unstable. In practice, most commercial circuits are designed to meet these conditions.

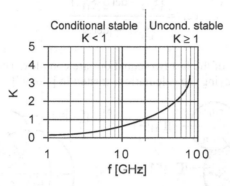

Fig. 4.2. Rollet's (K)-factor vs frequency

If K<1, the circuit is only conditional stable within a certain range of impedance terminations. To allow corresponding stabilisation strategies, it is important to know at which impedance terminations the circuit is stable and at which not. This investigation can be performed by means of the stability circle analysis. The analysis is based on the fact that boundary conditions for stability are given by $|\underline{\Gamma}_{11}| - 1$ and $|\underline{\Gamma}_{22}| - 1$. Recall that $\underline{\Gamma}_{11}$ and $\underline{\Gamma}_{22}$ represent $\underline{S}_{11}$ and $\underline{S}_{22}$ at any output and input termination, respectively. If $|\underline{\Gamma}_{11}|$ or/and $|\underline{\Gamma}_{22}| > 1$, the reflected signals would be stronger than the incoming signals clearly indicating instability. Beforehand, we investigate the impact of the output termination represented by $\underline{\Gamma}_L$ on the input reflection factor $\underline{\Gamma}_{11}$. Referring to Eq. (3.16), the input reflection coefficient can be calculated by

$$\left|\underline{\Gamma}_{11}\right| = \left|\underline{S}_{11} + \frac{\underline{S}_{21}\underline{S}_{12}\underline{\Gamma}_L}{1 - \underline{S}_{22}\underline{\Gamma}_L}\right|. \tag{4.3}$$

The values of $\underline{\Gamma}_L$ and subsequently $\underline{Z}_L$ can be drawn in a Smith chart indicating the status of stability together with Eq. (4.3). If $\underline{Z}_L$ is passive, it follows that $|\underline{\Gamma}_L| \leq 1$. Hence, $\underline{Z}_L$ does not exceed the Smith chart. Equation (4.3) can be transformed into a circle with radius of

$$ra_L = \frac{\left|S_{12}S_{21}\right|}{\left|S_{22}\right|^2 - \left|\det \underline{S}\right|^2} \tag{4.4}$$

describing the circle where $\left|\underline{\Gamma}_{11}\right| = 1$. The distance from the centre of the Z_L circle to the centre of the Smith chart is determined by

$$ce_L = \frac{\left(S_{22} - \det \underline{S} \cdot \underline{S}_{11}^{\;*}\right)^{*}}{\left|S_{22}\right|^2 - \left|\det \underline{S}\right|^2}. \tag{4.5}$$

Indeed, the derivations of the last two equations are complex. The reader may verify the relations by studying the calculations treated in [Gon97].

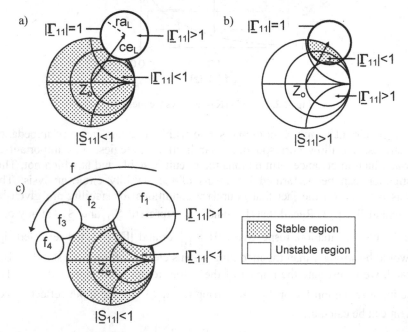

Fig. 4.3a–c. Input stability circles showing the boundary given by the input reflection coefficient $\underline{\Gamma}_{11}$ mapped on the output plane of $\underline{Z}_L$ with $\left|\underline{\Gamma}_L\right| < 1$: **a** $\left|\underline{S}_{11}\right| < 1 \Rightarrow \left|\underline{\Gamma}_{11}\right| < 1 \Rightarrow Z_0$ stable impedance; **b** $\left|\underline{S}_{11}\right| > 1 \Rightarrow \left|\underline{\Gamma}_{11}\right| > 1 \Rightarrow Z_0$ unstable impedance; **c** frequency dependency for $\left|\underline{S}_{11}\right| < 1$

The circle defines the boundary between stability and instability given by $\left|\underline{\Gamma}_{11}\right| = 1$. However, which area is stable? Is it the one inside the circle or the one outside the circle? This could be investigated by considering the phases in Eq. (4.5). A smart and fast approach is the feasibility check based on $Z_L=Z_0$ being the resistive reference impedance. At Z_0, $|\Gamma_L|=0$ and consequently $|\Gamma_{11}|=|S_{11}|$. Needless to say that this impedance equals the centre of the Smith chart. Now we can make the following conclusions: if $|S_{11}|<1$ as given by typical non-resonated transistors, we get $|\Gamma_{11}|<1$ yielding a stable point for $Z_L=Z_0$. Thus, according to Fig. 4.3a, the area outside the circle and inside the Smith chart is stable since it includes $Z_L=Z_0$. Circuits such as feedback amplifiers with high gain or oscillators could exhibit $|S_{11}|>1$. Consequently, we obtain $|\Gamma_{11}|>1$ yielding an unstable point for $Z_L=Z_0$. Thus, as illustrated in Fig. 4.3b, the intersection not including $Z_L=Z_0$ is stable. Figure 4.3c sketches the stability circuits for different frequencies demonstrating that the tendency for instability decreases with raised frequency. The presented characteristics hold for non-resonated transistors, where the gain falls with increased frequency.

In the discussions up to now, we have investigated the input reflection with respect to the output load. Analogically, we have to take into account the output reflection as a function of the input load. For stability, the requirements for both the input and output termination have to be fulfilled.

4.2 Power Gain

Power gain is an important performance measure for analogue circuits such as amplifiers. In this section, we will discuss the most important definitions for the power gain of 2-ports.

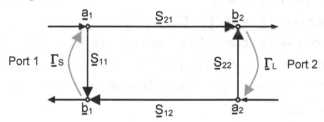

Fig. 4.4. Signal flow diagram of 2-port for calculation of transducer gain

4.2.1 Transducer Power Gain

The transducer gain is defined as

$$G_T = \frac{P_L}{P_S} \tag{4.6}$$

where P_L denotes the real power in the load and P_S is the available source power. To compute the transducer gain, we employ the signal flow diagram according to

Fig. 4.4 and the Mason rule. With a source voltage and impedance of $\underline{V}_S$ and $\underline{Z}_S$, the available source power is given by

$$P_S = \frac{\left|\underline{V}_S\right|^2}{4\,\mathrm{Re}\{\underline{Z}_S\}}. \tag{4.7}$$

The power transfer can be computed by means of the associated wave parameters, which according to Section 3.4, depend on the square-root of the power. Considering the reflections at the impedance interface, the real power transferred into the device is given by

$$\left|\underline{a}_1\right|^2 = P_S\left(1 - \left|\underline{\Gamma}_S\right|^2\right). \tag{4.8}$$

The real power transferred to the load amounts

$$P_L = \left|\underline{b}_2\right|^2\left(1 - \left|\underline{\Gamma}_L\right|^2\right). \tag{4.9}$$

Hence, the transducer gain yields

$$G_T = \left|\frac{\underline{b}_2}{\underline{a}_1}\right|^2\left(1 - \left|\underline{\Gamma}_S\right|^2\right)\left(1 - \left|\underline{\Gamma}_L\right|^2\right). \tag{4.10}$$

On basis of the Mason rule represented by Eq. (3.13), the $\dfrac{\underline{b}_2}{\underline{a}_1}$ relation can be calculated. We can identify the following terms:

One single forward path: $\underline{P}_1 = \underline{S}_{21}$

First order loop terms: $\underline{S}_{21}\underline{\Gamma}_L\underline{S}_{12}\underline{\Gamma}_S$, $\underline{\Gamma}_S\underline{S}_{11}$, $\underline{\Gamma}_L\underline{S}_{22}$

One second order loop product of 2 non-touching loops: $\underline{\Gamma}_S\underline{S}_{11}\underline{\Gamma}_L\underline{S}_{22}$

Overall determinant: $\underline{\Delta} = 1 - \underline{S}_{21}\underline{\Gamma}_L\underline{S}_{12}\underline{\Gamma}_S - \underline{\Gamma}_S\underline{S}_{11} - \underline{\Gamma}_L\underline{S}_{22} + \underline{\Gamma}_S\underline{S}_{11}\underline{\Gamma}_L\underline{S}_{22}$

Specific determinant remaining if $\underline{P}_1$ and all touching paths are removed: $\underline{\Delta}_1 = 1$

Consequently, with

$$\frac{\underline{b}_2}{\underline{a}_1} = \frac{\underline{S}_{21}}{\underline{\Delta}} = \frac{\underline{S}_{21}}{1 - \underline{S}_{21}\underline{\Gamma}_L\underline{S}_{12}\underline{\Gamma}_S - \underline{\Gamma}_S\underline{S}_{11} - \underline{\Gamma}_L\underline{S}_{22} + \underline{\Gamma}_S\underline{S}_{11}\underline{\Gamma}_L\underline{S}_{22}}, \tag{4.11}$$

we finally obtain

$$G_T = \frac{\left|\underline{S}_{21}\right|^2\left(1 - \left|\underline{\Gamma}_S\right|^2\right)\left(1 - \left|\underline{\Gamma}_L\right|^2\right)}{\left|1 - \underline{S}_{11}\underline{\Gamma}_S - \underline{S}_{22}\underline{\Gamma}_L + \det\underline{S}\cdot\underline{\Gamma}_S\underline{\Gamma}_L\right|^2} \tag{4.12}$$

where $\det\underline{S}$ has been defined in Eq. (4.2).

4.2.2 Unilateral Transducer Power Gain

The forward gain of active devices is much larger than the reverse gain, therefore $\underline{S}_{21} \gg \underline{S}_{12}$. For simplifications suppose that $\underline{S}_{12} = 0$, which means for example that the capacitive feedback in a common source amplifier is neglected. A reverse gain of zero could also be achieved by external compensation. This procedure is referred to as unilateralisation. Based on Eq. (4.12), the unilateral transducer gain is given by

$$G_{TU} = \frac{|\underline{S}_{21}|^2 \left(1-|\underline{\Gamma}_S|^2\right)\left(1-|\underline{\Gamma}_L|^2\right)}{\left|\left(1-\underline{S}_{11}\underline{\Gamma}_S\right)\left(1-\underline{S}_{22}\underline{\Gamma}_L\right)\right|^2} = G_S \cdot G_0 \cdot G_L \qquad (4.13)$$

with

$$G_S = \frac{1-|\underline{\Gamma}_S|^2}{|1-\underline{S}_{11}\underline{\Gamma}_S|^2} \quad \text{and} \quad G_L = \frac{1-|\underline{\Gamma}_L|^2}{|1-\underline{S}_{22}\underline{\Gamma}_L|^2} \quad \text{representing the impact of the input and}$$

output matching on the gain, and $G_0 = |\underline{S}_{21}|^2$ defining the intrinsic gain of the device. The factors G_S and G_L show that the gain magnitudes with constant values are located on circles and dependent on the values of $|\underline{\Gamma}_S|$ and $|\underline{\Gamma}_L|$, respectively.

These circles illustrate how sensitive the gain is versus impedance variations. Corresponding insights are valuable for the design of broadband amplifiers and amplifiers with variable load. Moreover, the designer can easily evaluate how the gain matching can be traded off with the matching for optimum noise, which will be discussed in Sect. 4.3.6. In Fig. 4.5, an example of the gain contribution of the input matching of a typical FET transistor is illustrated in a Smith chart. Please bring back to mind that circles in linear presentation remain circles in the Smith chart. We observe that minimum noise would lower the gain by 1.5 dB.

Maximum unilateral gain is achieved at conjugate match with $\underline{\Gamma}_S = \underline{\Gamma}_{S,opt} = \underline{S}_{11}^*$ and $\underline{\Gamma}_L = \underline{\Gamma}_{L,opt} = \underline{S}_{22}^*$ yielding

$$G_{TU\,max} = \frac{|\underline{S}_{21}|^2}{\left(1-|\underline{S}_{11}|^2\right)\left(1-|\underline{S}_{22}|^2\right)} \qquad (4.14)$$

with

$$G_S = \frac{1}{1-|\underline{S}_{11}^2|} \quad \text{and} \quad G_L = \frac{1}{1-|\underline{S}_{22}^2|}. \text{ In our example the input impedance transfor-}$$

mation yields $G_S = 2$ or equivalent 3 dB gain. For $\underline{S}_{11} = \underline{S}_{22} = \pm 1$ arising at purely reactive device impedances we would obtain a theoretical gain of infinity. In RF practice, purely reactive impedances do not exist. Furthermore, the complex impedance transformation would introduce significant losses. Given that

$\underline{S}_{11} = \underline{S}_{22} = 0$, which is the case if device and reference impedances are equal, we get $G_L = 1$, $G_S = 1$, and $G_{TU\,max} = |\underline{S}_{21}|^2$. Recall that the unit of S_{21} depends on the square-root of the power. In logarithmic scale the values of S_{21} and G_{TUmax} are equal, e.g. 10 dB S_{21} and 10 dB G_{TUmax} are the same.

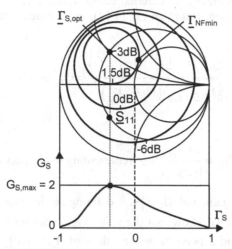

Fig. 4.5. Illustration of gain circles of typical input matching with $\underline{S}_{11} = 0.5 - 0.5j$. Moreover, the source reflection coefficient for minimum noise ($\underline{\Gamma}_{NF\,min}$) is included

4.2.3 Maximum Available Gain

In the equations for G_{TU}, we have assumed that $\underline{S}_{12} = 0$. However, according to Sect. 4.1.1 we know that S_{12} is very important for the stability. If $\underline{S}_{12} = 0$, a circuit could not become unstable since the K-factor would approach infinity. Let's go back to Eq. (4.12), where we have calculated the transducer power gain G_T without neglecting $\underline{S}_{12}$. Especially at matching for maximum gain, circuits can be unstable and G_{TUmax} may not be reached. The maximum G_T achieved at optimum source and load terminations, and under stable conditions is called maximum available gain (MAG). We obtain a relatively complex equation for the MAG, which can be simplified by incorporating the expression for the K-factor resulting in [Gonz97]

$$\text{MAG} = G_{T\,max}\big|_{K\geq 1} = \left|\frac{\underline{S}_{21}}{\underline{S}_{12}}\right|\left(K - \sqrt{K^2 - 1}\right). \qquad (4.15)$$

The MAG is defined for $K \geq 1$, which means that the circuit is stable at any source or load impedances. At high frequencies, the gain of non-resonated transistors is small enough to guarantee $K \geq 1$. The frequency characteristics of transistors are mainly determined by the dominant input pole acting as a first order low pass filter. Thus, the MAG falls with around 20 dB per decade or equivalent 6 dB per octave.

4.2.4 Maximum Stable Gain

At low frequencies, the gain of transistors can be very high and consequently the circuit can be unstable with K<1 demanding for stabilisation. At optimum stable terminations, a maximum stable gain of [Gon97]

$$MSG = G_{T\,max}\big|_{K<1} = \left|\frac{\underline{S}_{21}}{\underline{S}_{12}}\right| \tag{4.16}$$

can be achieved. At a certain frequency, the gain is moderate enough to generate a K-factor of unity. In this case, the MSG converts into the MAG. Due to the stabilisation, the gain increase toward low frequencies is lower than for the MAG. Typically, we find a slope in the order of 3 dB per octave.

4.2.5 Summary

The characteristics of the key power gain measures are summarised in Table 4.1 and are illustrated in Fig. 4.6 vs frequency.

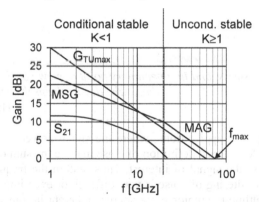

Fig. 4.6. Comparison of key gain measures vs frequency for a typical RF transistor

Table 4.1. Summary of power gain definitions

Name	Abbreviation and equation	Relations
Forward transmission parameter	$\underline{S}_{21} = \dfrac{\underline{b}_2}{\underline{a}_1}$	Equals $\sqrt{G_{TU\,max}}$ if device is matched to reference impedance
Transducer gain	$G_T = \dfrac{\left\|\underline{S}_{21}\right\|^2 \left(1-\left\|\underline{\Gamma}_S\right\|^2\right)\left(1-\left\|\underline{\Gamma}_L\right\|^2\right)}{\left\|1-\underline{S}_{11}\underline{\Gamma}_S - \underline{S}_{22}\underline{\Gamma}_L + \det\underline{S}\cdot\underline{\Gamma}_S\underline{\Gamma}_L\right\|^2}$	Considers impact of terminations
Unilateral Transducer gain	$G_{TU} = \dfrac{\left\|\underline{S}_{21}\right\|^2 \left(1-\left\|\underline{\Gamma}_S\right\|^2\right)\left(1-\left\|\underline{\Gamma}_L\right\|^2\right)}{\left\|\left(1-\underline{S}_{11}\underline{\Gamma}_S\right)\left(1-\underline{S}_{22}\underline{\Gamma}_L\right)\right\|^2}$	G_{TU} is special case of G_T for $\underline{S}_{12} = 0$
Maximum unilateral transducer gain	$G_{TU\,max} = \dfrac{\left\|\underline{S}_{21}\right\|^2}{\left(1-\left\|\underline{S}_{11}\right\|^2\right)\left(1-\left\|\underline{S}_{22}\right\|^2\right)}$	Maximum G_{TU} with $\underline{\Gamma}_S = \underline{S}_{11}^*$ and $\underline{\Gamma}_L = \underline{S}_{22}^*$
Maximum available power gain	$MAG = \dfrac{\left\|\underline{S}_{21}\right\|}{\left\|\underline{S}_{12}\right\|}\left(K - \sqrt{K^2 - 1}\right)$	G_T with optimum $\underline{\Gamma}_L$ and $\underline{\Gamma}_S$, defined for $K > 1$
Maximum stable power gain	$MSG = \dfrac{\left\|\underline{S}_{21}\right\|}{\left\|\underline{S}_{12}\right\|}$	G_T with optimum $\underline{\Gamma}_L$ and $\underline{\Gamma}_S$, valid for $K \leq 1$, $MSG = MAG$ for $K = 1$

4.3 Noise

Science is a harsh mistress, and in a pecuniary point of view poorly rewarding to those who devote themselves to her service.
Humphry Davy

Especially, for low power applications, where the signal power is limited, noise plays an important role. A loose definition of noise may be the sum of any random signals uncorrelated to the signal of interest. Thus, within the frequency band of the desired signal, the filtering of noise is very difficult or even impossible. In the following, we will outline the major noise sources inherent in transistors. Further discussions regarding the impact of noise on circuits and systems will follow in Sect. 5.4.7 and Chap. 8.

4.3.1 Thermal Noise

Every ohmic resistor[1] exhibits thermal noise caused by charge carriers generating a randomly varied current. As expected due to its thermal origin, the noise raises with increased temperature T. Thermal noise has broadband characteristics with equal distribution versus frequency. Thus, by intuition, it may be clear that the noise contribution increases proportionally with raised bandwidth Δf. Experiments have revealed that the maximum noise power, which can be transferred from a resistor to a load, is given by

$$P_n = kT\Delta f , \qquad (4.17)$$

where k denotes the Boltzmann constant. Since the associated noise sources have random characteristics, it is not possible to define a noise source at specific time. Attributed to the fact that thermal noise follows a Gaussian amplitude distribution, we are able to apply statistic measures such as the mean square values.

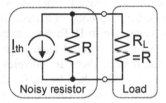

Fig. 4.7. Illustration of current noise source $\underline{I}_{th}^2$ associated with thermal noise of resistor R

Maximum noise power transfer is achieved at conjugate match where the source and load resistors are equal. Hence, according to Fig. 4.7, we get $P_n = \dfrac{1}{4} \cdot \overline{\underline{I}_{th}^2} \cdot R$ leading to a mean square noise current density of

$$\overline{\underline{I}_{th}^2} = 4 \cdot kT \cdot \frac{1}{R} \cdot \Delta f , \qquad (4.18)$$

According to Ohm's Law, a current is always associated with a voltage. Thus, optionally, the noise of the resistor can also be modelled by means of a voltage noise source. Thermal noise does not only appear in passive devices but also in active devices such as FETs and BJTs. Bias-depend characteristics have to be taken into account. Corresponding discussions will follow in the transistor sections.

It is self-explanatory that the impact of the thermal noise can be reduced by cooling. This is done for systems requiring highest performances, e.g. for space

---

[1] Please note that there are also non-ohmic resistors, which are operands rather than physical resistors. An example is the small signal conductance $\dfrac{1}{r_{ds}} = \dfrac{dI_d}{dV_{ds}}$, which does not generate any noise.

applications where only low signal power can be received due the high transmit distances from space to earth or vice versa.

4.3.2 Shot Noise

Shot noise is generated if current flows through a potential barrier such as a pn-junction illustrated in Fig. 5.3. The current is generated by electrons jumping from one terminal to the other. Since the duration an electron spends in one terminal changes arbitrarily, noise is generated. The square root of the shot noise current can be described by

$$\overline{I_{sh}^2} = 2 \cdot q \cdot I_{dc} \cdot \Delta f \qquad (4.19)$$

with q as the electron charge. As expected, the shot noise increases with DC current I_{dc} since it determines the number of available carriers. Thus, shot noise can be minimised by reducing the DC current. Unfortunately, as a considerable drawback, a reduced DC current may decrease the maximum possible gain and large signal properties of transistors. Consequently, a tradeoff has to be found.

Shot noise plays an important role in BJTs since they consist of pn-junctions. This holds especially for the forward biased base emitter junction. Usually, the shot noise of FETs is very small since there are no relevant pn-junctions, and the current flowing through them is weaker than for BJTs. However, the aggressively scaling of MOSFETs can introduce a significant current from the gate to the channel, which may generate shot noise. In contradiction to thermal noise, shot noise does not occur in an ideal resistor.

4.3.3 Flicker Noise

Flicker noise is caused by random charging and de-charging of traps, which are impure spots in devices. Most of the impurities are located at the surfaces of semiconductors, since it is challenging to fabricate clean transitions between materials. In this context, we have to review the possible types of current flow with respect to the surfaces. According to Fig. 4.8, we can distinguish between lateral and vertical current flow. The probability that the carriers interact with traps is much higher in the event of lateral current flow. The reason is that the lateral current flows for a relatively long time along the surface traps, whereas the vertical current flows only during a relatively short duration through the surface traps. Since the FET is based on lateral current flow, whereas the bipolar transistor is mainly based on vertical flow, the bipolar transistor usually exhibits lower flicker noise than the FET.

The probability of the trap charging increases with the potential travelling time of a carrier. Thus, the flicker noise current is anti-proportional to the frequency. This is the reason why the flicker noise is also referred to as 1/f noise. With $\overline{I_{fl0}^2}$

being the noise contribution given at a reference frequency f_r, the flicker noise current density can be estimated by

$$\overline{I_{fl}^2} = \overline{I_{fl0}^2} \cdot \frac{f_r}{f} .$$ (4.20)

Obviously, the flicker noise increases towards low frequencies. Consequently, at low frequencies, the flicker noise is usually much higher than the noise of other mechanisms. A useful measure for the $1/f$ performance of transistors is the $1/f$ corner frequency, which is the frequency, where the contribution of flicker noise equals the one of the thermal noise. The lower the $1/f$ corner frequency, the better the flicker noise performance. The corresponding $1/f$ corner frequencies are typically in the order of 0.1–1 MHz for FETs and 1–10 kHz for bipolar transistors. In this context, refer to Figs. 5.43b and 5.20b.

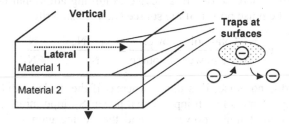

Fig. 4.8. Illustration of vertical (compare bipolar transistor) and lateral (compare FET) current flow with respect to surface traps

Flicker noise has a significant impact on the noise properties of two kinds of circuits and systems. First, for systems operating at very low frequencies such as the direct conversion or low-IF receivers, where the RF is directly mixed to DC or a low frequency, respectively. Second, for all large signal circuits exhibiting nonlinearities, generating desired or non-desired mixing products. These mixing products can include DC or low frequency components with strong flicker noise. Thus, although operating at RF frequencies, the signal performance of such circuits can be significantly degraded by the flicker noise originally allocated at low frequencies. Examples where we observe this mixing effect are oscillators and mixers. For small signal circuits such as RF low noise amplifier, the flicker noise is negligible.

Now let us compare the ability for the up-conversion of the flicker noise. With this regard, bipolar transistors have inferior prerequisites since they generally produce stronger nonlinearities than FETs. The reason is that the output current of BJTs follows an exponential function versus the input voltage, whereas the FET exhibits a quadratic dependency and therefore a lower level of nonlinearities. This may compensate the advantages of the bipolar transistor in terms of lower flicker noise. However, for a properly designed and biased BJT circuit, the advantage of the low $1/f$ corner frequency still outperforms the disadvantage concerning stronger nonlinearities.

4.3.4 Noise Figure

Reality does not take care about our wishes.
Bertrand Russel

In the wishes of a communication engineer, there is no noise. However, in real applications, undesired signals have to be taken into account. An important figure of merit for the signal quality in receivers is the signal to noise power ratio given by

$$SNR = \frac{Signal}{Noise} = \frac{S}{N}. \tag{4.21}$$

This factor expresses that, at same signal quality, the signal level can be decreased if noise is lowered, or noise can be increases if the signal is raised. To determine the impact of a device within a system, we define F (Noise Factor). This factor determines the noise increase due to a device exhibiting additional noise power N_a with respect to the noise power of the source N_1. We can write

$$F = \frac{Total\ noise\ power}{Noise\ power\ of\ source} = \frac{N_1 + N_a}{N_1} = 1 + \frac{N_a}{N_1}. \tag{4.22}$$

If the device adds no noise, F is unity, whereas if the device exhibits noise, F is larger than unity. A model with input signal power S_1, input noise power N_1, output signal S_2, output noise power N_2, and the device gain G is illustrated in Fig. 4.9. By definition, the noise is considered at the input of the devices. We will verify later that this definition makes sense.

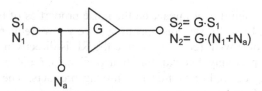

Fig. 4.9. Model of noisy device

It is worth knowing that the noise factor also expresses the relation between the SNR at the input and the output as shown in the following relation:

$$F = \frac{SNR_1}{SNR_2} = \frac{\dfrac{S_1}{N_1}}{\dfrac{S_2}{N_2}} = \frac{S_1 G(N_1 + N_a)}{N_1 G S_1} = 1 + \frac{N_a}{N_1}. \tag{4.23}$$

In other words, F describes the degradation of the SNR due to a device. According to Eq. (4.17), the thermal noise power is proportional to T. Thus, we can write for Eq. (4.23):

$$F = 1 + \frac{T_a}{T_1}, \tag{4.24}$$

where T_a and T_1 are the equivalent temperatures for the device and the source, respectively. Frequently, the noise factor is expressed in logarithmic representation with unit dB. In this case, we refer to the NF (Noise Figure) given by

$$NF = 10 \log_{10} F. \tag{4.25}$$

4.3.5 Noise Figure of Cascaded Stages

In systems such as the receiver shown in Fig. 2.2, several stages with different noise figures and gains are cascaded. In the following, we investigate the behaviour of two cascaded stages as illustrated in Fig. 4.10a. To compute the system noise figure, we apply the equivalent circuit depicted in Fig. 4.10b. By setting the overall noise figures at the outputs equal, we obtain

$$F_1 N_1 G_1 G_2 + (F_2 - 1) N_1 G_2 = F_{21} N_1 G_1 G_2, \tag{4.26}$$

resulting in

$$F_{21} = F_1 + \frac{F_2}{G_1} \frac{1}{}. \tag{4.27}$$

Similarly, we can show for n-stages that the overall noise figure is given by

$$F_{n1} = F_1 + \frac{F_2 - 1}{G_1} + \frac{F_3 - 1}{G_1 G_2} + \ldots + \frac{F_n - 1}{G_1 G_2 \ldots G_{n-1}} \tag{4.28}$$

or equivalent

$$F_{n1} = F_1 + \sum_{k=2}^{n} \frac{F_k - 1}{\prod_{l=1}^{k-1} G_l}. \tag{4.29}$$

According to its explorer, this famous equation is also called formula of Friis [Fri44].

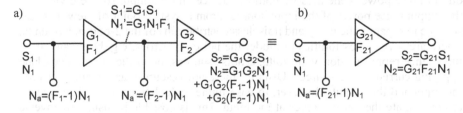

Fig. 4.10a,b. Two cascaded stages: **a** separated; **b** equivalent circuit

From the equation of Friis, we can gain the important insight that the noise of the first stage is dominant if the gain of the first stage is high. Thus, the noise figure of a receiver is particularly determined by the noise figure of the LNA located in front of the receiver. Given that the LNA gain is high enough, the noise contribution of the following stages such as the mixer or the IF amplifier is small. Let us make a quantitative review for a typical scenario according to Fig. 4.11. It becomes clear that the noise figure of the switch and the LNA directly adds to the total noise, whereas the noise figures of the mixer and the filter are not significant since the components are located after the LNA.

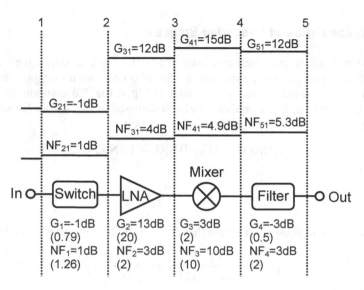

Fig. 4.11. Noise figure and gain of cascaded stages in typical receiver, numbers in brackets denote values in non-logarithmic scale

4.3.6 Minimum Noise Figure and Optimum Source Impedance

We calculate now the minimum noise figure of a 2-port and identify the major parameters impacting the noise. According to Eq. (4.22), F is defined as the total output noise power generated by both the device and the input source divided by the output noise power of the input source. From a system point of view it is reasonable to separate the noisy and noiseless contributions of the 2-port in two units. There are several possibilities to model the noise properties, e.g. by means of an impedance representation with voltage noise sources, or in the admittance plane applying current noise sources. Usually, these noise sources are required at both the input and the output. However, to ease further calculations, it is advantageous to concentrate the noise sources at the input. This is possible by using the inverse chain (ABCD) representation, which is based on descriptions of the input current and voltage. Consequently, referring to Fig. 4.12, the noise of the 2-port can be considered by adding a current noise source $\underline{I}_n$ and a voltage noise source $\underline{V}_n$ prior to the 2-port now being noiseless. The noise of the source is represented by the noise current $\underline{i}_s$ and the admittance $\underline{Y}_s = G_s + jB_s$.

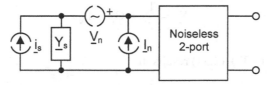

Fig. 4.12. Equivalent circuit for calculation of the noise figure of a noisy 2-port

Bearing the associated noise powers in mind, we can first consider noise optimisation. If the voltage noise is dominant, a high source impedance will minimise the noise, whereas if the current noise is dominant, a low source impedance will be advantageous. In practice, there are contributions from both noise sources. Thus, there must be an optimum noise source admittance

$$\underline{Y}_{opt} = G_{opt} + jB_{opt},\qquad(4.30)$$

where minimum noise figure F_{min} is achieved. Now, we calculate F_{min}, $\underline{Y}_{opt}$, and F at any source admittance $\underline{Y}_s$. Referring to Eq. (4.22) and by applying superposition of noise powers, we get

$$F = \frac{P_{is} + P_{in,en}}{P_{is}},\qquad(4.31)$$

with P_{is} the noise power due to the input noise source, and $P_{in,en}$ the noise power caused by $\underline{I}_n$ and $\underline{V}_n$. Since P_{is} and $P_{in,en}$ are based on different origins, no correlation between these noise power has to be expected. However, $\underline{I}_n$ and $\underline{V}_n$ are correlated, which can be considered by the mean square value of the sum of both. Since all powers have to be related on an arbitrary output impedance, which is cancelled out in the following equation, we can directly write the noise figure with respect to the mean square values of the currents yielding

$$F = \frac{\overline{I_s^2} + \overline{\left|\underline{I}_n + \underline{Y}_s\underline{V}_n\right|^2}}{\overline{I_s^2}}.\qquad(4.32)$$

Only a part of the noise power with respect to $\underline{I}_n$ and $\underline{V}_n$ is correlated. Thus, it makes sense to divide the total $\underline{I}_n$ into a correlated and an uncorrelated part denoted by $\underline{I}_c$ and $\underline{I}_u$, respectively:

$$\underline{I}_n = \underline{I}_c + \underline{I}_u . \tag{4.33}$$

To represent the correlation between $\underline{I}_n$ and $\underline{V}_n$, we can introduce the correlation admittance

$$\underline{Y}_c = \frac{\underline{I}_c}{\underline{V}_n} = G_c + jB_c . \tag{4.34}$$

Combining Eqs. (4.32)–(4.34) results in

$$F = 1 + \frac{\overline{I_u^2} + \left|\underline{Y}_c + \underline{Y}_s\right|^2 \overline{V_n^2}}{\overline{I_s^2}} . \tag{4.35}$$

Equation (4.35) contains three noise sources associated with $\underline{I}_s$, $\underline{I}_u$ and $\underline{V}_n$, which are determined by their equivalent thermal resistances and conductances:

$$G_s = \frac{\overline{I_s^2}}{4kT\Delta f} , \tag{4.36}$$

$$G_u = \frac{\overline{I_u^2}}{4kT\Delta f} \tag{4.37}$$

and

$$R_n = \frac{\overline{V_n^2}}{4kT\Delta f} . \tag{4.38}$$

With Eqs. (4.35)–(4.38), we can now write

$$F = 1 + \frac{G_u + \left|\underline{Y}_c + \underline{Y}_s\right|^2 R_n}{G_s} = 1 + \frac{G_u + \left[\left(G_c + G_s\right)^2 + \left(B_c + B_s\right)^2\right] \cdot R_n}{G_s} . \tag{4.39}$$

The optimum source admittance for minimum noise can be obtained by derivation of Eq. (4.39) resulting in

$$B_{opt} = -B_c , \tag{4.40}$$

meaning that the reactive part of the source impedance must be conjugate complex with respect to the correlated part of the reactive device impedance. Furthermore, for the real part of the source impedance the following identity holds for noise matching:

$$G_{opt} = \sqrt{\frac{G_u}{R_n} + G_c^2} \; . \tag{4.41}$$

The parameter G_{opt} for gain matching does not consider the term G_u/R_n. This imposes that at the same time, gain and noise matching are only possible if there is full correlation with $G_u = 0$, $B_c = B_s$ and $G_c = G_s$. The latter conditions are not feasible for transistors but for ideal resistors. Hence, for typical integrated circuits using transistors, ideal gain and noise matching is not possible simultaneously. With Eqs. (4.39)-(4.41), the minimum noise figure can be computed:

$$F_{min} = 1 + 2R_n \left[G_{opt} + G_c \right] = 1 + 2R_n \left[\sqrt{\frac{G_u}{R_n} + G_c^2} + G_c \right]. \tag{4.42}$$

Now we can modify Eq. (4.39) into

$$F = F_{min} + \frac{R_n}{G_s} \left[\left(G_s - G_{opt} \right)^2 + \left(B_s - B_{opt} \right)^2 \right]. \tag{4.43}$$

In Fig. 4.13, the three-dimensional characteristics of the latter relation are illustrated. The second term of the equation describes the increase of noise when the values of $\underline{Y}_s$ move away from $\underline{Y}_{opt}$. For specific $\underline{Y}_s$ located along concentric circles, we can identify noise figures with constant value, since the second term of the equation represents non-overlapping circles in the admittance plane as shown in Fig. 4.13a. In other words, the same value of F can be obtained for different combinations of G_s and B_s. The factor R_n/G_s determines how rapidly the noise figure degrades as $\underline{Y}_s$ moves away from $\underline{Y}_{opt}$. G_s is defined by the source, whereas R_n represents an important device characteristic. In Fig. 4.13b, it is illustrated that the lower R_n, the less the sensitivity of F on a non-optimum source impedance. According to Fig. 4.13c, we can find circles with constant F in the admittance plane. Since the Möbius transformation preserve circles, these circles can also be found in the Smith chart.

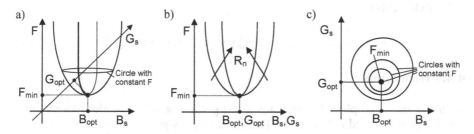

Fig. 4.13a–c. Noise factor: **a** three-dimensional representation vs complex source impedance $\underline{Y}_s = G_s + jB_s$; **b** impact of R_n on G_s or B_s; **c** circles with constant noise figure in admittance plane

We have seen that the optimum source impedance has a significant impact on the circuit noise properties. However, in practice, the source impedance is given by system constraints, e.g. by the output impedance of a preceding stage, which often exhibits an impedance of 50 Ω. Consequently, this impedance has to be transformed to the optimum impedance. Thus, the source impedance and the impedance seen by the device are usually different.

4.4 Linearity and Compression

The integral of the intelligence over the number of people on a planet is constant and the population is growing.

Unknown

Due to the high importance for wireless communication, the subject of linearity and compression has been intensively treated in the literature [Raz98, Maa03]. As a loose definition, linearity may be described as the signal level at desired frequency with respect to the signal level associated with undesired frequencies.

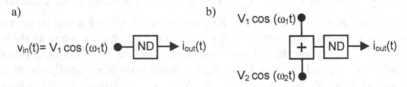

Fig. 4.14a,b. Feeding of a nonlinear device (ND): **a** single-tone input for generation of harmonics; **b** two-tone input for generation of intermodulation

Memory-less, time-invariant nonlinear devices can be represented by Taylor representations. An example is the transistor acting as a current source, where the output current $i_{out}(t)$ depends on the input voltage $v_{in}(t)$ and device specific amplitude coefficients c_n:

$$i_{out}(t) = c_0 + c_1 v_{in}(t) + c_2 v_{in}^2(t) + c_3 v_{in}^3(t) ... + c_n v_{in}^n(t) . \qquad (4.44)$$

4.4.1 Harmonics

A simple model employed to investigate the generation of harmonics in a nonlinear device is depicted in Fig. 4.14a. Feeding an input signal with voltage

$$v_{in}(t) = V_1 \cos(\omega_1 t) \qquad (4.45)$$

and single frequency yields an output current according to Eq. (4.44) of

$$i_{out}(t) = c_0 + c_1 \cdot V_1 \cos(\omega_1 t) + c_2 \cdot V_1^2 \cos^2(\omega_1 t) + c_3 \cdot V_1^3 \cos^3(\omega_1 t) \dots \quad (4.46)$$

As simplification, in the following, we consider only the contributions up to the 3rd order. Using trigonometric identities [Bro91] results in

$$i_{out}(t) =$$

$$\underbrace{c_0 + \frac{c_2 \cdot V_1^2}{2}}_{DC} + \underbrace{\left(c_1 \cdot V_1 + \frac{3 \cdot c_3 \cdot V_1^3}{4} \right) \cos(\omega_1 t)}_{\text{1st harmonic (fundamental)}} + \underbrace{\frac{c_2 \cdot V_1^2}{2} \cos(2\omega_1 t)}_{\text{2nd harmonic}} + \underbrace{\frac{c_3 \cdot V_1^3}{4} \cos(3\omega_1 t)}_{\text{3rd harmonic}}$$

$$(4.47)$$

demonstrating that in addition to the original fundamental frequency, the harmonic frequency components $\omega_n = n \cdot \omega$ are generated.

4.4.2 1 dB Compression Point

In the linear case, the fundamental frequency is determined by the term $c_1 V_1$. According to Eq. (4.47), in the event of nonlinearities, the amplitude of the fundamental is impacted by the terms $c_1 V_1$ and $\frac{3c_3}{4} V_1^3$. Depending on the device properties and the phases between c_1 and c_3, the amplitude of the fundamental component can either increase or decrease. Usually, the gain is decreasing with increasing input power due to saturation effects. We will focus on the latter case, where the terms $A_1 = c_1 V_1$ and $A_3 = \frac{3c_3}{4} V_1^3$ have opposite signs. A reasonable measure is the 1 dB compression point, where the gain of the fundamental frequency is decreased by 1 dB in comparison to the linear case. Figure 4.15 sketches the corresponding extrapolation. We can calculate the 1 dB compression point by means of

$$\frac{A_1 - A_3}{A_1} = 1 - \frac{3c_3}{4c_1} V_{1dB}^2 = 10^{-1/20}, \quad (4.48)$$

yielding

$$V_{1dB} \approx \sqrt{0.145 \left| \frac{c_1}{c_3} \right|}. \quad (4.49)$$

The stronger the compression, the higher the signal power transferred to the harmonics with n>1. The compression point can be referenced to both the input and output, which are related by means of the gain. Usually, for power applications, one refers to the output as a measure for the maximum output power, whereas the input representation is frequently employed for low power circuits indicating the maximum power, which can be fed.

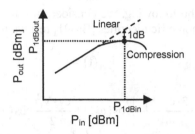

Fig. 4.15. Extrapolation of the 1 dB compression point

4.4.3 Intermodulation

Multiple signal frequencies are required to generate intermodulation. We consider the least complex case with two signals exhibiting different frequencies ω_1 and ω_2, and amplitudes V_1 and V_2. Suppose that the nonlinear circuit acts like a signal adder as illustrated in Fig. 4.14b yielding an input signal of

$$v_{in}(t) = V_1 \cos(\omega_1 t) + V_2 \cos(\omega_2 t) \tag{4.50}$$

By putting Eq. (4.50) in Eq. (4.44) we obtain

$$i_{out}(t) = c_0 + c_1 \left[V_1 \cos(\omega_1 t) + V_2 \cos(\omega_2 t) \right] +$$

$$c_2 \left[\begin{array}{l} \dfrac{V_1^2}{2}\left(1+\cos(2\omega_1 t)\right)+\dfrac{V_2^2}{2}\left(1+\cos(2\omega_2 t)\right) \\ +V_1 V_2 \left(\cos(\omega_1+\omega_2)t+\cos(\omega_1-\omega_2)t\right) \end{array} \right] +$$

$$c_3 \left[\begin{array}{l} \left(\dfrac{3}{4}V_1^3+\dfrac{3}{2}V_1 V_2^2\right)\cos(\omega_1 t)+\left(\dfrac{3}{4}V_2^3+\dfrac{3}{2}V_2 V_1^2\right)\cos(\omega_2 t) \\[2mm] +\dfrac{3V_1^2 V_2}{4}\left(\cos(2\omega_1+\omega_2)t\right)+\dfrac{3V_1^2 V_2}{4}\left(\cos(2\omega_1-\omega_2)t\right) \\[2mm] +\dfrac{3V_2^2 V_1}{4}\left(\cos(2\omega_2+\omega_1)t\right)+\dfrac{3V_2^2 V_1}{4}\left(\cos(2\omega_2-\omega_1)t\right) \\[2mm] +\left(\dfrac{3}{4}V_1^3\right)\cos(3\omega_1 t)+\left(\dfrac{3}{4}V_2^3\right)\cos(3\omega_2 t) \end{array} \right] \tag{4.51}$$

$$+...$$

The latter equation clearly demonstrates that, in addition to the fundamental frequencies and harmonics, intermodulation products are generated with frequencies of

$$n \cdot \omega_1 \pm m \cdot \omega_2 , \qquad (4.52)$$

where n and m are integer values. The frequency components and the associated amplitude coefficients can be presented in a pyramid:

Amplitudes			Associated Frequencies		
c_0			0		
c_1		ω_1		ω_2	
c_2	$2\omega_1$		$\omega_1 \pm \omega_2$		$2\omega_2$
c_3	$3\omega_1$	$2\omega_1 \pm \omega_2$		$\omega_1 \pm 2\omega_2$	$3\omega_2$

. . . .

The corresponding spectrum is illustrated in Fig. 4.16 up to a limited order. The third order intermodulation products IM3 yield frequencies of

$$2\omega_1 - \omega_2 \qquad (4.53)$$

and

$$2\omega_2 - \omega_1 . \qquad (4.54)$$

Given that the fundamental frequencies ω_1 and ω_2 are nearby, the IM3 products are close to the fundamental frequencies. Due to the small distance to the desired frequency, it is very difficult to filter the IM3 products. By inspection of Eq. (4.51) and considering common values for the amplitude coefficients, we can observe that the magnitude of the intermodulation products decrease with raised values of n and m. Since the latter parameters exhibit small values for the IM3 components, the corresponding amplitudes are relatively large. Consequently, the linearity of circuits is mainly determined by the IM3 products, which can be represented by the third order intercept point IP3 serving as linearity measure.

The IP3 is defined as the intersection point between the extrapolated power curves of the fundamental frequency and the IM3 products as shown in Fig. 4.17. Obviously, in the linear case, the output power of the fundamental frequency increases as much as the input power yielding a slope of unity. For low input power, the amplitude of the IM3 product is very low since the generated nonlinearities are weak. However, since the amplitude is proportional to V^3, it exhibits a strong slope of 3 in the double logarithmic scale. Thus, the IP3 at the input can be calculated by using a simple geometric rule:

$$IIP3 = P_{in,f} + \frac{\Delta P}{2} . \qquad (4.55)$$

$$= P_{out,f} + \frac{P_{out,f} - P_{out,IM3}}{2} = P_{in,f} + \frac{P_{in,f} - P_{in,IM3}}{2} = \frac{3P_{in,f} - P_{in,IM3}}{2}. \quad (4.56)$$

The outputs and the inputs are related by the gain: $P_{out,f} = P_{in,f} + G$ and $P_{out,IM3} = P_{in,IM3} + G$. In this paragraph, we use the logarithmic dBm notation. The indices out and in are related to the output and input measures, f and IM3 denote the fundamental and third order components, respectively.

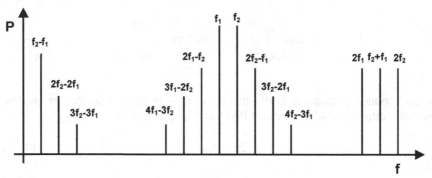

Fig. 4.16. Schematic spectrum showing two signals with frequencies f_1 and f_2 and their intermodulation products

Based on Eq. (4.51), we can estimate the IIP3. By definition, the amplitude of the two tones are equal with $V_1 = V_2$. For the magnitude of the third order terms we can identify $\frac{3}{4}c_3V_1^3$, whereas the amplitudes of the fundamental components are given by $c_1V_1 + \frac{3}{4}c_3V_1^3 + \frac{3}{2}c_3V_1^3$. Assuming that the first term in the last equation is dominant, we get

$$V_1 = V_{IIP3} = \sqrt{\frac{4}{3}\left|\frac{c_1}{c_3}\right|}. \quad (4.57)$$

With Eqs. (4.49) and (4.57) we can find a relation between V_{1dB} and V_{IIP3} of

$$\frac{V_{1dB}}{V_{IIP3}} \approx \sqrt{\frac{0.145}{\frac{4}{3}}} \approx 0.333 \quad (4.58)$$

revealing that the 1 dB compression point is approximately 9.6 dB lower than the third order intercept point. Although the calculation was based on coarse simplifications, this relation is quite well verified in practice.

a)   b)

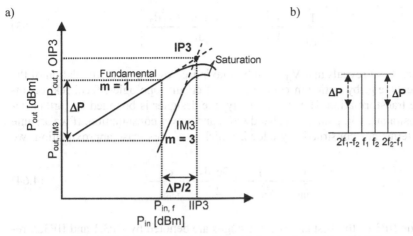

Fig. 4.17a,b. Extrapolation of the third order intercept point IP3

4.4.4 Cascaded Nonlinear Stages

As for the noise figure, the overall system performance has to be considered based on the linearities of the individual stages. Let's calculate the overall IIP3 of a system with two cascaded stages as illustrated in Fig. 4.18. We assume a filtering of the DC components and consider only the contributions up to the 3rd order leading to the following relations:

$$i_{out1}(t) = c_1 v_{in}(t) + c_2 v_{in}^2(t) + c_3 v_{in}^3(t) \tag{4.59}$$

and

$$i_{out2}(t) = d_1 i_{out1}(t) + d_2 i_{out1}^2(t) + d_3 i_{out1}^3(t) . \tag{4.60}$$

In turn we get

$$i_{out2}(t) = d_1 \left[c_1 v_{in}(t) + c_2 v_{in}^2(t) + c_3 v_{in}^3(t) \right]$$
$$+ d_2 \left[c_1 v_{in}(t) + c_2 v_{in}^2(t) + c_3 v_{in}^3(t) \right]^2 + d_3 \left[c_1 v_{in}(t) + c_2 v_{in}^2(t) + c_3 v_{in}^3(t) \right]^3 .$$

$$\tag{4.61}$$

The most important terms with respect to the IP3 are the first and third order terms. If we neglect all other contributions we get

$$i_{out2}(t) = \left[c_1 d_1 \right] v_{in}(t) + \left[c_3 d_1 + 2c_1 c_2 d_2 + c_1^3 d_3 \right] v_{in}^3(t) . \tag{4.62}$$

By comparison of the coefficients leading to Eq. (4.57) we can achieve an equivalent term of

$$\frac{1}{V_{IIP3}^2} = \frac{3}{4} \cdot \left[\frac{c_3 d_1 + 2c_1 c_2 d_2 + c_1^3 d_3}{c_1 d_1} \right].$$

(4.63)

Equation (4.63) reveals that V_{IIP3} can be maximised by proper manipulation of the coefficients, e.g. by optimum combination of circuit topologies, feedback, or impedance transformation. However, mostly, the designer is bounded by further design constraints, e.g. gain, noise, bandwidth and power consumption. If we assume that all terms are constructively added, which is a worst-case approximation, we obtain

$$\frac{1}{V_{IIP3}^2} = \frac{1}{V_{IIP3,1}^2} + \frac{3c_2 d_2}{2d_1} + \frac{c_1^2}{V_{IIP3,2}^2}.$$

(4.64)

where the IIP3 of the first and second stages are denoted by IIP3,1 and IIP3,2, respectively. In the majority of cases, the second term of Eq. (4.64) can be neglected because the product of $c_2 d_2$ is small since it is associated with frequencies far away from the fundamental frequency. Thus, it can be strongly attenuated by filters. Until now we have specifically discussed a nonlinear system with two stages. Similar calculations show that a general expression for nonlinear systems with n stages may be approximated by

$$\frac{1}{V_{IIP3}^2} \approx \frac{1}{V_{IIP3,1}^2} + \frac{c_1^2}{V_{IIP3,2}^2} + \frac{c_1^2 d_1^2}{V_{IIP3,3}^2} + \cdots$$

(4.65)

Equation (4.65) reveals that the stage having the lowest linearity has the dominant impact on the overall linearity. The gain of the preceding stages has a strong influence on the last stage. If the gain is high, the IIP3 of the last stage is degraded. On the other hand, low noise figure demands for a high gain of the input stages. Hence, a reasonable tradeoff has to be found.

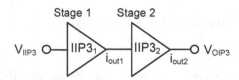

Fig. 4.18. Two cascaded nonlinear stages

4.5 Dynamic Range

Circuits such as amplifiers exhibit a lower and an upper signal limitation. The dynamic range represents the difference between these two limitations. The minimum detectable signal power is determined by the noise of the signal source and the required SNR. At high signal power, the dynamic range is limited by undesired signals such as intermodulation products or harmonics.

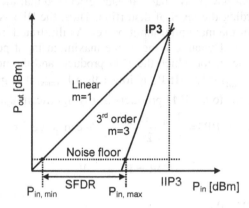

Fig. 4.19. Definition and extraction of the spurious free dynamic range (SFDR).

Referring back to Sect. 4.3.1, the resistive source impedance generates a noise power of

$$P_n = kT \cdot \Delta f . \tag{4.66}$$

At room temperature, we obtain a value $-174\text{dBm}/\text{Hz} \cdot \Delta f$. According to Sect. 4.3.4, it follows that

$$F = \frac{SNR_{in}}{SNR_{out}} = \frac{P_{in}/(kT \cdot \Delta f)}{SNR_{out}}, \tag{4.67}$$

where SNR_{in} and SNR_{out} denote the SNR at the input and the output, respectively. A minimum input power $P_{in,min}$ is required to achieve a certain SNR_{out}. Hence, from Eq. (4.67), we get

$$P_{in,min} = kT \cdot \Delta f \cdot F \cdot SNR_{out} . \tag{4.68}$$

Obviously, the noise can be lowered by minimising of Δf. However, at low Δf, the potential data rate is also low. We switch now to logarithmic notation:

$$P_{in,min}\big|_{dBm} = kT\big|_{dBm/Hz} + 10\log\Delta f + NF\big|_{dB} + SNR_{out}\big|_{dB} . \tag{4.69}$$

The part of $P_{in,min}$ excluding SNR_{out} is the total system noise. Referred to as the noise floor it amounts

$$P_{nf}\big|_{dBm} = kT\big|_{dBm/Hz} + 10\log\Delta f + NF\big|_{dB} = P_{in,min}\big|_{dBm} - SNR_{out}\big|_{dB} \,. \quad (4.70)$$

For sake of simplicity, in the following, we will not explicitly include the logarithmic notation.

The maximum possible input power can be defined in several ways, e.g. by means of the level of undesired intermodulation products or harmonics, which increase at rising input power. We have already discussed that the IIP3 is an important measure regarding the level of distortion. Thus, the IIP3 is applied as a reasonable measure for the maximum input power. As illustrated in Fig. 4.19, for the SFDR (Spurious Free Dynamic Range), the maximum input power $P_{in,max}$ is defined as the signal level for which the IM3 products and the noise floor become equal leading to $P_{in,IM3} \equiv P_{nf}$. If P_{in} is lower than $P_{in,max}$, it is guaranteed that the signal degradation due to the IM3 products is less significant than that of the noise floor. Eq. (4.56) yields $IIP3 = \dfrac{3P_{in} - P_{in,IM3}}{2}$. We can solve for

$$SFDR = P_{in,max} - P_{in,min} \,. \quad (4.71)$$

With $P_{in,max} = \dfrac{2IIP3 + P_{nf}}{3}$ we obtain

$$SFDR = \frac{2IIP3 + P_{nf}}{3} - \left(P_{nf} + SNR_{out}\right) = \frac{2\left(IIP3 - P_{nf}\right)}{3} - SNR_{out} \,. \quad (4.72)$$

Specified at a certain SNR_{out}, the SFDR describes the difference between the lowest detectable and the maximum possible input power on condition that the impact of the generated distortion is below that of the noise. Typical values of the SFDR of receivers are around 40–80 dB.

4.6 Tutorials

1. How is conditional and unconditional stability defined? What happens if a circuit is unstable? How can we investigate stability? Discuss the influence of the S-parameters on the stability factor.
2. What does the transducer gain describe? Which assumption is made for the unilateral transducer gain? What is the benefit? When is the unilateral transducer gain maximum? How can we separate the influence of the input, the device and the output? Which insights do we get? Can we always reach the maximum unilateral gain? Which other gain definitions are reasonable to describe a circuit?
3. How are the MSG and the MAG defined? Illustrate their characteristics versus frequency.
4. What are the main noise mechanisms? How are they determined? What are the physical sources?
5. Which transistor has lower flicker noise, the FET or the BJT? Why?
6. Does flicker noise have an impact at RF frequencies? Why? In this context, review the properties of FETs and BJTs. Which one has higher $1/f$ noise?
7. What does the SNR express?
8. Derive F and NF.
9. Calculate the overall NF and the gain of the system according to Fig. 4.20. Which NF is important, the one of the input or the one of the output stage? Which requirement has to be fulfilled to make the NF of the second amplifier unimportant? What happens if there are more stages?
10. Which parameters determine F_{min}? Comment on the role of the source impedance.
11. What does the 1 dB compression point represent? For which measure can it be used? What happens with the signal power when the saturation increases?
12. What does linearity or non-linearity mean?
13. Compared to the generation of harmonics, which additional requirement has to be met for intermodulation products? Consider the sources.
14. What are harmonics and intermodulation products? Which kind of devices generates them? How can we approximate a nonlinear device? Calculate the harmonics up to the second order.
15. Illustrate the spectrum of harmonics and intermodulation frequencies up to a reasonable order considering two input frequencies.
16. Comment qualitatively on the amplitude coefficient of harmonics and intermodulation products.
17. Two nearby frequencies are fed into a nonlinear device. What is the problem? How could we solve it?
18. What is the third order intercept point? How can we determine it graphically? Derive the corresponding equation. Why is it a good measure for linearity?
19. How are the 1 dB compression point and the third order intercept point related?

20. What happens if we cascade nonlinear circuits? What are the main impacts? Which stage dominates the properties?
21. What does the dynamic range describe and how is it defined?
22. Which kind of definition is used for the SFDR? Why is the definition reasonable? Calculate the SFDR for a receiver with NF=6 dB, IIP3=−10 dBm, Δf=100 kHz and SNR_{min}=10 dB.

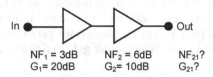

$$NF_1 = 3dB \qquad NF_2 = 6dB \qquad NF_{21}?$$
$$G_1 = 20dB \qquad G_2 = 10dB \qquad G_{21}?$$

Fig. 4.20. Cascaded system

References

[Bro91]   I. N. Bronstein, K. A. Semedjajew, Taschenbuch der Mathematik, Teubner Verlag Stuttgart, 1991.

[Fri44]   H. T. Friis, "Noise figure of radio receivers," Proceedings on IRE, Vol. 32, No. 7, pp. 419–422, July 1944.

[Gon97]   G. Gonzales, "Microwave transistor amplifiers", Prentice Hall, 1997.

[Maa03]   S. A. Maas, Nonlinear Microwave and RF Circuits, Artech House, 2003.

[Raz98]   B. Razavi, RF Microelectronics, Prentice Hall, Upper Saddle River, 1998.

[Rol62]   J. Rollet, "Stability and power gain invariants of linear two ports", IRE Transactions on Circuit Theory, CT-9, pp. 29–32, March 1962.

5 Transistors and Technologies

The transistor was probably the most important invention of the 20th century, and the story behind the invention is one of clashing egos and top-secret research.

Ira Flatow [Tra05]

5.1 Invention of Transistor and Integrated Circuit

Indeed, the history associated with the transistor development is very exciting and could serve as excellent basis for a screenplay. At the beginnings of the twentieth century the aim was to develop devices capable of switching and amplifying signals. One of the major tasks was the enlargement of the coverage range of wired telephone lines for intercontinental services. The first invention making amplification possible was the vacuum tube, also called triode, developed by De Forest. Unfortunately, the vacuum tube exhibits high power consumption and size, and provides only poor reliability, limiting the complexity and demanding costly maintenance.

In 1926, the first concept of a semiconductor transistor was reported and patented by Julius Lilienfeld. However, due to significant problems with the practical implementation of the idea, no working device was demonstrated. It took over 10 years until the research director of the Bell laboratories, Mervin Kelly, recognised that such a device could be a promising replacement for the vacuum tube. In 1945, he attracted excellent scientists and engineers to develop a semiconductor device that was more compact and reliable than the vacuum tube. The team was lead by Walter Shockley, who was a very creative visionary. Important members of this group were John Bardeen, an excellent theoretician, and Walter Brattain, an outstanding experimentalist. A photo of these guys is depicted in Fig. 5.1a. In the beginning, the working atmosphere within this group was very fruitful and motivating.

In 1947, John Bardeen and Walter Brattain experimentally demonstrated the first working semiconductor transistor, the point to contact transistor as shown in Fig. 5.1b. This device was similar to the bipolar junction transistor known today and exhibited a remarkable DC current gain of 330. Several years have been required to develop proper operation. The main reason was the insufficient understanding of the behaviour of charge carriers at the semiconductor junctions and surfaces. It took quite some time to discover that these material discontinuities generate a potential barrier at the surface impacting the control of the current flow.

By using proper material combinations and optimising the dimensions, they successfully manipulated this potential barrier. It was possible to control the channel current by means of a terminal isolated by a pn-junction – the transistor was born.

Shockley was furious that he was not directly involved in the invention, which in his opinion was based on his ideas. He believed that the others had betrayed him. Consequently, the excellent collaboration within the team was suddenly interrupted. Claims in terms of inventor rights and patents changed the cooperative atmosphere into a highly competitive one.

To preserve his standing concerning the achievements of Bardeen and Brattain, Shockley worked on his own on a further promising principle based on the field effect, which he has already considered in 1945. In a burst of creativity and anger, Shockley developed the concept of the field effect junction transistor in a hotel room in Chicago. This transistor was successfully implemented in 1950 and provided significantly higher reliability than that designed by Bardeen and Brattain.

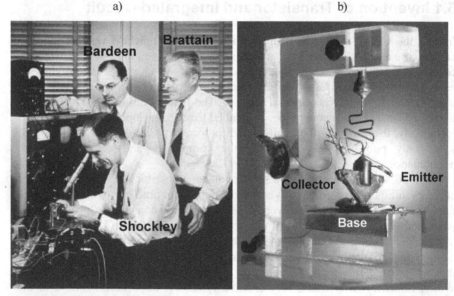

Fig. 5.1a,b. The first transistor, photos from 1947, courtesy of Bell/Lucent laboratories: **a** inventors; **b** realisation of the point to contact transistor

In 1956, Bardeen, Brattain and Shockley received the joined Nobel Price for their genius contributions. However, none of them managed to exploit their ideas commercially. Recognising the potential of the transistor, Shockley left the Bell Laboratories and founded his own company. Due to mismanagement and personal problems, the company was not able to benefit from the promising inventions. Eminent people such as Gordon Moore and Robert Noyce left Shockley's company to co-found Intel cooperation. While his company went bankrupt, Shockley had to watch the successful growth of many other companies creating the well-known Silicon Valley located in the area around Santa Clara in California [Léc05].

A further invention, namely the integrated circuit (IC), was an important step in forcing the commercialising of the transistor. Integrated circuits allow the implementation and connection of a high number of devices with various functionalities in one single chip. At the same time both increased complexity and lowered costs are feasible. In this context, we can refer to the contributions of Jack Kilby and Robert Noyce, who developed a suitable fabrication process in 1948 [Ber05]. By using lithography with high resolution it was possible to implement multiple layers featuring specific properties.

However, a simple and easily scalable transistor with highest yield was required to exploit the possibilities of the integrated circuit today incorporating billions of devices. The corresponding solution was the MOSFET (Metal Oxide Semiconductor Field Effect Transistor), which has primarily been developed by Bell's scientist John Atalla around 1960. It is worthwhile mentioning that the MOSFET is similar to Shockley's junction transistor, since it also uses the field effect previously elaborated by Shockley.

One remarkable fact swept under the carpet in most historical essays about the invention of the transistor has to be added. At the same time and independently from Bardeen and Brattain, the German scientists Herbert Mataré and Heinrich Welker invented the transistron. It was a working transistor as well and nearly identical to the one developed by the Bell Labs [Dor04]. Hence, it is not clear who really invented the first transistor. A radio transceiver was successfully demonstrated using these transistors. Mataré founded the company Intermetall, launching transistor based products including broadcast radios, which may have been the first commercial available products of this kind. This was more than one year before Texas Instruments claimed that milestone.

For further reading concerning the enthralling history of the transistor, the reader may refer to the corresponding literature [Tra05, Mill83, Sho72].

5.2 Charge Transport in Transistors

Transistors play a dominant role in integrated circuits [Sze88]. Their tasks are the control of signals, e.g. the switching or amplification. They mainly consist of a signal channel with input and output terminal, and an electrode capable of controlling the carrier transport in the channel. This controlled transport is mainly based on drift or diffusion. To maximise the efficiency of transistors, it is aimed to control a strong channel current by means of a weak control signal. We will see that ideal transistors behave like a voltage controlled current source.

5.2.1 Drift

In Fig. 5.2a, the drift transport is illustrated. Due to an applied voltage and the associated electrical field E, carriers, e.g. electrons, can be transported from the input to the output of the channel. In a first approximation, a model with variable resistance R can be employed to describe the channel properties and the resulting current

$$I = f(R) = f(\rho \cdot \frac{1}{w \cdot d}) = f(V_c), \tag{5.1}$$

where ρ denotes the channel resistivity and d the channel thickness. We assume that the length l and width w are fixed for a fabricated device. Which parameters can be impacted by means of the control voltage V_c? The following techniques are efficiently used:

1. By impacting the carrier density N, we can determine ρ.
2. A high-ohmic barrier can influence the effective d.
3. Mix of a. and b.

Since a high electrical field is required for the controlled drift transport, the devices are called FETs (Field Effect Transistors). According to Fig. 5.2b, FETs can be segmented into MOSFETs, MESFETs (Metal Semiconductor FETs) and HEMTs (High Electron Mobility Transistors). Since the type of carrier and basic channel charge are equal, we talk about majority carrier transport, e.g. n-carriers in n-channel material or p-carriers in p-channel material.

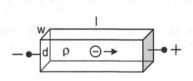

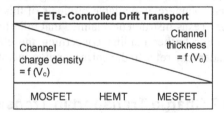

Fig. 5.2a,b. Drift transport: **a** illustration; **b** control principles and associated FET types

5.2.2 Diffusion

The second transport method is based on pn-junctions as shown in Fig. 5.3. By applying a forward bias voltage determined by V_c, we generate a diffusion current aiming at compensating the imbalance of the carriers. Free electrons and holes travel to the p- and n-terminals, respectively. In contradiction to the drift transport, the free carriers are transported in a channel material with opposite intrinsic charge. Thus, they are referred to as minority carriers. According to common literature treating device physics [Sze88, Bäc02], the diffusion current of the electrons through the p-material can be approximated by

$$I_n = \frac{\left[n_p(x_p) - n_{po}\right]}{L_n} \cdot D_n \cdot A \cdot q \tag{5.2}$$

with $\dfrac{\left[n_p(x_p) - n_{po}\right]}{L_n}$ as the gradient of the minority n-carriers at the junction. L_n

denotes the diffusion length. The diffusion constant D_n describes the ability of the minority carriers to travel in the majority material, A is the area of the junction surface and q is the elementary charge. Increasing V_c raises the imbalance of the carriers forcing the diffusion current to grow. Based on Eq. (5.2), we can derive the relation between diffusion current and V_c:

$$I_n = \frac{D_n \cdot A \cdot q \cdot n_{po}}{L_n} \cdot \left(e^{V_c/V_T} - 1\right). \tag{5.3}$$

The temperature voltage $V_T = \dfrac{kT}{q}$ considers the temperature T and the Boltzmann

constant k. Remarkable is the exponential dependence of the diode current vs the applied control voltage. To be shown later, the BJT (Bipolar Junction Transistor) is based on this control mechanism.

It is worthwhile mentioning that the drift transport may also impact diffusion-controlled devices since the drift transport is used for the transport to the extrinsic terminals of the devices.

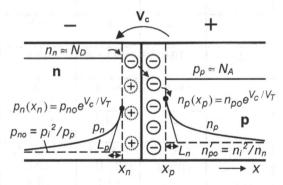

Fig. 5.3. Controlled diffusion of electrons in a pn-junction, which is forward biased by the control voltage V_c. There is a similar diffusion of holes from the p- to the n-terminal. N_D: donator doping, N_A: acceptor doping, n_n: n-majority carrier density=n-carriers in n-terminal, p_p: p-majority carrier density=p-carriers in p-terminal, L_n: diffusion lengths of n-carriers, L_p: diffusion lengths of p-carriers, n_p: n-minority carrier density=n-carrier in p-terminal determining n-diffusion current, p_n: p-minority carrier density=p-carrier in n-terminal determining p-diffusion current, n_i: intrinsic carrier density

5.3 Materials

Today, most integrated circuits are implemented in silicon (Si), silicon germanium (SiGe), gallium arsenide (GaAs) or indium phosphide (InP). Silicon carbide and gallium nitride technologies are considered as promising candidates for the future. The key characteristics of the basic materials are listed in Table 5.1 and described in the following sections.

Table 5.1. Key characteristics and comparison of MMIC technologies

	Silicon	Silicon carbide	InP	GaAs	Gallium nitride
Electron mobility at 300K [cm^2/Vs]	1500	700	5400	8500	1000-2000
Hole mobility at 300K [cm^2/Vs]	450	n.a.	150	400	n.a.
Peak/saturated electron velocity [10^7 cm/s]	1.0/1.0	2.0/2.0	2.0/2.0	2.1/n.a.	2.1/1.3
Peak/saturated hole velocity [10^7 cm/s]	1.0/1.0	n.a.	n.a.	n.a.	n.a
Bandgap [eV]	1.1	3.26	1.35	1.42	3.49
Critical breakdown field [MV/cm]	0.3	3.0	0.5	0.4	3.0
Thermal conductivity [W/cm·K]	1.5	4.5	0.7	0.5	>1.5
Relative dielectric constant	11.8	10.0	12.5	12.8	9
Substrate resistance [Ωcm]	1–20	1–20	>1000	>1000	>1000
Number of transistors in IC	> 1 billion	<200	<500	<1000	<50
Transistors	MOSFET, Bipolar, HBT	MESFET, HEMT	MESFET, HEMT, HBT	MESFET, HEMT, HBT	MESFET, HEMT
Costs prototype/ mass fabrication	High/low	Very high/ n.a.	High/very high	Low/high	Very high/ n.a.
Ecological compatibility	Good	Good	Bad	Bad	Bad

T=300 K, part of data taken from [Eas02]

5.3.1 Speed

We have discussed the fact that the drift transport is important for both drift and diffusion devices. The carrier drift velocity gives first insights regarding the speed. According to Fig. 5.4, the velocity is a function of the applied electrical field. We can summarise the following properties:

- At low to moderate electrical fields as typically applied, the velocity of electrons is higher than that of holes. This difference is very significant for III/V[1] based devices. Consequently, electrons are superior for high-speed applications.

- The peak electron velocity of GaAs is approximately two times higher than that of silicon. Furthermore, compared to GaAs, more electrical field is required in silicon to reach the peak velocity making the design of low-power consuming RF circuits in silicon challenging.

- The characteristics differ when comparing the velocities of holes. Compared to III/V technologies, the hole velocity is higher in silicon. In addition, the speed gap between electrons and holes is much smaller in silicon than in III/V technologies. Thus, if complementary technologies featuring both types of carriers are applied, III/V technologies do not pay off any more in terms of speed. In this case, silicon based processes are superior due to the lower costs.

- A promising performance is observed for silicon carbide yielding peak velocities similar to III/V technologies. However, an extremely high electrical field is required to reach this velocity. Thus, the realistic velocity is much lower.

- In practice, the electrical field along the channel does not have a constant field due to interferences with other potentials, e.g. the voltage controlling the conductivity in the channel of a FET. Consequently, the available electrical field at the input terminal and the associated velocity are lower than those at the output node.

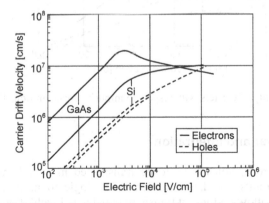

Fig. 5.4. Drift velocity of electrons vs the electrical field, T=300 K

---

[1] Materials of the 3rd and 5th main group of the elementary material system.

A further important indicator for the speed of a current is the mobility given by

$$\mu = \frac{dv}{dE}.$$ (5.4)

The mobility describes how fast the carrier velocity and the associated current can be varied with respect to an applied electrical field E. Hence, the mobility determines the time required to approach the maximum velocity. This is important for switches or RF circuits in general, where fast control of currents is a key issue. Figure 5.5, depicts the mobilities of electrons and holes in silicon and GaAs versus the level of impurities. Let us review the most important observations:

- Compared to the III/V counterparts, the electron mobilities of silicon-based technologies are lower.
- The mobility of electrons is generally higher than that of holes, e.g. 2.5 times higher in silicon and more than 10 times higher in GaAs and InP. Consequently, for high-speed applications, electrons rather than holes are applied as carriers.
- Similar to the values for the velocities, the gap between the electron and hole mobilities is much lower in silicon than in III/V technologies. The silicon hole mobility is even higher than that in GaAs and InP.

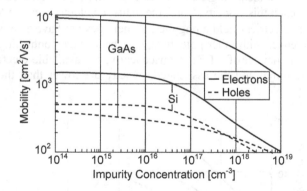

Fig. 5.5. Mobility of carriers with respect to impurity concentration, T=300k

5.3.2 Supply Power and Integration

In addition to electrons, holes have to be transported in complementary logic. Complex digital circuits mandate complementary logic to minimise the static power consumption allowing highest transistor density and yield. Recall that in the complementary logic either the n-or p-channel transistor is switched off if no dynamic signal is applied. Thus, ideally, complementary logic does not consume any static power. Without this mechanism, the required DC power in VLSI circuits would be very high, leading to insoluble heating problems. Consequently, in III/V technologies, the integration complexity is low since no adequate complementary logic can be implemented. One major reason is the low hole mobility. To get rid of the heating, the thermal conductivity should be as high as possible. The thermal conductivity of silicon is approximately twice as high as that of GaAs.

An Intel Itanium processor incorporates 1.7 billion transistors in one CMOS IC, whereas integration in III/V chips is limited to a transistor count of around 1000. In addition to thermal problems, GaAs suffers from limited fabrication yield. As a result of commercial interests very high efforts have been spent on the fabrication of silicon ICs. This is one reason why the yield of silicon based circuits is higher than that of III/V based semiconductors. Moreover, silicon oxide junctions can be manufactured with very high reproducibility. We can conclude that silicon based ICs are the first choice for low-cost VLSI applications.

5.3.3 Costs

In 2000, the IC market was a vendor market with a high demand and a low supply. Today, the market is determined by strong competition. Thus, it can be classified as a consumer market.

Usually, III/V based technologies have lower fixed costs than silicon based technologies since less processing steps are required. Hence, low-volume prototyping costs are less significant for III/V based technologies. The fixed costs are mainly determined by the costs for the lithographic masks.

In mass fabrication, costs can be scaled down more significant for silicon-based technologies due to lower variable fabrication costs per IC. Reasons are the high yield making large wafer sizes possible, and the lower material costs. Silicon material, which is based on quartz sand, is quite cheap. In Table 5.2, the typical sizes of silicon, GaAs and InP wafers are listed. Figure 5.6a shows the silicon wafer size generations versus time.

The mask costs for typical silicon technology are illustrated in Fig. 5.6b versus the minimum possible dimensions of devices. Together with variable cost, wafer costs can easily exceed € 100,000. The typical costs per mm^2 of III/V and silicon based processes are listed in Table 5.3 for both prototyping and mass fabrication.

Table 5.2. Typical wafer sizes

Parameter	Silicon	GaAs	InP
Size	20–30cm	10–20cm	8–10cm
Area normalised to silicon	~1	~1/4	~1/9

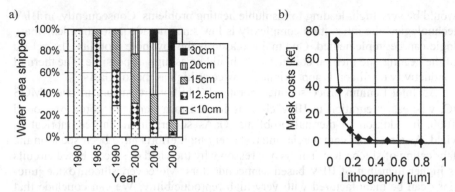

Fig. 5.6a,b. Data CMOS wafer: **a** semiconductor wafer size versus time, source IC Insights [Ici05]; **b** average costs per fabrication mask 2005, source IC Knowledge [Ick05]

Table 5.3. Approximate IC fabrication costs for prototyping and mass fabrication excluding testing and packaging

Technology/ speed	Prototyping (MPW)			Mass fabrication (100 wafer)	
	Run	IC with 1 mm^2	Total area	IC with 1 mm^2	Total IC area
0.2 μm GaAs PHEMT/ f$_t$=80GHz	€ 12k	€ 80	Approximately 15 repetition pieces each having 10 mm^2	< € 1	10 cm wafer ~0.8 Mio. mm^2
90 nm CMOS/ f$_t$=110GHz	€ 50k	€ 330		< € 0.1	30 cm wafer ~7.1 Mio. mm^2

MPW: Multi Project Wafer, numbers estimated in 2008

5.3.4 RF Output Power

The speed of CMOS technology is achieved by aggressively downscaling of the dimensions. At lowered dimensions, the voltages have to be scaled down as well to keep the electrical field below a critical value. Hence, as a significant drawback, the maximum possible RF output power decreases.

Due to the advantages in terms of speed, III/V based technologies do not have to be scaled as aggressive as the silicon counterparts. Thus, generally, the possible supply voltage and the associated RF output power of III/V technologies are larger. That is the major reason why III/V based power amplifier ICs are still used in mobile phones.

In the future, we also have to consider the materials silicon carbide and gallium nitride. Being superior for high power applications they can handle very high electrical fields due to their large band-gap. There are still a lot of processing challenges for these novel materials, which have to be solved prior to commercial exploitation. However, with enhanced processing technology, these materials may play a growing role in the future. Their superior thermal conductivity is advantageous as well.

5.3.5 Ecologic Compatibility

What about the ecologic compatibility of the technologies? Consider all the mobile phones and computers not used any more. They have to be trashed or, if possible, recycled. The amount of harmful substances produced during IC fabrication may be comparable for the different technologies. However, as far as IC trash is concerned, the ecological compatibility of silicon-based products is better than for GaAs or InP, since silicon is based on the natural material quartz sand. GaAs and InP contain very poisonous materials, e.g. arsine.

5.3.6 Conclusions

Despite the superior prerequisites for III/V technologies in terms of speed, silicon CMOS technology plays the most important role for today's IC market. The major reasons are the low costs in mass-fabrication and the excellent ability for highest level of integration. In this context, we have to take into account the very low static power consumption of the complementary logic in silicon. Due to their low hole mobilities, complementary logic cannot be successfully applied in III/V technologies. More and more circuits are implemented in the digital domain. Thus, there is the trend to optimise low-cost technologies for digital rather than analogue applications. As a consequence, the design of analogue circuits in digital VLSI technologies is challenging. In terms of RF output power, there will be a niche for III/V technologies featuring high breakdown voltages. At the same speed, III/V technologies do not require a similar level of downscaling. Because of their very high breakdown voltages and good thermal conductivities, novel technologies such as silicon carbide and gallium nitride may serve as promising alternatives in the future.

5.4 MOSFET

If we will not make it on CMOS – somebody else will do it.
Unknown

MOSFETs (Metal Oxide Semiconductor Field Effect Transistor) are fabricated in silicon, and feature both n- and p-channel devices. Hence, they are suited for complementary logic with the aforementioned advantages in terms of static power consumption and circuit complexity. Recall that the gap between the electron and hole mobilities of silicon based devices is relatively small. Nevertheless, n-channel FETs are still the first choice for circuits demanding highest speed.

5.4.1 Functional Principle

MOSFETs, as well as other FET types, are based on the drift transport of majority carriers within a channel terminated by two ports called source and drain. The properties of the channel can be controlled by the gate terminal, which is capacitively coupled to the channel. This capacitance eliminates the required DC control power and decouples DC-wise the control gate and the channel. In case of the MOSFET, control of the channel conductance is mainly performed by means of the carrier density.

In the following sections we focus on the description of the n-channel FET. However, the functionality of the p-channel FET is very similar. The only major difference is that the signs for the voltages and currents are inverted.

Equal to all other FET types, MOSFETs can be realised as D (Depletion)- or E (Enhancement)-FETs. In case of the n-channel FET, the threshold voltages are negative and positive, respectively. Hence, the D-FET is conductive for gate source voltages V_{gs} above the negative threshold voltage (V_{th}), whereas the E-FET is conductive for V_{gs} larger than the positive V_{th}. At $V_{gs}=0$ V, the D-FET is conductive, whereas the E-FET is not. Due to the inverted signs, p-channel FETs are conductive for $V_{gs}<V_{th}$.

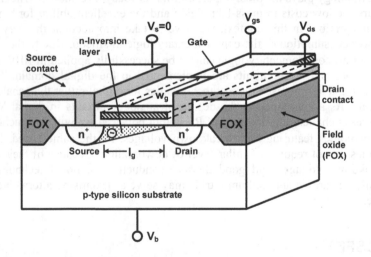

Fig. 5.7. Cross-section of an enhancement n-channel bulk MOSFET, FOX: Field Oxide Layer

5.4.2 DC Characteristics

Let's qualitatively explain the basic principle of a MOSET based on a typical n-channel E-FET as illustrated in Fig. 5.7. Heavily doped source and drain regions terminate the channel, which is embedded into a lightly doped p-substrate. Given that $V_{gs} > V_{th}$, the holes are strongly depleted whereas electrons are accumulated as shown in the band diagram depicted in Fig. 5.8. The impact of the gate source voltage is illustrated in Fig. 5.9a, implicating that the electron density and the channel conductance raise with increased V_{gs}.

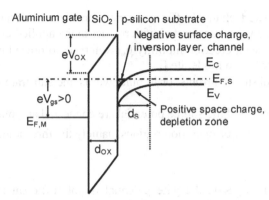

Fig. 5.8. Band diagram of the n-channel MOSFET in inversion. Due to the positive gate potential, negative surface charge is accumulated in the p-silicon substrate forming an n-channel, E_C: conductance band, E_V: valance band, $E_{F,S}$: Fermi level semiconductor, $E_{F,M}$: Fermi level metal, d_{ox}: oxide thickness, d_s: thickness of depletion zone

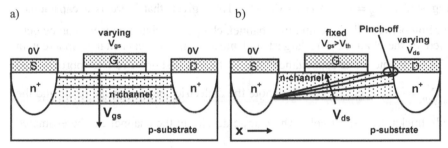

Fig. 5.9a,b. Enhancement n-channel MOSFET: **a** variation of V_{gs}; **b** variation of V_{ds}, increasing of V_{ds} above $V_{ds,sat}$ pinches off the channel and keeps I_D approximately constant

Since the isolation of the silicon substrate is weak, the DC potential of the substrate (V_b) has to be defined since it influences, e.g. the threshold properties. Thus, in contrast to other transistor types such as MESFETs, the conventional MOSFET has an additional port called bulk. In practice, this port is frequently connected to ground.

To activate the current flow directed from source to drain, a drain source voltage V_{ds} has to be applied. According to Fig. 5.9b, the charge density along the channel is not constant any more. We observe that the effective gate potential is also a function of the drain source voltage, which has an increasing impact towards the drain where it is maximal. The effective gate potential may be approximated by

$$V_g = V_{gs} - V(x) - V_{th} .$$ (5.5)

From source to drain $V(x)$ increases from zero to V_{ds} where x describes the location along the channel. For further calculations, we make the following assumptions:

- The so-called short channel effects are not considered. We assume that the electron velocity v is not saturated with respect to the applied electrical field E leading to a constant electron mobility μ_n. Referring to preceding insights this is only valid for low to moderate E.
- There are no substrate currents, or in other words, the substrate is an ideal isolator.
- The bulk potential and the source potential are connected and put on ground.

We can identify four major operation regions, namely the threshold, triode, linear[2] and saturation region.

I. Threshold Region

If $V_g = V_{gs} - V(x) - V_{th} \leq 0$, the gate potential is not sufficient to generate the conductive inversion layer. Consequently, the channel conductance is very low resulting in $I_d \approx 0$. Or in other words, the device is switched off.

II. Triode Region

Suggest that $V_g = V_{gs} - V(x) - V_{th} > 0$. Thus, given that there is a capacitance coupling between the gate and the channel, charge associated with V_{gs} can be generated in the channel. The largest and most significant capacitance inherent in FETs is the gate oxide capacitance[3] $C_{ox}^{"}$, which in first order can be approximated by $C_{ox}^{"} = \dfrac{\varepsilon_{ox}}{t_{ox}}$ where ε_{ox} and t_{ox} denote the dielectric constant of the oxide and the oxide thickness, respectively. The charge density in the channel can be estimated by

$$Q_n(x) = C_{ox}^{"} \left[V_{gs} - V(x) - V_{th} \right].$$ (5.6)

The resulting channel current is determined by the channel charge density times the velocity of the electronics with respect to the channel width w_g:

$$I_d = w_g \cdot Q_n(x) \cdot v(x) .$$ (5.7)

Together with Eq. (5.6),

$$v(x) = \mu_n \cdot E(x) ,$$ (5.8)

---

[2] Please note that the two terms linear region, and linearity with respect to device large signal performance do not exhibit any relations.
[3] Capacitance per area indicated by $"$.

and the fact that the electrical field is the derivation of voltage with respect to x, we get for the channel current

$$I_d = \mu_n \cdot w_g \cdot C''_{ox} \cdot \left[V_{gs} - V(x) - V_{th} \right] \cdot \frac{dV}{dx} . \tag{5.9}$$

The integration of the current along the channel yields

$$\int_0^L I_d \cdot dx = \int_0^{V_{ds}} \mu_n \cdot w_g \cdot C''_{ox} \cdot \left[V_{gs} - V(x) - V_{th} \right] \cdot dV . \tag{5.10}$$

Finally we get

$$I_d = k_n \cdot \left[\left(V_{gs} - V_{th} \right) V_{ds} - \frac{V_{ds}^2}{2} \right] \tag{5.11}$$

with $k_n = \dfrac{w_g}{l_g} \cdot \mu_n \cdot C''_{ox}$. By means of Eq. (5.11), it can be shown that up to the

drain source saturation voltage

$$V_{ds,sat} = V_{gs} - V_{th} , \tag{5.12}$$

I_d raises with increased V_{ds}. The relation $V_{gs} - V_{th}$ is also known as the gate overdrive voltage. Equation (5.11) is only valid for $V_{ds} \leq V_{ds,sat}$ and is not valid for $V_{ds} > V_{ds,sat}$ since in this case, along an increasing part of the channel length, the charge density would drop to zero and the channel resistance would approach infinity, which cannot be observed in reality.

III. Linear Region
The linear region is a sub-region of the triode region and is valid for $V_{ds} \ll V_{gs} - V_{th}$. In this case, from inspection of Eq. (5.11), we can approximate

$$I_d = k_n \cdot \left[V_{gs} - V_{th} \right] \cdot V_{ds} \tag{5.13}$$

resulting in a linear dependence of V_{ds} versus I_d. That is the reason why this region is called linear region.

IV. Saturation Region
According to Eq. (5.11), in the triode region, maximum I_d is obtained at $V_{ds,sat}$. In the saturation region, which is defined as the region with $V_{ds} > V_{ds,sat}$, I_d stays constant due to the so called pinch-off effect. With increased channel potential towards the drain, the effective gate to channel potential gets smaller. Thus, the conductive inversion is reduced towards the drain. Due to the high resistivity in this non-inverted region, a voltage drop is created preventing the current from rising significantly. Hence, the channel current stays more or less constant if $V_{ds,sat}$ is

approached. Thus, for $V_{ds} > V_{ds,sat}$, we can substitute V_{ds} with $V_{ds,sat}$. With Eqs. (5.11) and (5.12), we obtain

$$I_d = \frac{k_n}{2} \cdot \left(V_{gs} - V_{th}\right)^2 . \tag{5.14}$$

This equation reveals that the ideal transistor behaves like a voltage controlled current source with V_{gs} as the control voltage. There is no dependency on V_{ds}. Recall that the performance of current sources can be described by the transconductance

$$g_m = \frac{dI_d}{dV_{gs}} = k_n \cdot \left(V_{gs} - V_{th}\right) . \tag{5.15}$$

Rearrangement on the basis of Eq. (5.14) yields

$$g_m = \sqrt{2 \cdot I_d \cdot k_n} = \sqrt{2 \cdot I_d \cdot \frac{w_g}{l_g} \cdot \mu_n \cdot C_{ox}^{"}} . \tag{5.16}$$

As we will see later, g_m is one of the major small signal device and design parameters for active circuits. It greatly impacts the voltage, current and power gain of a circuit.

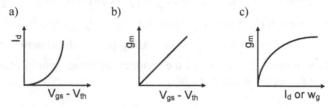

Fig. 5.10a–c. FET characteristics of channel current and transconductance vs key design parameters assuming that the other parameters are kept constant

From Eqs. (5.14)–(5.16), we can derive some important basic insight, which are valuable for the circuit designer. All parameters not listed are assumed to be constant.

- $I_d \sim \left(V_{gs} - V_{th}\right)^2$, given that $V_{gs} > V_{th}$; see also Fig. 5.10a.

- $g_m \sim V_{gs} - V_{th}$; refer to Fig. 5.10b.

- $g_m \sim \sqrt{I_d}$ if w_g is kept constant; compare Fig. 5.10c.

- $g_m \sim \sqrt{w_g}$ if I_d is kept constant; see Fig. 5.10c.

- $g_m \sim \sqrt{\mu_n}$.

- $g_m \sim \sqrt{C_{ox}^{"}}$.

- $g_m \sim w_g$ if w_g and I_d are scaled at the same time meaning that the current density is kept constant while scaling w_g.

- $g_m \sim \dfrac{1}{\sqrt{l_g}}$ if I_d is kept constant.

- If the dependency of I_d with respect to l_g (approximation $I_d \sim 1/l_g$) is taken into account we get $g_m \sim \dfrac{1}{l_g}$.

The last two observations show that the smallest l_g available in the used technology is favourable for most applications. Given by the used material, the parameters μ_n and $C_{ox}^{"}$ cannot be impacted by the designer. Thus, the remaining design parameter, which can be optimised from circuit perspective, is w_g.

a)   b)

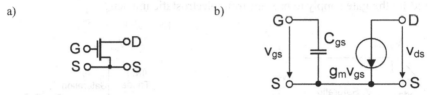

Fig. 5.11a,b. MOSFET: **a** common source topology; **b** first order small signal model including current source and gate capacitance

Now, we derive a first order small signal FET model consisting of the two major elements, which are the voltage controlled current source and the capacitance of the device. To do this, a common reference potential has to be defined. For modelling proposes, the common source topology as shown in Fig. 5.11a is applied, where the source is defined as reference potential. If not otherwise mentioned, throughout this book, we use the common source topology. By reviewing the charge profile in Fig. 5.7, we observe that the capacitance of the transistor is mainly determined by the contribution between the gate and the source. Hence, in first order, we may represent the capacitive behaviour of the device by means of the gate source capacitance C_{gs}. The corresponding equivalent small signal model is depicted in Fig. 5.11b. We find a relation of $C_{gs} \approx C_{ox}^{"} \cdot w_g \cdot l_g \cdot k_{ox}$ with $k_{ox} \leq 1$ as a device and bias dependent parameter. For typical CMOS devices operated in saturation k_{ox} amounts to approximately $\dfrac{2}{3}$.

In Fig. 5.12a,b, the current voltage (IV) curves and the corresponding operation regions are illustrated.

So far, the calculations have been eased to deduce simple insights. The accuracy is sufficient for many circuit evaluations. More realistic IV curves are plotted in Fig. 5.12c showing that in reality, I_d increases even in the saturation region. Moreover, the maximum voltages are limited by breakdown effects. The DC characteristics excluding the breakdown region of a 90-nm n-channel MOSFET are shown in Fig. 5.12d.

V. Drain Break-down Region

At large V_{ds}, we approach the drain breakdown voltage leading to a very high channel current, which can damage the device. Since the drain breakdown voltage scales down with l_g, only small drain voltages can be applied for aggressively scaled devices.

VI. Gate Break-down Region

We have learned that increased V_{gs} raises the channel current. At a certain maximum channel current, the transistor can be damaged limiting the maximum V_{gs}. Furthermore, excessive gate voltage can lead to tunnelling effects short-circuiting the oxide. The gate oxide thickness has to be scaled together with l_g. Thus, the maximum V_{gs} is limited by scaling. As for the drain voltage, protection circuits are required for the gate supply to prevent from electrostatic impacts.

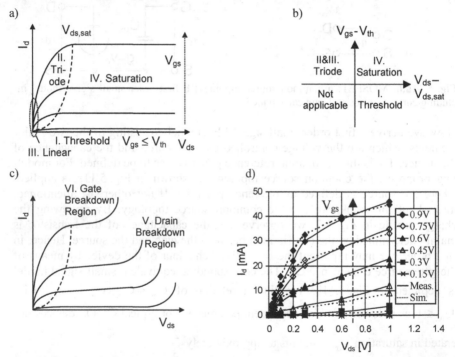

Fig. 5.12a–d. Current voltage characteristics of FET: **a** first order simplification; **b** illustration of most important operation regions; **c** realistic; **d** characteristics of 90-nm n-channel SOI FET with w_g=64 μm

5.4.3 MOSFET Types

Since the n- and p-channel devices can be realised as E- or D-FETs, four different FET types are possible. Compared to n-channel FETs, signs for currents and voltages are inverted for p-channel FETs. With respect to the applied V_{gs}, D-FETs conduct at lower V_{gs} than E-FETs. To define the V_{th}, the substrate doping can be adapted. Since the D-FET requires an earlier inversion effect than the E-FET it should exhibit a lower substrate doping. Further details are listed in Table 5.4. Obviously, compared to E-FETs, D-FETs have higher maximum gate overdrive voltage $V_{gs}-V_{th}$ since the magnitudes of both voltages are added, whereas they are subtracted for E-FETs. The maximum possible V_{gs} determined by breakdown effects are approximately equal for both device types. Thus, the D-FET provides higher gate voltage swing and consequently higher channel current, thereby being favourable for power applications.

Table 5.4. Different types of MOSFETs

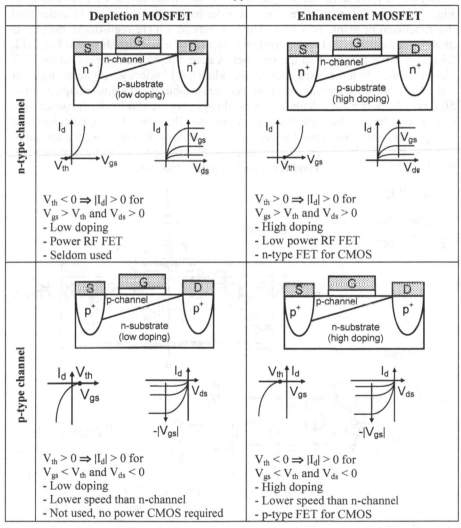

	Depletion MOSFET	Enhancement MOSFET
n-type channel	$V_{th} < 0 \Rightarrow |I_d| > 0$ for $V_{gs} > V_{th}$ and $V_{ds} > 0$ - Low doping - Power RF FET - Seldom used	$V_{th} > 0 \Rightarrow |I_d| > 0$ for $V_{gs} > V_{th}$ and $V_{ds} > 0$ - High doping - Low power RF FET - n-type FET for CMOS
p-type channel	$V_{th} > 0 \Rightarrow |I_d| > 0$ for $V_{gs} < V_{th}$ and $V_{ds} < 0$ - Low doping - Lower speed than n-channel - Not used, no power CMOS required	$V_{th} < 0 \Rightarrow |I_d| > 0$ for $V_{gs} < V_{th}$ and $V_{ds} < 0$ - High doping - Lower speed than n-channel - p-type FET for CMOS

Compared to n-channel FETs, the signs for currents and voltages are inverted for p-channel FETs.

At given f_t or f_{max}, the enhancement FET has lower power consumption, which is an important prerequisite for digital circuits with high complexity. For this reason, most low-cost technologies feature only enhancement devices. CMOS IC technologies are dictated by the needs of digital rather than analogue RF circuits.

The reader is referred to a useful interactive tutorial tool concerned with the modelling of the MOSFET [Edu05].

5.4.4 Small Signal Modelling

Hitherto, we have used the simple small signal model of the FET according to Fig. 5.11b, which is well suited for first order hand calculations and to understand the functional principle of a FET. The structure of a more realistic MOSFET, its associated parasitics and the equivalent small signal model are shown in Fig. 5.13. Listed in Table 5.5, we find the parameters according to their impact on the FET characteristics. In most cases, models including the 1st priority parameters together with the 2nd priority parameters show very reasonable small signal accuracy up to 50 GHz. To ease hand calculations it is always very fruitful to check, which specific parameter can be neglected and which not. However, for the calculation of complex networks we should apply CAD simulations. In this case, we can additionally include the 3rd priority parameters thereby improving the accuracy towards highest frequencies and for specific circuit problems.

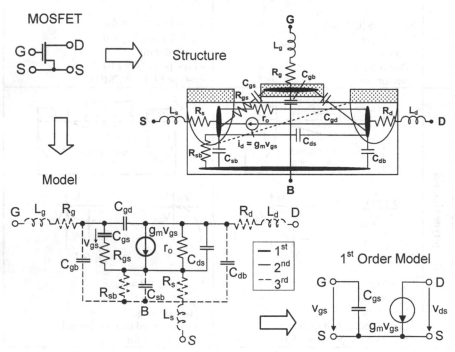

Fig. 5.13. Small signal equivalent circuit of the MOSFET in common source configuration

Table 5.5. FET small signal model parameters according to their priority with respect to impact on one hand and complexity on the other

Parameter		Origin and impact	Related S-parameters
1st priority			
Transconductance	g_m	Current source properties: $g_m = \dfrac{dI_d}{dV_{gs}} = \dfrac{i_d}{v_{gs}}$	$\underline{S}_{21}$
Gate source capacitance	C_{gs}	Largest capacitance, $C_{gs} > C_{gd} > C_{ds}$, mainly based on oxide capacitance	Imaginary part of $\underline{S}_{11}$
2nd priority			
Output resistance	r_o	$r_o = \left(\dfrac{dI_d}{dV_{ds}}\right)^{-1} = \left(\dfrac{i_d}{v_{ds}}\right)^{-1} = r_{ds} \| R_{ds}$, small signal output impedance extrapolated from large signal characteristics r_{ds}: small signal operand, no ohmic resistance R_{ds}: parasitic ohmic resistance due to channel and substrate leakage, decreasing value and dominate role for a. low V_{ds} (triode region) b. aggressively scaled devices c. towards high frequencies	Real part of $\underline{S}_{22}$
Gate source resistance	R_{gs}	Considers losses of oxide capacitance including connection to source. Usually, $R_{gs} > R_g$	Real part of $\underline{S}_{11}$
Gate drain capacitance	C_{gd}	Input to output feedback, introduces large virtual input capacitance at high voltage gain due to Miller effect (refer to Sect. 7.1.1.6).	Imaginary part of $\underline{S}_{12}$
Drain source capacitance	C_{ds}	Relative small capacitance, has to be considered for fine tuning of output impedance matching, e.g. for power amplifiers	Imaginary part of $\underline{S}_{22}$
3rd priority			
Gate contact resistance	R_g	At optimum layout, very small values in the order of 1 Ω, difficult to extract, analytical approximations yield reasonable results	
Drain contact resistance	R_d		
Source resistance	R_s		
Gate contact inductance	L_g	At optimum layout, values are small, assuming 0.7 nH/mm and 20 μm contact lengths yields 15 pH showing impact only for very high frequencies. However, L_s can be significant for power FETs with low source impedances, since strong series feedback may be generated at high frequencies. Given that source is put on ground wide and thick connections should be used to minimise L_s	
Drain contact inductance	L_d		
Source inductance	L_s		
Gate bulk capacitance	C_{gb}	Can be absorbed in C_{gs} as reasonable approximation. Due to low substrate resistivity, capacitance has strong resistive losses	
Drain bulk capacitance	C_{db}	Can be absorbed in C_{ds} as reasonable approximation. Due to low substrate resistivity, capacitance has strong resistive losses	
Source bulk capacitance	C_{sb}	Can be deactivated by connection of source and bulk. Due to low substrate resistivity, capacitance has strong resistive losses	
Source bulk resistance	R_{sb}		

Much more complex models are available in today's circuit simulators such as Cadence Virtuoso Spectre and ADS. These models can also describe the large signal properties of devices – an important feature required for power and nonlinear investigations.

In the small signal approach, we assume that the input power and the associated signal swing in the IV curves are very small. The latter property does not hold under large signal conditions. Large signal means that the model parameters are a function of the time variant input signal, e.g. $g_m = f(V_{gs}) = f(t)$. Thus, large signal models can be treated as bias dependent array of small signal parameters. For further information about large signal models the reader is referred to the specific literature [Tro03].

5.4.5 S-Parameter Extraction and Fitting

After definition of the MOSFET small signal model and the discussion of the tradeoff between complexity and accuracy, we determine the values of the small signal model parameters. Different extraction methods have been discussed in literature [Maa03, Fuk79, Yan86, Hol86, Dam88, Ber90, Ror96]. The element values can be obtained by fitting of measured S-parameters. Especially for the important 1st and 2nd order parameters, adequate start values are helpful. In this context, the proper separation of each parameter with respect to available measurable data is important. In a certain range of observation one can obtain similar fits, e.g. for S_{11} with high C_{gs} and low C_{gd}, or also for low C_{gs} and high C_{gd}. We present a method, which allows simple extractions by means of Y-parameters, which can be converted from S-parameter measurements as discussed in Sect. 3.5. C_{gd} can be easily determined from

$$\underline{Y}_{12} = -j\omega C_{gd} .\qquad(5.17)$$

C_{gd} and r_0 can be found from the complex value of

$$\underline{Y}_{22} = \frac{1}{r_o} + j\omega C_{ds} + j\omega C_{gd} .\qquad(5.18)$$

C_{gs} and R_{gs} can be extracted from

$$\underline{Y}_{11} = j\omega C_{gd} + \frac{j\omega C_{gs}}{1 + j\omega C_{gs} R_{gs}} .\qquad(5.19)$$

With

$$\underline{Y}_{21} = \frac{g_m}{1 + j\omega C_{gs} R_{gs}} - j\omega C_{gd} ,\qquad(5.20)$$

we can deduce g_m. It is noted that $g_m = \dfrac{dI_d}{dV_{gs}}$ can also easily be determined by

DC measurements. Based on the start values, one may accomplish fine-tuning of the model parameters with respect to measurements. This can be performed on basis of an optimisation routine in a CAD tool, where an error function describing the mismatch between modelled and measured results is minimised. Since we have four S-parameters each having a real and imaginary part, we have eight individual error functions, e.g.

$$e_{Re\{\underline{S}_{11}\}} = \sqrt{\left(Re\{\underline{S}_{11measured}\}\right)^2 - \left(Re\{\underline{S}_{11modelled}\}\right)^2}. \qquad (5.21)$$

To allow overall optimisation, a global error relation including all individual error functions may be defined. According to their individual relevance, e.g. $\underline{S}_{21}$ more relevant than $\underline{S}_{12}$, weighting factors k can be applied yielding an error function of

$$e_{overall} = k_{11r} \cdot e_{Re\{\underline{S}_{11}\}} + k_{11i} \cdot e_{Im\{\underline{S}_{11}\}} + k_{22r} \cdot e_{Re\{\underline{S}_{22}\}} + k_{22i} \cdot e_{Im\{\underline{S}_{22}\}}$$

$$+ k_{21r} \cdot e_{Re\{\underline{S}_{21}\}} + k_{21i} \cdot e_{Im\{\underline{S}_{21}\}} + k_{12r} \cdot e_{Re\{\underline{S}_{12}\}} + k_{12i} \cdot e_{Im\{\underline{S}_{12}\}}. \qquad (5.22)$$

Fitting with respect to magnitude and phases of the S-parameters is possible as well.

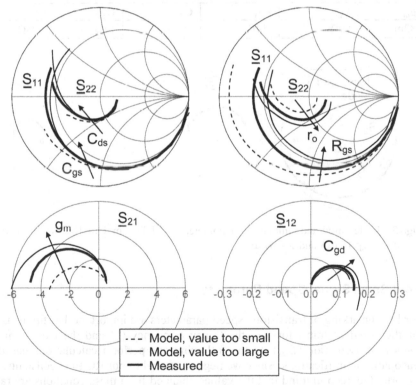

Fig. 5.14. Influence of model parameter on S-parameter. Measurements represent data of a 90-nm SOI n-channel E-FET in a typical class-A bias point with V_{gs}=0.5 V, V_{ds}=1 V, I_d=17 mA and w_g=64 μm from 2–100 GHz

Optionally, we can also match the parameters manually. This has a strong learning effect regarding the modelling. A Smith chart proves to be very useful for the comparison of the S-parameters $\underline{S}_{11}$ and $\underline{S}_{22}$. The transmission coefficients $\underline{S}_{21}$ and $\underline{S}_{12}$ are not defined for a Smith chart and can, e.g. be plotted in a polar diagram.

In Fig. 5.14, the characteristics of the S-parameters are illustrated for variations of the 1st and 2nd order parameters. The measured S-parameters represent data of a typical 90-nm SOI n-channel E-FET with a gate width of 64 μm. In Table 5.6, the values obtained by fitting are listed. According to the rule of thumb for inductances per lengths, we have determined layout dependant values for L_g, L_d and L_s of 30 pH, 30 pH and 10 pH, respectively. Based on the geometries and the specific sheet resistances of the metals we have calculated values of around 1 Ω for R_g and R_d, and 0.5 Ω for R_s. In Fig. 5.15, the measured and simulated S-parameters of the FET are plotted. Good agreement is achieved up to 100 GHz verifying the model.

Table 5.6. Key model parameters of 90-nm SOI n-channel E-FET

g_m	C_{gs}	C_{gd}	C_{ds}	R_{gs}	r_0
82mS	90fF	30fF	15fF	14Ω	75Ω

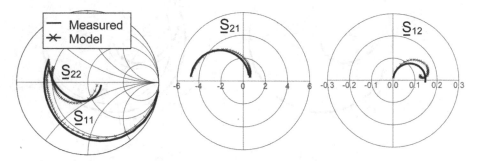

Fig. 5.15. Measured and simulated S-parameters of FET from 1 to 100 GHz at V_{ds}=1 V, V_{gs}=0.5 V, I_d=17 mA and w_g=64 μm

5.4.6 Small Signal Speed Parameter

For benchmarking of transistors, speed parameters are important. Useful measures for the speed of transistors are the transit frequency (f_t) and the maximum frequency of oscillation (f_{max}). In the following, we perform calculations based on reasonable simplifications. Since we neglect several minor RC time constants, it is important to keep in mind that the values obtained from these equations are rather optimistic. In reality, the values may be up to 25% lower.

5.4.6.1 Transit Frequency

The transit frequency is defined as the frequency where the current gain $h_{21}(\omega)$ equals unity. Thus, f_t serves a useful speed measure for current switches. By definition, the current gain is calculated by short-circuiting of the output port. Since the highest output current is obtained with this short, f_t is a rather optimistic value. In reality, where non-zero impedances have to be driven, lower output current, current gain and subsequently f_t has to be expected. We apply the first order equivalent circuit according to Fig. 5.13 yielding

$$h_{21}(\omega_t) = \left|\frac{I_d}{I_g}\right| = \frac{g_m V_{gs}}{\omega_t C_{gs} V_{gs}} = 1. \tag{5.23}$$

Solving for f_t results in

$$f_t = \frac{g_m}{2\pi \cdot C_{gs}}. \tag{5.24}$$

From the latter equation, and on the condition that the current saturation associated with V_{gs} and short channel effects can be neglected, we can draw the following conclusions:

- Considering the quadratic characteristics of I_d vs V_{gs}, $g_m = dI_d/V_{gs}$ and f_t can be raised by increase of the supply current.
- Since g_m and C_{gs} are both proportional to the gate width w_g, f_t is independent of w_g.
- g_m is inversely proportional to l_g, whereas C_{gs} is proportional to l_g. Thus, in idealised theory, f_t increases with $1/l_g^2$. We will see later that due to short channel effects such as the velocity saturation caused by the high electrical drain source field, the accuracy of this conclusion is limited.

It is promising to know that the scaling of l_g improves the speed of FETs. This can also be explained by the fact that a lower channel length decreases the travel time from one terminal to the other. Micro- and nano-technologies exploit this property.

In Sect. 7.1.1.6 we will discuss the Miller effect, which increases the effective input capacitance in case of a high voltage gain a_v. To improve the accuracy of Eq. (5.24), we may substitute C_{gs} with $C_{in} = C_{gs} + C_{gd}(1 - a_v)$.

Figure 5.16 shows the f_t performances of several bulk n-channel MOSFET generations featuring different l_g vs bias. Due to the g_m increase vs bias, f_t raises up to saturation observed at high bias currents, where no further g_m improvement is possible due to decreased mobility associated with the high electrical field. A maximum f_t of around 110 GHz is yielded by the 90-nm technology.

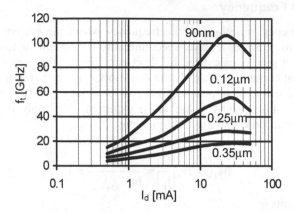

Fig. 5.16. Measured dependence of the transit frequency versus bias current for different bulk n-MOSFET generations featuring different l_g, w_g=100 µm, V_{ds}=1V

5.4.6.2 Maximum Frequency of Oscillation

The maximum frequency of oscillation is defined as the frequency, where the MAG (Maximum Available Power Gain) equals one. The MAG is achieved at conjugate matching, where the parasitic reactive impedances of the transistor are compensated by reactive matching networks. Consequently, maximum real power is transferred into the load.

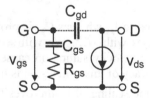

Fig. 5.17. Equivalent circuit for the calculation of f_{max}

For the determination of f_{max}, the consideration of the resistive contributions at the input and output terminals is mandatory. Otherwise we would not be able to calculate the required relations between the real powers. Thus, in addition to the first order model used for the calculation of f_t, we have to consider the gate resistance R_{gs} and the output impedance of the transistor r_{out}. The latter parameter is obtained by small signal calculations. The calculation of f_{max} is quite difficult. Hence, we have to make some simplifications. For the calculation of the input impedance we neglect the feedback associated with C_{gd}. We may verify this simplification by the fact that the voltage gain, the Miller capacitance (refer to Sect. 7.1.1.6), and consequently the impact of C_{gd} is small towards f_{max} since the losses increase towards highest frequencies. However, for the calculation of the output impedance we consider the feedback associated with C_{gd}, which has a significant impact on the real part of the output impedance r_{out}. We neglect the parasitic channel resistance[4] R_{ds}. The parasitic channel capacitance C_{ds} is compensated. A corresponding equivalent network is depicted in Fig. 5.17. We get

$$MAG = \frac{P_{out}}{P_{in}} = \frac{1}{4} \cdot \left| \frac{I_d^2}{I_g^2} \right| \cdot \frac{r_{out}}{R_{gs}} \; . \qquad (5.25)$$

Referring to Eq. (5.23), around f_t, the current gain is inversely proportional to the frequency. Usually, the values of f_t and f_{max} are not far away from each other. Hence, around f_{max}, we may apply the following linearisation:

$$\left| \frac{I_d}{I_g} \right| \approx \frac{f_t}{f} \; . \qquad (5.26)$$

Now, we can write for the maximum available power gain

$$MAG = \frac{P_{out}}{P_{in}} \approx \frac{1}{4} \cdot \left(\frac{f_t}{f} \right)^2 \frac{r_{out}}{R_{gs}} \; . \qquad (5.27)$$

Recall that f_{max} represents the frequency where MAG=1. It is straightforward to rearrange the last equation into

$$f_{max} = \frac{f_t}{2} \sqrt{\frac{r_{out}}{R_{gs}}} \; . \qquad (5.28)$$

This equation reveals that the resistive losses in the equivalent circuit of the FET have a strong impact on f_{max}. For $r_{out} \to \infty$ and $R_{gs} \to 0$, f_{max} would approach ∞. For frequencies well below f_t, R_{gs} may be neglected with respect to C_{gs} since $1/\omega\, C_{gs} > R_{gs}$. With

$$\underline{V}_{ds} = \underline{V}_{gs} + \underline{V}_{dg} = \underline{V}_{gs} \left(1 + \frac{C_{gs}}{C_{gd}} \right) \qquad (5.29)$$

we may deduce

$$r_{out} = \frac{V_{ds}}{I_d} = \frac{1}{g_m} \left(1 + \frac{C_{gs}}{C_{gd}} \right) \; . \qquad (5.30)$$

On condition that R_{ds} and C_{ds} are neglected, Eq. (5.30) may also be verified by Eq. (7.34) and the corresponding sub-section C. By taking into account Eq. (5.24) and $C_{gs} > C_{gd}$, we can modify Eq. (5.30) into

---

[4] This assumption may not be reasonable for aggressively scaled MOSFET devices, where the R_{ds} scales down with gate length thereby playing a considerable role. Due to the additional losses, in reality, the MAG and f_{max} may be lower.

$$r_{out} \approx \frac{1}{2\pi f_t \cdot C_{gd}} . \tag{5.31}$$

With Eqs. (5.29)–(5.30), we get

$$f_{max} \approx \sqrt{\frac{f_t}{8\pi \cdot R_{gs} \cdot C_{gd}}} . \tag{5.32}$$

The latter equation reveals that f_{max} can be increased by minimising R_{gs} and C_{gd}. In this context, the proper layout of the device is important. Gates with multiple fingers help to decrease R_{gs}. On the other hand, the connection capacitances associated with multiple fingers raise C_{gd}. Hence, an optimum layout exists. Given a low R_{gs}, for MOSFETs, the f_{max} can exceed the f_t. The input and output capacitances have no impact on f_{max} since they can be tuned out by an inductance. The maximum frequency of oscillation is a key parameter for the design of analogue circuits such as amplifier and oscillators. To compensate losses and to achieve adequate gain, f_{max} has to be much higher than the operation frequency of a circuit. As a rule of thumb, for amplifiers, f_{max} should be approximately 2–10 times higher than the operation frequency in ideal case resulting in an MAG amounting to 6–20 dB.

5.4.6.3 Gate Width Scaling

In Table 5.7, the ideal characteristics of key model parameters are listed when w_g is scaled by a factor β and the current density is kept constant. By assuming that the relations between g_m/C_{gs} and $R_{gs} \cdot C_{gd}$ remain constant, the values of the speed parameters f_t and f_{max} are constant as well. These assumptions are verified quite well by measurements. It is worthwhile to repeat that the g_m/C_{gs} ratio can be improved by l_g scaling to improve both speed parameters. Details about l_g scaling will follow in Sect. 5.4.9.

Table 5.7. Effect of gate width scaling with scaling factor β on model parameters

w_g	Area	V	I	g_m	C	R	f_t	f_{max}
β	β	con.	β	β	β	$1/\beta$	con.	con.

V: voltages, I: currents, C: capacitances, R: resistances, con: constant

5.4.6.4 Performance of 90-nm CMOS Technology

In Fig. 5.18, the power- and current gains of 90-nm SOI (Silicon on Insulator) CMOS n-channel FETs are depicted in a typical class-A operating point. An f_{max} of 147 GHz and an f_t of 149 GHz are extracted from measurements [Ell1]. Slightly higher speed performance with f_{max} up to 160 GHz and f_t up to 150 GHz is achieved at increased supply currents. It is noted that, due to some process variations, the speed performance of this IC hardware was not optimum. For ideal IC hardware and DC bias, values of f_t and f_{max} up to 243 GHz and 208 GHz, respectively have been reported for the n-FETs of this technology at V_{gs}=0.7 V and V_{ds}=1.2 V [Plo05, Zam04]. As will be discussed in Sect. 5.4.10.1, SOI technology provides slightly better performance than conventional CMOS bulk technology.

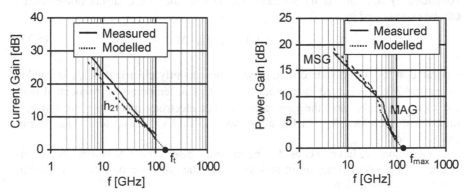

Fig. 5.18. Power and current gain of 90-nm SOI CMOS n-channel FETs with w_g=64 μm at V_{gs}=0.5 V, V_{ds}=1 V and I_d=17 mA vs frequency, and extraction of f_t and f_{max}

5.4.7 Noise Modelling

The noise modelling of transistors is complex since different effects influence each other. We follow the theories presented in literature [Lee04, Zie86, Sch03] and focus on n-channel devices if not otherwise mentioned. To allow meaningful analytical derivations and didactic insights it is reasonable to minimise the complexity of the equivalent circuit. The small signal equivalent circuits sketched in Fig. 5.19 provide a good compromise between complexity and accuracy. The capacitive input and the current source are considered.

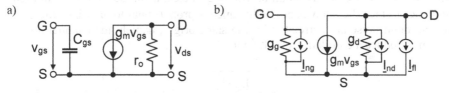

Fig. 5.19a,b. Simplified small signal models of the MOSFET in common source configuration: **a** typical FET model; **b** noise model with I_{ng} : thermal noise current source at gate, I_{nd} : thermal noise current at drain, I_{fl} : flicker noise, g_g: equivalent ohmic gate conductance, g_d: ohmic channel conductance

The resistive parasitics of the channel may be modelled by a voltage dependent resistor with r_o=r_{ds}||R_{ds}. However, r_{ds} does not generate any noise since it is not associated with any noise process. We assume that the ohmic nature of the output

impedance dominates and that all ohmic contributions are incorporated into the
ohmic channel conductance[5] g_d, which is mainly determined by R_{ds}. According to
Sect. 4.3.1 and Eqs. (4.17)–(4.18), g_d produces a thermal noise current. At zero
V_{ds}, we define a drain source admittance g_{d0}, which is a function of V_{gs}. The
dependency of a non-zero V_{ds} on g_d can be expressed by means of the drain noise
coefficient γ equalling unity for zero V_{ds}. Hence, the drain current noise source
can be represented by

$$\overline{I_{nd}^2} = 4 \cdot kT \cdot \gamma \cdot g_{d0} \cdot \Delta f . \tag{5.33}$$

Operated at low drain source voltage and subsequently electrical fields, a γ value
of around 2/3 is observed in saturation and at room temperature. Associated with
the electrical fields and the generation of hot electrons increasing the noise, γ
raises with increased V_{ds}. In literature, we can find values up to 1.5 for 90-nm
technologies.

   In addition, we have to consider the thermal noise at the gate caused by capacitive coupling between the drain and the gate. The corresponding noise contribution can be modelled by

$$\overline{I_{ng}^2} = 4 \cdot kT \cdot \delta \cdot g_g \cdot \Delta f , \tag{5.34}$$

where δ is the gate noise coefficient and

$$g_g \approx \frac{\omega^2 C_{gs}^2}{5 g_{d0}} , \tag{5.35}$$

which is a frequency dependent equivalent ohmic conductance. The derivation of
Eq. (5.35) can be gleaned in [Zie86]. Similar to γ, the value of δ depends on V_{ds}. δ
values in the range of 4/3 to 2 can be observed for $V_{ds} \geq 0$. Due to the coupling to
the channel, it is clear that the gate noise increases with raised electrical field.
While the gate noise is very small at low frequencies, it can be significant at high
frequencies since the equivalent conductance is proportional to ω^2. Since the gate
and the drain noise are correlated by the same origin, namely the channel noise,
we can introduce a correlation coefficient

$$\underline{c} = \frac{\overline{I_{ng} \cdot I_{nd}^*}}{\sqrt{\overline{I_{ng}^2} \cdot \overline{I_{nd}^2}}} . \tag{5.36}$$

---

[5] Real ohmic resistances are typically denoted in capital letters. However, for noise modelling, the
conductance may be an equivalent ohmic conductance rather than a conductance existing in reality. We
will see later that this applies for the gate conductance g_g. Consequently, for noise modelling, the conductances are denoted in small letters as it is common practice in literature.

Dominated by the impact of C_{gs}, the correlation coefficient is particularly reactive. A value[6] of approximately $-j0.4$ can be found in the literature for typical long channel MOSFETs [Lee04]. Referring back to Sect. 4.3.6, with given noise source impedance $\underline{Y}_s = G_s + jB_s$,

the noise factor $F = 1 + \dfrac{G_u + \left[(G_c + G_s)^2 + (B_c + B_s)^2 \right] \cdot R_n}{G_s}$ of a 2-port can be

characterised by the following device parameters:

- Equivalent noise resistance $R_n = \dfrac{\overline{V_n^2}}{4kT\Delta f}$, where $\underline{V}_n$ denotes the equivalent input noise voltage.

- Complex correlation admittance $\underline{Y}_c = \dfrac{\underline{I}_c}{\underline{V}_n} = G_c + jB_c$ with $\underline{I}_c$ as the total input noise current correlated with $\underline{V}_n$.

- Conductance $G_u = \dfrac{\overline{I_u^2}}{4kT\Delta f}$, where $\underline{I}_u$ considers the input noise current not exhibiting any correlation with $\underline{V}_n$.

Let's now derive equations for the latter 4 parameters (R_n, G_c, B_c and G_u) leading to relations for the optimum source impedance and F_{min}. The drain noise current associated with $\underline{V}_n$ amounts $\underline{I}_{nd} = g_m \cdot \underline{V}_n$. Hence, from Eqs. (4.38, 5.33) we get:

$$R_n = \frac{\gamma g_{d0}}{g_m^2}. \tag{5.37}$$

The input noise voltage does not account for the full drain noise current. In this context, the input noise current associated with $\underline{V}_n$ and the input admittance must be considered, which amounts

$$\overline{I_{n1}^2} = \overline{V_n^2} \left(j\omega C_{gs} \right)^2. \tag{5.38}$$

Again we have assumed that the input impedance is purely capacitive. The correlation admittance is defined as the total correlated input current divided by $\underline{V}_n$. The total correlated input current consists of $\underline{I}_{n1}$ and the correlated noise current $\underline{I}_{ngc}$. It follows that:

---

[6] The sign depends on the reference direction. Accordingly, in literature, we can find also values with positive sign.

$$\underline{Y}_c = \frac{\underline{I}_{n1} + \underline{I}_{ngc}}{\underline{V}_n} = j\omega C_{gs} + \frac{\underline{I}_{ngc}}{\underline{V}_n} = j\omega C_{gs} + g_m \cdot \frac{\underline{I}_{ngc}}{\underline{I}_{nd}} = j\omega C_{gs} + g_m \cdot \frac{\overline{\underline{I}_{ng} \cdot \underline{I}_{nd}^*}}{\overline{\underline{I}_{nd}^2}}. \quad (5.39)$$

The last term can be rearranged together with the correlation coefficient $\underline{c}$ resulting in

$$\underline{Y}_c = j\omega C_{gs} + g_m \cdot \underline{c} \cdot \sqrt{\frac{\overline{\underline{I}_{ng}^2}}{\overline{\underline{I}_{nd}^2}}}. \quad (5.40)$$

With Eqs. (5.33)–(5.35) and (5.40), and $\underline{c}$ assumed to be purely imaginary we get

$$G_c = 0, \quad (5.41)$$

and in turn

$$\underline{Y}_c = j\omega C_{gs}\left(1 - \frac{g_m}{g_{d0}} \cdot |\underline{c}| \cdot \sqrt{\frac{\delta}{5\gamma}}\right). \quad (5.42)$$

The optimum susceptance of the source admittance $\underline{Y}_s = G_s + jB_s$ can be obtained with Eqs. (4.34), (4.40), and (5.42) yielding an optimum B_s of

$$B_{opt} = -\omega C_{gs}\left(1 - \frac{g_m}{g_{d0}} \cdot |\underline{c}| \cdot \sqrt{\frac{\delta}{5\gamma}}\right). \quad (5.43)$$

With Eqs. (4.34) and (4.40), it can be deduced that

$$B_c = \omega C_{gs}\left(1 - \frac{g_m}{g_{d0}} \cdot |\underline{c}| \cdot \sqrt{\frac{\delta}{5\gamma}}\right). \quad (5.44)$$

The uncorrelated part of the gate noise current is given by

$$\overline{\underline{I}_u^2} = \overline{\underline{I}_{nd}^2} \cdot \left(1 - |\underline{c}|^2\right). \quad (5.45)$$

Hence, with Eqs. (5.34), (5.35), (4.37) and (5.45) we find

$$G_u = \frac{\delta\omega^2 C_{gs}^2\left(1 - |\underline{c}|^2\right)}{5g_{d0}}. \quad (5.46)$$

Equation (4.41), (5.37) and (5.46) lead to an optimum source admittance G_s of

$$G_{opt} = \frac{g_m}{g_{d0}} \cdot \omega \cdot C_{gs} \sqrt{\frac{\delta}{5\gamma}\left(1-|\underline{c}|^2\right)} . \qquad (5.47)$$

Finally, by Eqs. (4.41) and (5.47), the minimum noise figure yields

$$F_{min} = 1 + \frac{2}{\sqrt{5}}\frac{\omega}{\omega_t}\sqrt{\gamma\delta\left(1-|\underline{c}|^2\right)} . \qquad (5.48)$$

Indeed, the derivations of these equations are not trivial. To keep the overview, the key parameter are summarised in Table 5.8.

Table 5.8. Summary of FET noise parameters

General parameters	Parameters for minimum noise
$R_n = \dfrac{\gamma g_{d0}}{g_m^2}$	
$G_c = 0$	
$B_c = \omega C_{gs}\left(1-\dfrac{g_m}{g_{d0}}\cdot|\underline{c}|\cdot\sqrt{\dfrac{\delta}{5\gamma}}\right)$	$B_{opt} = -\omega C_{gs}\left(1-\dfrac{g_m}{g_{d0}}\cdot|\underline{c}|\cdot\sqrt{\dfrac{\delta}{5\gamma}}\right)$
$G_u = \dfrac{\delta\omega^2 C_{gs}^2\left(1-|\underline{c}|^2\right)}{5g_{d0}}$	$G_{opt} = \dfrac{g_m}{g_{d0}}\cdot\omega\cdot C_{gs}\sqrt{\dfrac{\delta}{5\gamma}\left(1-|\underline{c}|^2\right)}$
Result	
$F = 1 + \dfrac{G_u + \left[\left(G_c + G_s\right)^2 + \left(B_c + B_s\right)^2\right]\cdot R_n}{G_s}$	$F_{min} = 1 + \dfrac{2}{\sqrt{5}}\dfrac{\omega}{\omega_t}\sqrt{\gamma\delta\left(1-|\underline{c}|^2\right)}$

$$\underline{c} \approx -j0,4$$

From Eq. (5.48) we can derive some important insights:

- F_{min} can be improved by lowering of the $\dfrac{\omega}{\omega_t}$ ratio. As expected F_{min} raises with frequency. F_{min} can be decreased by increasing of ω_t, which in turn depends on l_g and the applied supply current.

- F_{min} is proportional to $\sqrt{\gamma\delta}$. Remember that, due to the raised electrical field and hot electron effects, typically γ and δ become larger with decreased l_g. However, concerning scaling we observe that the impact of the noise source factors γ and δ is less significant than the benefit associated with the $\dfrac{\omega}{\omega_t}$ ratio.

- F_{min} is not directly dependent on w_g since the parameters ω_t, γ and δ remain constant vs w_g. However, with respect to the matching of a specific source impedance, w_g can play a remarkable role when considering the associated losses of the impedance matching network.

- Although we have assumed a purely capacitive input impedance, we observe that the imaginary part of the optimum source admittance B_{opt} depends not only on C_{gs} but also on other parameters. Why that? B_{opt} considers also the compensation of the correlated noise power contributions associated with the gate and the drain. From this fact we can deduce that noise and power/return loss matching is not possible simultaneously. Thus, a tradeoff between the matching approaches has to be made.

To allow analytical calculations, the presented noise model is based on many simplifications. However, it provides sufficient accuracy for many applications and gives fruitful insights in terms of major noise effects inherent in FETs. For more precise noise predictions, we can perform computer simulations based on much more complex equivalent circuit models. It is straightforward that all further resistances such as the contact resistances R_s, R_d and R_g also exhibit thermal noise and can be considered using an equivalent noise source according to Eq. (4.18). Aggressively scaled technologies exhibit significant gate leakage, which can be considered by a thermal noise source in parallel to C_{gs}.

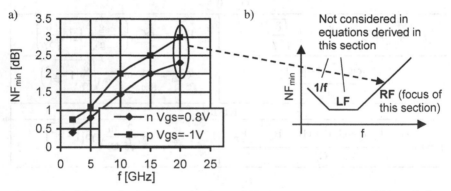

Fig. 5.20a,b. Measured minimum noise figure (NF_{min}): **a** measurements of 90-nm bulk n-channel and p-channel MOSFETs versus frequency at V_{ds} of 1 V and -1 V, and I_D of 17 mA and 9.5 mA, respectively, w_g=32×2 µm; **b** illustration versus full frequency range including the RF zone according to Fig. 5.20a, and additionally low frequency (LF) and flicker noise range

In Fig. 5.20a, the NF_{min} of n-channel and p-channel transistors of typical 90-nm CMOS technology is shown. As expected with respect to Eq. (5.48), the noise increases with frequency. At 20 GHz and optimum low noise bias, the n-channel and p-channel transistors yield an NF_{min} of 2.3 dB and 3 dB, respectively. It can be concluded that n-channel devices provide better performances at the costs of higher current consumption. One reason is the higher electron mobility of n-channel devices, which rises the channel conductance and the channel current. Consequently, ω_t increases and according to Eq. (5.48) F_{min} falls. Due to their superior properties in terms of noise, most low noise circuits apply n-channel transistors.

In addition to the discussed noise sources, we have to consider the flicker noise. Referring to Eq. (4.20), the associated noise power is inversely proportional to frequency. Thus, 1/f noise is the dominant noise source at frequencies below the 1/f corner frequencies, which are typically in the range of 0.1–1 MHz. The corresponding noise source can be included between the drain and the source as been shown in Fig. 5.19b. Since the trap charging process associated with the flicker noise is very slow, the contribution of the flicker noise is low at high frequencies. Flicker noise may have a significant contribution at high frequencies when up-converted by nonlinear effects. The corresponding behaviour will be explained in the section concerned with oscillators.

At moderate frequencies (let's say from the sub-MHz to the lower GHz range), one typically observes a region with nearly constant NF_{min} vs frequency. This behaviour is attributed to the effect of impedances with constant real part and negligible reactive contribution. The resulting performance ranging from DC to RF is illustrated in Fig. 5.20b.

5.4.8 Scaling

If nano research is the Mt. Everest, we have barely reached the base camp!
Charles M. Lieber, Harvard University

One of the most significant speed limitation of MOSFETs is the carrier transit time through the channel below the gate. As discussed before, the speed can be improved by decreasing of l_g. To meet the market demands of VLSI (Very Large Scale Integration) circuits in terms of chip density and size all other dimensions have to be scaled as well as illustrated in Fig. 5.21. Moreover, this overall scaling is mandatory to prevent from excessive short channel effects. Just four years after the first planar integrated circuit has been discovered Gordon Moore made a famous observation in 1965. The press called it "Moore's Law" and the name has stuck. Moore observed that the number of transistors per integrated circuit doubles every year, and predicted that this trend would continue [Moo65, Bon98]. This is verified in Fig. 5.22a up to date. It is remarkable that this law has held now for over 40 years. Moreover, the data density in integrated circuits doubles every 18 months. The question is how long this performance improvement is possible. A basic requirement to keep pace with the law is the continuous scaling of device dimensions. The required l_g scaling versus time is illustrated in Fig. 5.22b. Due to limits concerning the maximum reliable electric fields and current densities, the voltages and currents have to be scaled together with the geometrical dimensions. This case is called constant electrical field scaling.

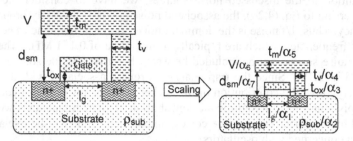

Fig. 5.21. VLSI scaling of CMOS technology, α_{1-7}: scaling factors, l_g: gate length, t_{ox}: thickness of gate oxide, t_m: metal thickness, t_v: metal via thickness, a_{sm}: distance from metal to active substrate, ρ_{sub}: substrate resistivity, V: applied voltage. For simplicity only one metal layer is shown

In the second column of Table 5.9, the resulting scaling values are summarised assuming that the scaling factors are equal. At the same time, an f_t improvement, and a circuit area and power consumption reduction of α^2 can be observed. However, due to the problems associated with the decreasing supply voltages, for circuits such as amplifiers, the supply voltages may not scale in the same amount as the other parameters do. In this case, this reader is referred to the third column of Table 5.9, where an additional electric field scaling parameter ε is considered.

Table 5.9. Ideal scaling rules

Parameter	Constant electrical field	Scaled electrical field
Electrical field	1	ε
Gate length	$1/\alpha$	$1/\alpha$
Wiring width	$1/\alpha$	$1/\alpha$
Voltages	$1/\alpha$	ε/α
Current	1	ε/α
Substrate doping	α	$\varepsilon\alpha$
Substrate resistivity	$1/\alpha$	$1/\alpha$
Area	$1/\alpha^2$	$1/\alpha^2$
Parasitic capacitances	$1/\alpha$	$1/\alpha$
Transconductance	α	α
Transit frequency	α^2	α^2/ε
Power dissipation	$1/\alpha^2$	ε^2/α^2
Power density	1	ε^2

$\alpha = \alpha_{1..7}$, two scaling cases are shown: 1. constant electrical field, and 2. non-constant electrical field represented by electrical field scaling factor ε

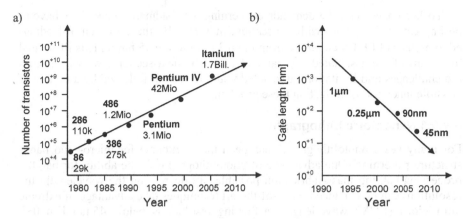

Fig. 5.22a,b. VLSI scaling: **a** number of integrated transistors per chip, verification of Moore's Law **b** gate length required to achieve performance

5.4.9 Scaling Performances and Limits

In May 2003, the CTO of IBM stated the following comment at the International Electronics Forum in Prague:

> *Scaling is already dead but nobody noticed it had stopped breathing*
> *and its lips had turned blue.*
> Bernard S. Meyerson

The latter comment points out that towards the nanometre range, it is not possible to achieve the ideal scaling characteristics by scaling alone.

Table 5.10. Key performance parameters for leading-edge n-channel FETs, V_{th}: threshold voltage, V_{dd}: drain dc voltage, t_{ox}: oxide capacitance, g_m: transconductance, g_{ds}: drain source channel conductance, compare [Pek04, Jag06]

Node	nm	350	250	180	130	90	65	45
l_g	nm	250	180	130	92	63	43	30
t_{ox}	nm	9	6.2	4.5	3.1	2.2	1.8	1.6
V_{dd}	V	3.3	2.5	1.8	1.5	1.2	1	0.8
V_{th}	V	0.6	0.44	0.43	0.34	0.36	0.24	0.2
$g_{m, peak}$	µS/µm	225	335	500	720	1060	1400	1900
g_{ds} at $g_{m,peak}$	µS/µm	12	22	40	65	100	230	400
g_m/g_{ds}	-		18	15.2	12.5	11.1	10.6	6.1	4.8
$f_{t, max}$	GHz	23	35	53	94	140	210	290

The most relevant limitations involve the resolution of the lithography, the maximum supply and operation voltage, the threshold voltage, the leakage through the gate oxide, the reduced intrinsic gain and the reliability.

To keep pace with the demands concerning the roadmap, innovations have to be implemented together with the scaling. In Tab. 5.10, the evolution of leading-edge n-channel FET technology from the 0.35 µm to the 45 nm node is illustrated by means of the associated key parameters. In the next sections, we will discuss the challenges and limitations associated with the scaling followed by a review of possible innovations mitigating these problems.

5.4.9.1 Nanometre Lithography

For many years photolithography has been the workhorse for micro- and nano-structure patterning. One advantage of photolithography is the ability for parallel processing making mass-fabrication possible. By decreasing the wavelength, the resolutions can be enhanced. State-of-the-art techniques take advantage of extreme ultra violet (EUV) wave-lengths achieving resolutions below 45 nm [End03]. Higher resolution can be achieved with electron beam lithography. Resolutions in the order of 10 nm have been demonstrated [Lan94]. The resolution is degraded by electron scattering, known as the proximity effect. Corresponding correction methods have been reported [Har00]. The disadvantage of electron beam lithography is that it allows only serial processing thereby significantly increasing the fabrication time. Thus, electron beam lithography is mainly used for prototyping and mask writing for photolithography.

5.4.9.2 Threshold and Supply Voltage

Referring to Tab. 5.10, we can identify the following trends:

- To minimise the power consumption, the FET off-current should be kept as low as possible. For most circuits, the off-voltage is defined by a voltage of zero. A decrease of the threshold V_{th} towards zero increases the off-current. This is not acceptable because of the increased power consumption. The impact due to process variations increases towards low V_{th}, thereby decreasing the yield and raising the costs. Consequently, to mitigate these undesired effects, towards the nanometre regime, V_{th} can't be scaled by a constant value.
- To prevent from electrical breakdown, V_{dc} has to be scaled. On one hand, the power consumption can be decreased, whereas on the other hand the maximum possible output voltage and power is reduced. This is a significant problem for analogue RF applications such as power amplifiers and drivers.

To preserve a sufficient V_{dc} to V_{th} margin, V_{dc} is less scaled than a constant electrical field would demand. This results in an increased electrical field represented by the electric field scaling factor ε as been outlined in the third column of Table 5.9.

5.4.9.3 Gate Oxide

As mentioned before, in the ideal case, all dimensions should be scaled to protect from excessive short channel effects. The most significant limit is given by the thickness of the thin gate isolation layer realised by silicon oxide ($Si0_2$). If it

would not be scaled as well, the gate terminal would be too far away to have sufficient impact on the channel leading of a reduction of g_m. In this case V_{th} would
approach unpractical values leading to conditions where the channel can't be controlled any more. Thus, a fairly constant t_{ox}/l_g ratio has to be maintained. In Tab.
5.10, typical values for the gate oxide thickness t_{ox} are illustrated vs l_g. Technologies with l_g around 65 nm feature already ultra low thicknesses of 2.2 nm corresponding to only a few atom layers when considering an atom diameter of around
0.1 nm. It is obvious that we are already very close to the oxide scaling limit.
Much thinner t_{ox} is not possible without significant degradation of the reliability
and the yield of the devices. Due to the associated quantum tunnelling, a strong
gate leakage is generated, which increases the noise and the power consumption.

5.4.9.4 Speed Performance

The f_t performances vs l_g are listed in Tab. 5.10. As discussed in Sect. 5.4.6.1, the
f_t is proportional to $1/l_g^2$ for ideal long channel devices when assuming a linear
scaling of both g_m and C_{gs}. This relation typically holds for devices down to the
0.5 µm node. However, for aggressively scaled devices, g_m increases less than expected due to saturation effects caused by the raised electrical field in the channel

leading to $f_t \sim \dfrac{1}{l_g^s}$ with $s < 2$ typically ranging from 1.6 to 1 for the sub-0.1 µm re-

gime. Further f_t degradations are due to scaling limits associated with t_{ox} and
extrinsic parasitics. To preserve $s > 1$, innovations as discussed in Section 5.4.10
have to be applied. An f_t up to 210 GHz has been demonstrated in 65 nm technology. However, considering typical connection parasitic in circuits, the f_t may be
25 % lower.

5.4.9.5 Gain

We have observed the speed improvement by means of l_g scaling. On the other
hand, for most circuit applications we need gain to exploit the speed. Referring to
Section 7.1.1.3, the magnitude of the voltage gain of the frequently used common

source circuit is bounded by $\left| \underline{a}_v \right| = \dfrac{g_m}{g_{ds}}$, where the relation $\dfrac{g_m}{g_{ds}}$ is referred to as in-

trinsic gain. Unfortunately, the channel conductance increases when l_g is scaled
down. Hence, the maximum possible impedance seen as the load decreases.
Moreover, recall that the g_m scaling is constrained by the high electrical field and
the nonideal t_{ox} scaling. Hence, the maximum possible intrinsic gain and conse-
quently $\left| \underline{a}_v \right|$ are limited. In Tab. 5.10, the intrinsic gain is listed versus technology
generation. In 65 nm technology, an intrinsic gain of less than 5 is feasible. Com-
pared to a value of 15 possible in 0.25 µm technology, this is a remarkable degradation. We can conclude that speed is improved by the expense of gain.

5.4.9.6 Passive Devices

The scaling strategy of digital CMOS technologies focuses on the optimisation of transistors rather than of passive devices. Device and metal dimensions are scaled down to increase the transistor speed and the integration density. Unfortunately, this degrades the performance of passive devices [Ell40]. Major reasons are:

- The coupling to the lossy substrate is increased due to the lowering of the distance between conductor metals and substrate.
- The series resistance of a simple metal line can be approximated by

$$R_s = \rho_m \cdot \frac{l_m}{w_m \cdot t_m} \propto \alpha, \tag{5.49}$$

where ρ_m is the metal resistivity, l_m is the metal length, w_m denotes the metal width and t_m is the metal thickness. From Eq. (5.49) it is obvious that R_s increases with the scaling factor α, which is anti-proportional to the dimensions as defined before. All dimensions are scaled but not ρ_m.
- To decrease V_{th} as mentioned in Sect. 5.4.9.2, the substrate doping has to be increased resulting in a lower resistivity. Unfortunately, this increases the parasitics associated with the substrate and significantly degrades the performance of area consuming passive elements such as inductors.
- Metal filling, typically up to 40%, is required for yield reasons. This raises the capacitive coupling to the lossy substrate. Filling exclusion is strongly limited or not allowed at all.
- For design rule and density reasons during fabrication, holes are made in metals thereby further increasing the parasitic series resistance of lines. This process step is called cheesing.
- Most low-cost VLSI processes do not provide optimised MIM (Metal Insulator Metal) capacitors with high dielectric constant and small distance between the metal plates. Thus, they require a large area. Consequently, they have a strong coupling to the lossy substrate and relatively high losses.

These effects have a significant impact on analogue circuits, where passive devices such as inductors, transmission lines, capacitors and pads with high Q are required. Circuit concepts, which do not require inductors have advantages since their influence on the mentioned drawbacks are reduced. However, such inductor-less topologies suffer from significantly lower performances in terms of operation frequency, noise and power consumption. Passive devices will be elaborated in Chap. 6.

5.4.10 Advanced CMOS Techniques

Any sufficiently advanced technology is indistinguishable from magic.
Arthur C. Clarke, science fiction author

The preceding sections have revealed that traditional scaling is not sufficient any more to comply with the demands of the semiconductor roadmap. This is illustrated in Fig. 5.23, where the progress by traditional scaling and the benefits of innovations are compared for IBM CMOS generations. Innovative ideas and the use of new transistor, material and fabrication concepts become mandatory. Some of the most promising innovations are SOI (Silicon on Insulator), low-k gate dielectrics, strained silicon, copper metal and low-k interconnect dielectrics. The potential of MEMS (Microelectrical Mechanical System) technologies and the operation at low temperatures have already been demonstrated. In the future, promising enhancements are expected by multi-gate FETs. These techniques are discussed in the following sections.

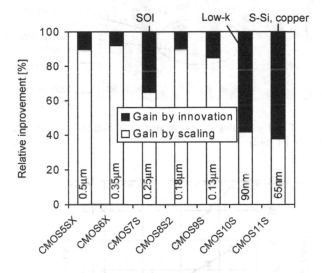

Fig. 5.23. Comparison of transistor speed improvement by scaling and innovation of IBM CMOS technology versus generation and gate length. Significant enhancements are based on the silicon on insulator (SOI), low gate dielectric (low-k) and strained silicon technology (S-Si). The performance of interconnects is improved using copper metal. Compare [Wal04]

5.4.10.1 Silicon on Insulator Technology

A solution for enhanced l_g scaling is the SOI technology. Referring to Fig. 5.24, an SOI isolation layer is implanted between the thin active device plane and the main substrate. Thus, the resistivity of the active layer determining the threshold voltage and the resistivity of the main substrate can be set independently. Consequently, relatively high substrate resistivities can be chosen for the main substrate without degrading the V_{th} properties. This results in reduced parasitics for active and passive devices. Improvements of f_t in the order of 30 % have been demonstrated [Ibm42].

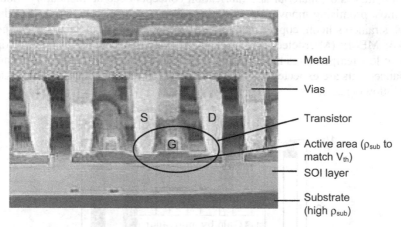

Fig. 5.24. Illustration of SOI technology: implementation of SOI layer to separate the active from the main substrate [Ibm42]

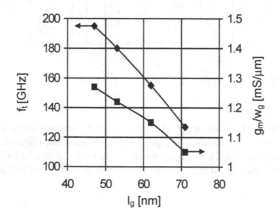

Fig. 5.25. Measured transit frequency (f_t) and transconductance (g_m) of IBM SOI n-channel MOSFETs versus effective gate length versus effective gate length [Zam02]

In addition, the isolation is significantly improved. This can be important for circuits being susceptible to undesired coupling, feedback or noise pickup. A silicon substrate behaves like a high-pass filter with a cut-off frequency, which may be approximated by

$$f_{c,sub} = \frac{1}{2\pi \rho_{sub} \varepsilon_{sub}},$$ (5.50)

where ρ_{sub} and ε_{sub} are the resistivity and the dielectric constant of the silicon substrate, respectively. At substrate resistivities of 1.5 Ωcm, 15 Ωcm and 150 Ωcm, the substrate cut-off frequencies are approximately 1 GHz, 10 GHz and 100 GHz, respectively [Plo03]. The higher $f_{c,sub}$ the higher the isolation. Due to the DC isolation effect of the SOI layer, no bulk contact is necessary.

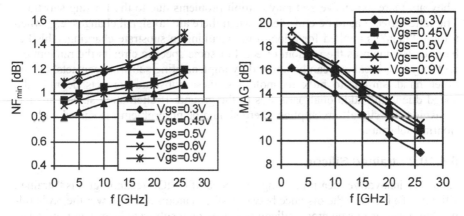

Fig. 5.26. Measured minimum noise figure and maximum available gain of 90-nm SOI n-channel MOSFETs vs frequency and V_{gs} at V_{ds}=1 V, w_g=64×1 μm

For fabrication, yield and cost reasons, the thickness of the SOI isolation layer is very thin. Typically, the thickness of VLSI SOI layers is below 0.1 μm. Therefore, the RF isolation effect of the SOI layer is very weak compared to the much thicker isolation layer between substrate and metal stack.

Figure 5.25 illustrates the characteristics of f_t and g_m versus l_g for typical SOI technology. For conventional bulk technology, the speed enhancement due to scaling is less efficient. Matching of the V_{th} requirements in bulk technologies worsens the isolation properties thereby partly compensating the beneficial effect of l_g scaling.

The measured NF_{min} and gain properties of SOI n-channel FETs vs gate bias at V_{ds}=1 V are illustrated in Fig. 5.26. At 20 GHz and optimum low noise bias of 17 mA, the transistor yields a very low NF_{min} of 1 dB and a MAG of 13 dB.

In Table 5.11 [Rob02], the noise performance of state-of-the-art III/V technologies are compared. The SOI technology outperforms the gallium arsenide (GaAs) high electron mobility transistor (HEMT) process and is only slightly worse than the indium phosphide (InP) HEMT technology. Due to the lower substrate losses, SOI technologies demonstrate also better performances for the passive devices. Further information can be found in Sect. 6.2.5.3.

Table 5.11. Comparison of minimum noise figures of CMOS and state-of-the-art III/V technologies around 20 GHz

SOI CMOS	Bulk CMOS	GaAs		InP	
N-FET	N-FET	PHEMT	MMHEMT	LMHEMT	PHEMT
1 dB	2.3 dB	1.6 dB	1.2 dB	0.4 dB	0.35 dB

Devices operate at optimum noise bias, PHEMT: pseudo-morphic HEMT, MMHEMT: metamorphic HEMT, LMHEMT: lattice-matched HEMT.

SOI processes exhibit two drawbacks compared to conventional bulk technology. They are more expensive and may exhibit problems due to the floating substrate. The latter effect has to be considered when there are no signal changes, e.g. when a logic "high" is applied for a long time resulting in substrate charging. This behaviour is known as the random effect. For some specific circuits the random effect may be critical and can be omitted by implementing substrate contacts. On the other hand, similar to bulk technologies, these substrate contacts degrade the speed due to the additional parasitics to the ground. If alternating signals such as sinusoidal RF signals are applied, there should be no problem associated with the floating substrate.

5.4.10.2 Strained Silicon

A further innovative idea improving the speed of CMOS technologies is "strained silicon". To increase the distance between silicon atoms and to lower the probability that a carrier hits an atom, silicon is placed on a substrate layer featuring a larger lattice constant, e.g. on SiGe (silicon germanium). Thus, the silicon has to "strain". As a consequence, the mobility in the channel is improved. This is illustrated in Fig. 5.27. By using silicon with a 15–20 % Ge portion for the substrate layer, the electron mobility in the silicon channel has been enhanced by 70% leading to a speed improvement in the order of 30 % [Rim01]. Especially at high Ge fractions, the hole mobility can also be improved with ratios up to 5. However, the enhancement for holes diminishes at high electrical fields [Rim02]. Due to the existing knowledge in terms of SiGe processing, corresponding technologies will probably be available on the market within a few years. The most difficult issue is to maintain the strain while manufacturing a thin channel.

Fig. 5.27. Comparison between common and strained silicon MOSFET technology [Ibm43]

5.4.10.3 Multi Gate MOSFETs

> *Finally, it is the spirit, which makes technology alive.*
> Johann Wolfgang von Goethe

A further prime candidate capable of enhancing the speed and the scaling proper-ties is the DG (Double Gate) FET [Tau97, Ieo02]. Many papers on this topic have been published over the last decade covering a broad range of investigations such as device design [Fra92, Won98], device fabrication [Wei04, Won97] and device physics including modelling [Tau01, Ieo01]. A simplified schematic of a DG MOSFET with planar aligned gates is illustrated in Fig. 5.28a. Two gates, one at the top and one at the back of the channel are implemented. These two gates are electrically connected so that they can both serve to modulate the channel. The following advantages can be observed with respect to conventional MOSFETs:

- Much better controllability of the channel is achieved since the field acts from both sides. Velocity degradation due to carriers travelling close to the lossy substrate is circumvented.

- In principle, two channels are connected in parallel. Hence, more current can flow at given gate source voltage.
- A steeper sub-threshold slope and subsequently lower gate leakage current is achieved.
- Due to volume-like definition of the gate potential, the impact of the pinch off effect is decreased. Thus, variations of V_{th} with respect to V_{ds} and l_g are smaller improving the yield and reliability.

Consequently, more aggressive scaling is possible with the double gate transistor. Monte Carlo device simulations and analytical calculations predict continuous improvement in device performance down to 20-30 nm l_g without degrading the performance with respect to undesired short channel effects [Won97].

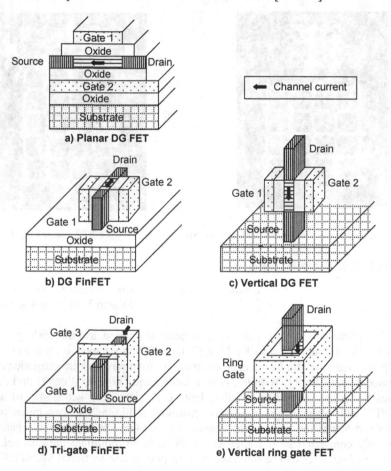

Fig. 5.28. Illustration of multi-gate MOSFETs

The fabrication of DG MOSFETs is challenging. In the horizontally aligned double gate MOSFET, the fabrication and low-loss contacting of the back gate is difficult. Both gates should preferably provide the same delays to modulate the channel in phase. Full symmetry can be obtained by applying laterally aligned gates. Corresponding devices can be realised as DG FinFETs[7] or vertical DG FETs as depicted in Fig. 5.28b,c, respectively. The idea behind the DG devices, namely the enhancement of the channel control can be further advanced by applying tri-gate or ring-gate FETs. Strongly simplified illustrations are sketched in Fig. 5.28d,e, respectively. First tri-gate devices have been realised showing promising performance [Doy03]. Ring-gate FETs will be subject to future research. Three-dimensional processing is required, which is challenging and costly.

5.4.10.4 High Barrier Gate Oxide

It has been mentioned in Sect. 5.4.9.3 that quantum tunnelling in thin oxide layers is a major limitation for CMOS scaling. According to Fig. 5.29, this leads to a gate leakage current that increases exponentially as t_{ox} is scaled down. Typically, MOSFETs with 20 nm l_g have leakage current densities in the order of 10^{-2}–10^{-1} A/cm^2. One possibility to mitigate the leakage is to use high dielectric gate materials with an increased barrier for the tunnelling [Cha03]. Promising candidates are high-k[8] materials such as hafnium oxide (HfO_2) and zirconium oxide (ZrO_2), as well as their silicates. According to Fig. 5.29, adding of hafnium decreases the gate leakage current by a factor of more than 10000 for a 20-nm device.

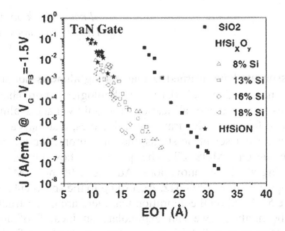

Fig. 5.29. Gate leakage current density vs effective oxide thickness (EOT) for devices with l_g=20 nm using silicon and hafnium gate oxides © 2004 IEEE [Akb04]

---

[7] Name related with the shape of transistor, Fin does not represent an abbreviation.

[8] k means value of dielectric constant.

The manufacturing of such high-k dielectrics is challenging since impurities and traps at the interfaces have to be avoided. One disadvantage is the increased parasitic gate capacitance caused by the high dielectric constant. However, for most digital VLSI applications, this degradation is less severe than the benefit gained by the reduced gate leakage.

5.4.10.5 Low Temperature

The potential to improve the performance of MOSFETs at low temperatures is known for a long time [Gae77]. Cooling systems for server systems have been developed and are on the market since several years [Sch00]. Major reasons for the enhancements are higher carrier mobility and lower interconnect resistance yielding performance gains in the order of 1.5-2 compared to the operation at room temperatures [Sun87]. As discussed in previous sections, the gate leakage is a significant problem for CMOS devices with small threshold voltages. In this context, low temperature operation provides an additional advantage. The sub-threshold slope of the leakage current versus the gate voltage significantly increases at low temperatures [Jae88]. This results in lower leakage currents mitigating the problem of scaled threshold voltages.

Nevertheless, cooling has considerable disadvantages. Mandating for additional supply power and significantly increased system size, the method is not suited for mobile and low-power applications. But maybe in Siberia …

5.4.10.6 Conclusions and Speed Extrapolation

I never think about the future.
The future comes soon enough.
Albert Einstein

The important lesson we have learned is that scaling alone is not sufficient any more to improve the speed of sub 0.1 µm technologies. Short channel effects, boundaries in terms of gate oxide fabrication, threshold voltage scaling limits and gate leakage demand for innovative transistor principles, materials and fabrication procedures. At the current scientific status, the most promising are SOI, strained silicon and the double gate MOSFET. The question is how far we can go when successfully applying all these innovations. Advanced 90 nm CMOS transistors provide an f_t of approximately 120 GHz. According to the speed enhancement factors listed in Table 5.12 and on the condition that there are no destructive dependencies between the methods, we can extrapolate an ideal f_t of approximately 400 GHz at room temperature. Additional cooling may lead to a further enhancement of around 0.8 THz. To minimise the leakage effects, high-k dielectrics with enhanced barrier properties have to be developed for the gate oxides. The author is convinced that technologies with such speeds will be available in research laboratories within the next 5 years and on the commercial market in the next 10 years. To keep pace with the advancing speed of the transistors, technologies such as low-k dielectrics, copper metals and MEMS have to be applied for the passive devices and interconnects.

Table 5.12. Improvement factors.

Innovation	SOI	Strained silicon	Double gate MOSFET	Cooling
Effect	Less parasitic capacitances	Higher electron mobility	Scaling down to 20nm without increasing short channel effects	Higher mobility and lower resistances
Enhancement	30%	30%	200%	200%
Reference	[Ibm42]	[Rim01]	[Won97]	[Sun87]

5.5 MESFET

Compared to the MOSFET, the MESFET (Metal Semiconductor Field Effect Transistor) has many similarities. To avoid redundancies, we will refer back to the MOSFET properties where appropriate. The MESFET is also based on the drift transport of majority carriers between the source and the drain forming the channel.

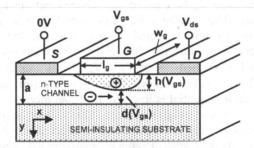

Fig. 5.30. Depletion MESFET, charge in channel with thickness a is controlled by adjusting thickness of depleted Schottky barrier zone thereby influencing channel thickness $d(V_{gs})$.

Let's point out the main differences between the MESFET and the MOSFET. In the case of the MOSFET, the capacitive coupling from the gate to the channel and the DC decoupling is performed by means of a MOS capacitor. For the same tasks, MESFETs feature a metal semiconductor Schottky diode. The control of the channel conductance in MESFETs is possible by variation of the channel thickness. In MOSFETs the carrier density is varied to perform this task. In the beginnings, silicon has been used for MESFET fabrication. Today, the most common material for MESFETs is GaAs exhibiting superior electron mobility compared to silicon. Occasionally, MESFETs are also fabricated in InP offering even higher electron mobility than in GaAs. However, since GaAs and InP have much lower hole mobility than silicon, it is not reasonable to fabricate complementary logic requiring p-channel devices in addition to the n-channel FETs. In comparison, complementary logic in silicon is faster, provides higher integration density, has lower static power consumption and is less expensive in mass-fabrication. By the way, the major reason why MOSFETs are not realised in GaAs is the fact that the required GaAs-oxides have a poor quality and yield. Much better oxides can be processed in silicon.

5.5.1 Functional Principle

Since the MESFET is fabricated as n-channel device only, and most MESFET applications are optimised for power applications, we focus on the discussion of the n-channel depletion MESFETs. Recall that the D-FET has a negative V_{th}[9] allowing for a high gate overdrive voltage and channel current. The gate source control voltage V_{gs} and the channel are isolated, since the depleted zone of the Schottky barrier is very high-ohmic and mainly capacitive. GaAs substrates offer very high resistivities. Thus, no bulk terminal as for the silicon bulk MOSFET is required. Fortunately, the current flowing through the isolating substrate is extremely small and therefore negligible. The channel resistance is controlled by adjusting the thickness of the depleted Schottky diode $h(V_{gs})$, thereby determining the effective channel thickness $d(V_{gs})$ bounded by the fixed channel thickness a as shown in Fig. 5.30. The impact of a varying V_{gs} at zero V_{ds} is illustrated in Fig. 5.31a.

A drain source voltage V_{ds} is required to motivate electrons to move from the source to the drain. The resulting Schottky barrier zones are sketched in Fig. 5.31b. With growing drain potential towards the drain, the effective gate to channel potential decreases. Thus, the channel thickness is decreasing towards the drain and the channel is pinched-off for $V_{ds} \geq V_{ds,sat}$. It is emphasised that similar to the MOSFET, the MESFET channel is never pinched-off totally. At $V_{ds} \geq V_{ds,sat}$, the effective drain potential is lowered due to an increased voltage drop in the pinch-off region exhibiting high resistance. This voltage drop reduces the effective channel voltage leading to a compensation effect. Thus, in first order estimation, the resulting channel current remains approximately constant for $V_{ds} \geq V_{ds,sat}$.

a)   b)

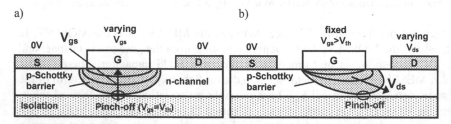

Fig. 5.31a,b. Depletion MESFET: **a** variation of V_{gs}, the channel is pinched-off at negative $V_{gs}=V_p$; **b** increasing of V_{ds} pinches off the channel, thus keeping I_d constant

---

[9] In literature, the negative threshold voltage of depletion MESFETs is frequently called pinch-off voltage. To maintain consistency with the MOSFET definitions and to minimise the potential for confusion, we call this voltage negative threshold voltage as for the MOSFETs.

5.5.2 DC Characteristics

After the qualitative considerations, we will now discuss the Shockley model, which allows first quantitative insights regarding the DC characteristics [Bäc02]. In Fig. 5.32, the simplified structure is shown. To simplify the calculations we make the following assumptions:

a. Gradual channel approach: $l_g \gg a$; thus short channel effects are not considered.
b. h is a linear function of x and is abruptly reduced to zero at edges of gate.
c. Channel is completely pinched off for $V_{ds} \geq V_{sat}$ yielding $h(x=l_g)=a$.
d. Substrate is an ideal isolator.
e. Mobility μ_n is constant.
f. Constant doping in channel.

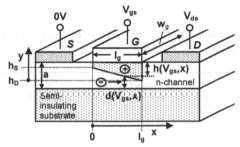

Fig. 5.32. Shockley model for the calculation of the MESFET IV characteristics

I. Threshold Region

The threshold is reached at the point where the channel is pinched off, h=a and subsequently d=0. Referring to common literature addressing metal to semiconductor junctions [Sze88, Bäc02], the depletion layer thickness can be expressed by

$$h(x) = \sqrt{\frac{2 \cdot \varepsilon \cdot V_{BP}}{q \cdot N_D}},$$ (5.51)

where the barrier potential is given by $V_{BP} = \phi_B - V_{gs} + V(x)$, q denotes the elementary charge, and N_D is the donator doping in the semiconductor. ϕ_B is the surface potential between the metal and the semiconductor, and V(x) is the potential along the channel associated with V_{ds}. By the way, the surface potentials have delayed the invention of the transistor. It took a while until the Bell Lab guys understood its existence and properties. Let us calculate now V_{BP} for $V_{ds}=0$ simply leading to V(x)=0. Pinch off is reached when h=a. In this case, as per definition, V_{BP} equals the pinch of voltage V_p. Using Eq. (5.51), we obtain

$$V_p = \frac{q \cdot N_D \cdot a^2}{2 \cdot \varepsilon}. \tag{5.52}$$

The threshold voltage is given by

$$V_{th} = V_{gs}(h = a) = \phi_B - V_p. \tag{5.53}$$

II. Triode Region
Now, we consider the general case with nonzero V(x) required for current flow in the channel. The drain current along the channel is not dependent on x. At any point x, the channel current is given by

$$I_d = \kappa d w_g E(x) = q N_D \mu_n (a - h) w_g \frac{dV(x)}{dh} \frac{dh}{dx} = \frac{q^2 N_D^2 \mu_n h}{\varepsilon}(a - h) w_g \frac{dh}{dx}. \tag{5.54}$$

In the last equation we have applied the relation of E(x)=dV(x)/dx. Integration of x from 0 to L, and h from h_S to h_D yields

$$\int_0^L I_d dx = \frac{q^2 N_D^2 \mu_n w_g}{\varepsilon} \int_{h_S}^{h_D} \left(ah \text{-} h^2\right) dh, \tag{5.55}$$

and in turn

$$I_d = I_{dss}\left[\frac{3}{a^2}\left(h_D^2 - h_S^2\right) - \frac{2}{a^3}\left(h_D^3 - h_S^3\right)\right] \tag{5.56}$$

including the saturation drain current

$$I_{dss} = \frac{q^2 N_D^2 \mu_n w_g}{6 \varepsilon l_g}. \tag{5.57}$$

By considering the boundaries V(x=0)=0 at h(x=0)=h_S, and V(x=L)=V_{ds} at h(x=l_g)=h_D, we can calculate the drain current in the triode region based on Eqs. (5.51) and (5.56):

$$I_d = I_{dss}\left[\frac{3V_{ds}}{V_p} - \frac{2}{V_p^{3/2}}\left(\left(\phi_B - V_{gs} + V_{ds}\right)^{3/2} - \left(\phi_B - V_{gs}\right)^{3/2}\right)\right]. \tag{5.58}$$

III. Saturation Region
Suppose that the channel is pinched off in the saturation region. However, the reader should keep in mind that in reality, the channel is never pinched off completely. With the additional boundary of h_d=a, we can calculate with Eq. (5.56) a drain current of

$$I_d = I_{dss} \left[1 - 3\frac{\phi_B - V_{gs}}{V_p} + 2\left(\frac{\phi_B - V_{gs}}{V_p}\right)^{3/2} \right]. \tag{5.59}$$

If $V_{gs} = \phi_B$ we find by observation of the last equation that $I_d = I_{dss}$. Equations (5.52) and (5.59) reveal that

$$I_d = I_{dss} \left[-2 + 3\frac{V_{gs} - V_{th}}{V_p} + 2\left(1 - \frac{V_{gs} - V_{th}}{V_p}\right)^{3/2} \right]. \tag{5.60}$$

As illustrated in Fig. 5.33, Eq. (5.60) may be approximated by

$$I_d \approx I_{dss} \left(\frac{V_{gs} - V_{th}}{V_p}\right)^2 \tag{5.61}$$

demonstrating quadratic dependence of I_d vs V_{gs}. This is very similar to the MOSFET.

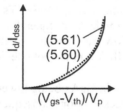

Fig. 5.33. Simplified depletion MESFET IV characteristics

IV. Drain Break-down Region
The breakdown considerations are similar to those of MOSFETs. However, due to the material properties of GaAs, MESFETs can handle higher breakdown fields. In addition, for a given f_t, the l_g of the GaAs FET does not have to be scaled by the same amount as the silicon MOSFET. Consequently, higher supply voltages can be applied for GaAs MESFETs. This is favourable for power applications.

V. Gate Break-down Region
As expected, also the gate break-down region shows similarities with the MOSFET. One significant advantage of the MESFET is that we do not have the problem of gate oxide scaling since a Schottky diode instead of a MOS capacitance is employed.

We can conclude that the DC characteristics of MESFETs are similar to those of MOSFETs. In this context we refer back to the plots shown in Fig. 5.12.

5.5.3 Small Signal and Noise Modelling compared to MOSFET

Since both MOSFETs and MESFETs are based on the same principle, namely the field effect, also the basic properties in terms of small signal and noise modelling are analogical. Thus, the equivalent circuits and equations for f_t, f_{max} and F_{min} derived for the MOSFET can also be applied for the MESFET. To minimise redundancies, we would like to point out the differences rather than to repeat material already discussed for the MOSFET. Some major differences can be highlighted:

*   Due to the high substrate isolation of III/V technologies used for MESFETs, we can neglect bulk effects and the associated parasitics. Consequently, at given l_g, speed parameters such as f_t, f_{max}, and F_{min} are superior for MESFETs.
*   The MESFET can be described as pure 3-port transistor, whereas for precise modelling, the MOSFET has to be considered as 4-port device including the bulk contact.
*   The C_{gs} of a MESFET is based on a Schottky diode and not on a MOS diode. Nevertheless, the corresponding capacitance per w_g is in the same order as for the MOSFET.
*   For aggressively scaled l_g, we have to consider a remarkable gate leakage current for the MOSFET due to the very thin gate oxide. This leakage can be represented by a resistor in parallel to the gate source capacitance. Such a resistor has a strong impact at low frequencies, whereas at high frequencies it is shorted by C_{gs}. At comparable speed, MESFETs have less gate leakage since they do not require such an oxide. Consequently, we do not find such a significant F_{min} increase towards low frequencies as exhibited by aggressively scaled MOSFETs. Because of the flicker noise both devices show $1/f$ noise characteristics towards DC.
*   Due to the low hole mobility in GaAs being lower than that in silicon, the performance of p-channel devices is poor. Consequently, they are not used. Thus, no complementary MESFET technologies comparable to CMOS are available. However, there is some scientific works going on in this area aiming at mitigating this problem by using enhanced material systems.

Further information about large and small signal modelling of MESFETs can be found in [Ell53.]

5.5.4 MESFET Types

The value of V_{th} and the associated D- or E-FET characteristics are defined by the channel thickness a and the channel doping. Corresponding information and illustrations are provided in Table 5.13. Obviously, because of the high channel thickness together with the high possible voltage swing and the associated maximum possible channel current, D-FETs have significant advantages in terms of maximum output power and linearity. Because of the thin channel, the E-FETs have a relative low DC current thereby yielding a high f_t to DC power ratio. Thus, D-FETs are used for power amplifiers, whereas E-FETs are preferred for low-power consuming receiver applications. Process variations of V_{th} can have a strong impact on the device performance. Usually, D-FETs have lower sensitivity since V_{th} mismatches are related with the large gate overdrive voltage. Thus, D-FETs are frequently used for stable current sources.

Table 5.13. Different types of MESFETs

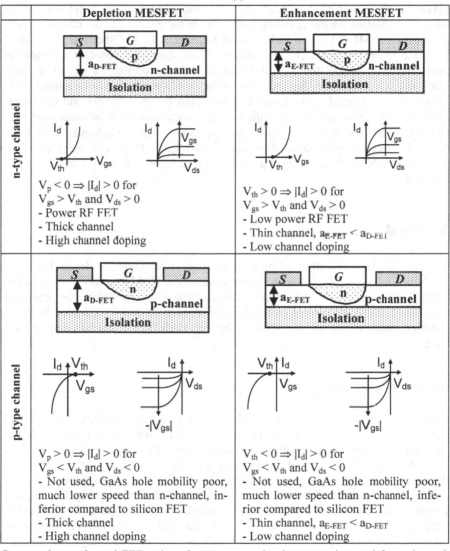

	Depletion MESFET	Enhancement MESFET
n-type channel	$V_p < 0 \Rightarrow |I_d| > 0$ for $V_{gs} > V_{th}$ and $V_{ds} > 0$ - Power RF FET - Thick channel - High channel doping	$V_{th} > 0 \Rightarrow |I_d| > 0$ for $V_{gs} > V_{th}$ and $V_{ds} > 0$ - Low power RF FET - Thin channel, $a_{E\text{-}FET} < a_{D\text{-}FET}$ - Low channel doping
p-type channel	$V_p > 0 \Rightarrow |I_d| > 0$ for $V_{gs} < V_{th}$ and $V_{ds} < 0$ - Not used, GaAs hole mobility poor, much lower speed than n-channel, inferior compared to silicon FET - Thick channel - High channel doping	$V_{th} < 0 \Rightarrow |I_d| > 0$ for $V_{gs} < V_{th}$ and $V_{ds} < 0$ - Not used, GaAs hole mobility poor, much lower speed than n-channel, inferior compared to silicon FET - Thin channel, $a_{E\text{-}FET} < a_{D\text{-}FET}$ - Low channel doping

Compared to n-channel FETs, signs for currents and voltages are inverted for p-channel FETs

5.5.5 Limitations

It is clear that the speed of MESFETs can also be improved by l_g scaling. However, we have to live with velocity saturation effects and decreasing breakdown voltages. For several reasons, the impact is not as severe as for the MOSFET. The electron saturation velocity and mobility of MESFETs are higher. Hence, for a given f_t, we do not have to scale the l_g by the same amount as for the silicon MOSFET. Similar to the gate oxide scaling for the MOSFET, the channel thickness a of the MESFET has to be scaled together with l_g. Otherwise, we would generate undesired short channel effects. If no somewhat constant l_g/a ratio is maintained, the associated V_{th} conditions are not appropriate for circuit applications, e.g. a V_{th} with very high negative value would result for the n-channel FET. Due to the scaling of a, we lower the maximum channel current and the corresponding RF output power. Nevertheless, due to the superior aforementioned prerequisites, the best RF power to speed figures of merit are achieved with III/V MESFET technologies. State-of-the-art GaAs MESFETs provide an f_t of up to 168 GHz, an NF_{min} below 1.2 dB up to 20 GHz and a V_{ds} DC supply of a least 2 V [Nis96].

5.6 HEMT

The HEMT (High Electron Mobility Transistor) is an advanced device based on the MESFET principle. Thus, in many respects, the HEMT is very similar to the MESFET. Like the other FET types, it has a channel terminated by the drain and the source. The gate control terminal is DC decoupled and RF coupled by a capacitive metal semiconductor Schottky barrier. Both E- and D-FETs can be realised.

5.6.1 Conventional HEMT

Conventional HEMTs are usually fabricated on GaAs substrates. The channel doping has a strong impact on the properties of FETs. We want to have a high carrier density, channel conductance and current demanding for high channel doping. As a drawback, high channel doping decreases the mobility since the probability that carriers hit each other rises degrading the speed. This problem can be solved by implementing a heterojunction along the channel. Referring to Fig. 5.34a,b, this junction may consist of an n-AlGaAs (Aluminium GaAs) layer under the gate, a very thin undoped AlGaAs spacer layer and an undoped GaAs layer serving as channel. Since undoped, the channel yields a very high mobility and superior speed even at moderate l_g. With increasing V_{gs} and due to the potential characteristics of the abrupt heterostructure, a quasi two-dimensional electron gas is generated in the undoped layer forming a very thin channel with thickness a. Aggressively scaling is feasible while maintaining an l_g/a ratio suitable for efficient control of the channel by means of the gate voltage. A drawback of the thin channel is the limited current handling leading to early channel current saturation. Let's summarise the points making the HEMT to a very fast device:

1. Generally, the used III/V AlGaAs channel material allows high carrier velocities and mobilities.
2. The low channel doping makes further improvement of the mobility possible.
3. A high current density is achieved by the efficient accumulation of free carriers in the quasi two-dimensional channel.
4. The thin channel makes a low l_g feasible.

At high V_{gs}, a part of the channel current may flow in a parasitic MESFET-like sub-channel exhibiting low carrier mobility. To avoid corresponding speed deteriorations, the maximum V_{gs} and in turn the maximum output current and power are limited.

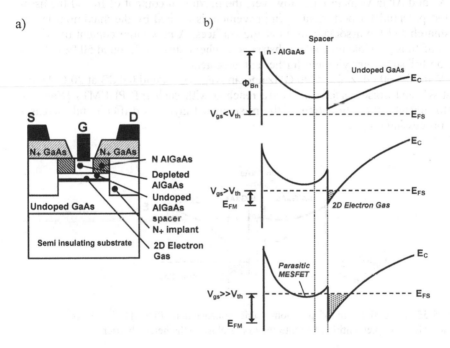

Fig. 5.34a,b. HEMT: **a** cross-section; **b** band diagram of GaAs enhancement device. Thin electron channel is generated if applied gate source voltage is higher than threshold voltage. At high gate source voltages, a parasitic MESFET may be activated

Due to the complex characteristics of the implemented heterostructure, quantitative modelling of HEMTs is more difficult than for MESFETs. However, from the perspective of a circuit designer, the IV curves, the small signal and noise model are analogical to those of the MESFET or MOSFET. In this respect, the reader can map the insights of the previous sections.

Last but not least a comment regarding the costs. The realisation of the hetero-junction is complex. Hence, the costs of HEMTs are beyond those of MESFETs. At very high volumes, the cost gap increases with respect to MOSFET technologies.

5.6.2 Pseudomorphic HEMT

Compared to conventional HEMTs, the performance can be improved by implementing In (Indium) into the GaAs channel as sketched in Fig. 5.35. Together with an AlInAs (Aluminium Indium Arsenide) spacer, the InGaAs channel generates a lattice mismatch, which increases the bandgap ΔE_c at the channel surface. Thus, the parasitic MESFET is better suppressed than for the conventional HEMT allowing for higher V_{gs} and channel carrier densities. Moreover, the InGaAs material further increases the electron mobility and the saturation velocity compared to GaAs and AlGaAs materials. However, the maximum content of In and the associated potential for performance improvement is limited by the maximum lattice mismatch and the associate strain at the surfaces. A maximum content of around 15% of In is possible on GaAs substrates. Higher content of around 50 % is feasible on InP substrates yielding further enhancements.

With f_t=560 GHz, f_{max}=300 GHz and an NF_{min} of around 0.4 dB at 20 GHz, the best RF and noise performances are reached with such InP PHEMTs [Yam02]. Unfortunately, the fabrication of the mismatched layers is difficult and increases the processing costs.

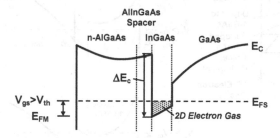

Fig. 5.35. Band diagram of pseudomorphic enhancement PHEMT on GaAs substrate, higher ΔE_c by implementing of In into the GaAs close to the hetero-barrier

5.6.3 Metamorphic HEMT

Due to the higher content of In, InP based HEMTs exhibit better performance than those realised on GaAs wafers. Unfortunately, InP is more expensive than GaAs. In this regard, the metamorphic GaAs HEMT is a good option. By introducing metamorphic buffer layers with stepped lattice constants, at similar surface stress, an In content of up to 50% is possible on GaAs substrates. Consequently, at lower material costs, the performance of metamorphic GaAs HEMTs is close to that of devices on InP substrates [Tes04].

5.7 Bipolar Transistor

With respect to FETs, BJTs (Bipolar Junction Transistors) apply a different concept for the control of the channel current. Instead of the drift mechanism, the properties of the diffusion transport are exploited. Similar to FETs, BJTs can be represented as 3-port devices. Control is performed by means of the base electrode located between the collector and emitter contacts forming the channel. These terminals can be compared with the gate, drain and source of a FET, respectively. For BJTs the channel current flows in vertical direction, whereas common FETs features lateral current transport. Recall that the direction of the current flow related to the material surfaces has an impact on the flicker noise. In this regard, BJTs are advantageous. For cost reasons, BJTs are mainly realised in silicon. If higher performance is required, the superior SiGe HBT (Heterojunction Bipolar Transistor) can be applied. Today, we find no commercial BJTs in GaAs or InP technologies. A significant difference between the transistor concepts concerning the channel carrier type can be identified. The current of FETs is mainly determined by majority carriers, whereas the charge transport in BJTs is based on minority carriers exhibiting opposite charge polarity with respect to those of the channel material, e.g. n-carriers are transported in p-material. BJTs can be realised as complementary devices with npn and pnp structures. Since npn devices apply electrons as charge carriers, npn devices are faster than pnp transistors. Recall that the electron mobility is higher than the hole mobility.

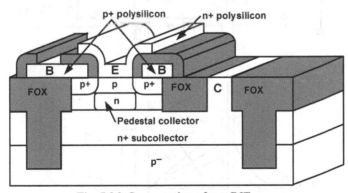

Fig. 5.36. Cross-section of npn BJT

5.7.1 Functional Principle

In Fig. 5.36, the cross-section of a common npn BJT is shown, which will be discussed in this section. Similar to FETs, the insights can be mapped for pnp devices by inverting the signs of all voltages and currents. According to Fig. 5.37, the transistor consists of two diodes connected through the same base. We focus on the so-called forward biased operation, where the base emitter diode is forward biased and the base collector diode is reverse biased. Due to its superior properties in terms of gain and noise, the forward biased operation is mostly applied for RF circuit applications. Two parallel base contacts may be used to decrease the base resistance resulting in improved RF performance. The forward biased emitter base junction injects electrons from the emitter into the base as indicated by I_{neb}, and holes from the base to the emitter represented by I_{pbe}. The first current is the desired channel current, whereas the latter one is a parasitic current, which must be minimised. Some of the injected electrons recombine with holes in the base creating the parasitic current I_{nEBr}. However, this recombination current is much weaker than the base current. After reaching the edge of the base collector region by diffusion, the remaining electrons are swept to the collector by means of the high electrical field across the reversed biased junction.

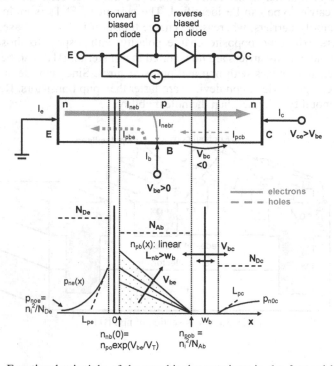

Fig. 5.37. Functional principle of the npn bipolar transistor in the forward biased mode. Electrons (minority carriers) pass the forward biased base emitter control diode, go through the p-base by diffusion, are swept away by the reverse biased base collector diode and finally reach the collector

The BJT is actually based on two transport mechanisms: In first priority, the diffusion for channel current control. Moreover, the drift transport of the carriers to the collector via an n+ doped sub-collector. If properly designed, the speed is mainly determined by the diffusion within the base rather than by the drift in the highly conductive collector region. However, for ultra high-speed HBTs featuring very fast diffusion, we observe that the drift delay becomes more and more to a limiting factor.

According to Sect. 5.2.2, the diffusion effect is the response to an imbalance of carriers. Carrier gradients determine the current flow and subsequently the channel conductivity. The control mechanism of bipolar transistors is based on this gradient. By intuition, we may understand that the desired I_{neb} is proportional to the gradient of the electrons in the base n_{pb}:

$$I_{neb} \sim \frac{dn_{pb}(x, V_{be})}{dx}.$$

(5.62)

From the diode theory we know that the gradient vs the control voltage V_{be} is exponential. The current is proportional to the material dependent diffusion constant of the electrons D_n, and inversely proportional to the base acceptor doping concentration N_{Ab}. To maximise the gradient and the associated diffusion current, the base has to be made as thin as possible. Moreover, due to a decrease of the transit time through the channel also the transistor speed can be improved by decreasing the base width w_b. Given that w_b is much smaller than the electron base diffusion length L_{nb}, the effective available L_{nb} is determined by w_b. By inspection of the diffusion tails, we observe that the gradient of the electron concentration $n_{pb}(x, V_{be})$ is approximately linear for $w_b \ll L_{nb}$. These insights may help to understand that the electron current is given by

$$I_{neb} = I_{Sneb} \cdot \exp\left(\frac{V_{be}}{V_T}\right).$$

(5.63)

where $I_{Sneb} = \dfrac{q \cdot A \cdot D_n \cdot n_i^2}{w_b \cdot N_{Ab}}$, with A as the junction area, n_i is the intrinsic carrier density and q is the elementary charge. The temperature voltage $V_T = \dfrac{kT}{q}$ considers the temperature T and the Boltzmann constant k. In the latter identity, we have neglected the contribution of the weak current I_{nebr}. The hole current I_{pbe} in the emitter can be approximated by

$$I_{pbe} \sim I_{Spbe} \cdot \exp\left(\frac{V_{be}}{V_T}\right),$$

(5.64)

where $I_{Speb} = \dfrac{q \cdot A \cdot D_p \cdot n_i^2}{L_{pe} \cdot N_{De}}$. D_p denotes the p diffusion constant and N_{De} is the emitter donator doping concentration. We should note that the illustrations in Fig. 5.37 do not represent the real dimensions regarding the base, emitter and collector widths. In reality, the base is much thinner than the collector and the emitter. In Eq. (5.64), the full emitter diffusion length L_{pe} has to be taken into account since the emitter width is larger than L_{pe}. The small hole current from the collector to the base arising due to thermal generation of carriers at the junction I_{pcb} can be neglected.

From the last equations we have observed a remarkable property regarding the collector current characteristics versus the control voltage. The relation is exponential for BJTs. Recall that it is quadratic for FETs. These properties impact the g_m representing the slope of the output current vs the input voltage. By assuming that the collector current I_c equals I_{neb}, the transconductance of the BJT can be approximated by

$$g_m = \frac{dI_c}{dV_{be}} = \frac{1}{V_T} \cdot I_{Sneb} \cdot \exp\left(\frac{V_{be}}{V_T}\right) = \frac{I_c}{V_T} . \qquad (5.65)$$

As expected, g_m increases with I_c. At room temperature, the factor $1/V_T$ exhibits a value around 40 V^{-1} resulting in a high g_m. A disadvantage of BJTs is the high sensitivity on the base emitter voltage and the threshold voltage, which can be impacted by system conditions and process variations, respectively. Corresponding compensation circuits using DC feedback are important in practice. Moreover, associated with the strong change of the output current vs the input voltage, the nonlinearities of BJTs are more significant than for FETs. For nonlinear circuits such as mixers requiring a high nonlinear conversion gain, this is favourable, whereas it is disadvantageous for linear amplifiers.

Complementary bipolar logic similar to CMOS can be realised in silicon since both npn and pnp transistors are feasible. However, we have observed that the BJT exhibits a significant base current. This is a remarkable disadvantage for logic circuits since power is required to control the gate. Recall that the gate control current of FETs can be neglected due to the gate DC decoupling. In this respect, we can identify a major difference compared to FETs. In ideal case, FETs do not exhibit any DC control current through the gate control terminal. Furthermore, the base current introduces noise. It becomes clear that is aimed to minimise the parasitic base current related to the desired channel current. Consequently, the current gain β must be maximised. With $I_{neb} \gg I_{nebr}$ and $I_{pbe} \gg I_{pcb}$ we can write

$$\beta \approx \frac{I_{neb}}{I_{pbe}} \approx \frac{D_n \cdot N_{De} \cdot L_{pe}}{D_p \cdot N_{Ab} \cdot w_b} . \qquad (5.66)$$

The following important conclusions can be drawn:

- To maximise the current gain, the emitter doping must be high, whereas the base doping has to be low. Unfortunately, low base doping increases the base resistance R_{bb} thereby lowering the maximum frequency of oscillation as will be shown in Eq. (5.68), and increases the noise. Consequently, a tradeoff has to be found. For the corresponding band diagrams and doping profile refer to Figs. 5.43a and 5.44a.

- Minimising w_b improves both β and f_t due the lowered transit time and base resistance. Consequently, w_b has to be made as thin as possible. However, a thin base requires a more advanced processing technology, decreases the yield and raises the costs.

5.7.2 DC Characteristics

The DC properties of BJTs are illustrated in Fig. 5.38. Despite the fact that the BJT employs a completely different control principle compared to the FET, we can find similarities compared to the DC characteristics of the FET illustrated in Fig. 5.12c. The major regions are described in the following paragraphs. For a helpful interactive illustration of the functional principle of bipolar transistors the reader is referred to [Edu05]

a)   b)

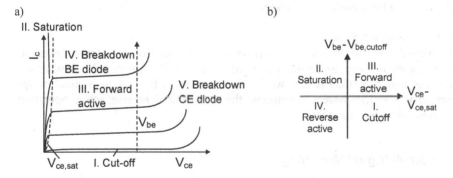

Fig. 5.38a,b. BJT DC characteristics: **a** IV curves; **b** illustration of major operation areas

I. Cutoff Region
Below a certain base emitter voltage, which is referred to as the cut-off voltage, no channel current flows into the collector. This voltage corresponds to the threshold voltage of FETs. Typically, the cut-off voltage is around 0.7 V. Above the cut-off voltage, the increase of the collector current is very strong due to the exponential relation.

II. Saturation Region
The collector current increases continuously with raised collector emitter voltage since a higher number of electrons provided by diffusion from the base to the

emitter can be swept into the collector. From global perspective, this behaviour is similar compared to that of the FET in the triode region. Typically, the saturation voltage of the BJT $V_{ce,sat}$ is smaller than that of the FET. By virtue of the naming, the saturation regions of the BJT and the FET are different.

III. Forward Active Region

Because of the good RF and large signal properties, the forward biased region is the superior operation region for most BJT circuit applications. It is similar to the saturation region of the FET. As previously explained, the base emitter diode is forward biased, whereas the base collector diode is reverse biased. At high collector emitter voltages, we experience also a saturation effect as for the FET. However, the reasons are different. Let us recapitulate that the saturation of the FET drain current is due to the pinch-off effect. In contradiction, the saturation of the collector current can be attributed to an increase of the width of the base collector barrier zone leading to a higher resistance. The resulting voltage drop decreases the effective base collector voltage. Hence, the collector current is more or less constant after $V_{ce,sat}$ is reached.

IV. and V. Diode Breakdown

At specific voltages, the diodes are driven into breakdown leading to a strong increase of the current and eventually to the death of the device. These breakdown voltages are generally lower for the forward biased emitter diode than for the reversed biased base collector junction.

VI. Reverse Active Region

At inverted signs of V_{be} and V_{ce}, a reverse active operation is possible based on the transport of holes. Since the doping of the diodes is optimised for forward active operation, the effect is very weak. If corresponding bias voltages could be applicable in specific system conditions, this non-desired reverse active operation has to be taken into account.

5.7.3 Small Signal Modelling

In Fig. 5.39, the equivalent small signal circuit of a bipolar transistor including the most important elements is shown. Details regarding the elements are compiled in Table 5.14. Compared to the FET, we can observe many similarities. The idealised bipolar transistor acts also as a voltage controlled current source with output current $i_c = g_m v_{be}$. Let us review major differences compared to the FET:

- The g_m of BJTs scales with I_c only ($g_m \sim I_c$), whereas the g_m of FETs scales with both I_d ($g_m \sim \sqrt{I_d}$, w_g constant) and w_g ($g_m \sim \sqrt{w_g}$, I_d constant) leading to $g_m \sim I_d$ if w_g is scaled at the same time.
- At typical DC bias and low to moderate w_g, FETs typically provide lower g_m than BJTs as illustrated in Fig. 5.40. In the event of very high w_g, the g_m of

FETs may be higher. However, such large devices are rarely used for RF applications.

- The significant RC constants mainly associated with the base emitter pn-diode lead to a moderate f_t, which limits the exploitation of the large g_m of BJTs towards highest frequencies.
- In contrast to the FET, the input of the BJT consists of a conducting diode, which can be modelled by a capacitance and a resistance connected in parallel. Thus, the input impedance has a strong resistive content and a relatively low impedance. For small devices as required for low power applications, impedance matching to 50 Ω is simplified. Due to the decreased sensitivity in terms of impedance matching, the R_n of BJTs is smaller than for FETs. Hence, BJTs are well suited for broadband matching.
- For circuits demanding huge transistors, e.g. power amplifiers, the low impedance can be problematic for 50 Ω matching.
- Because of the low doping within the base required for high current gain, a base series resistor R_{bb} has to be introduced in the equivalent model. Due to its high value it significantly affects the performance.

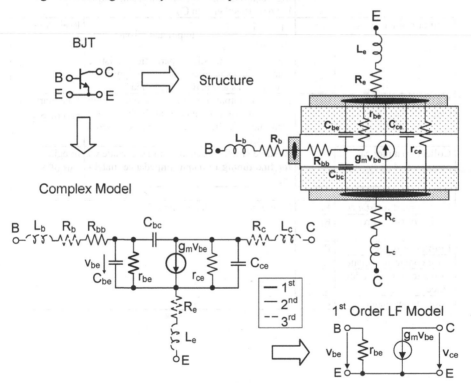

Fig. 5.39. Small signal equivalent circuit of BJT in common emitter configuration, LF: low frequency

Table 5.14. BJT small signal model parameters according to their priority for high frequency applications with respect to impact on one hand and complexity on the other hand

Parameter		Origin and impact	Related S-parameter
1st priority			
Transconductance	g_m	Current source properties: $g_m = \dfrac{dI_c}{dV_{be}} = \dfrac{i_c}{v_{be}}$	$\underline{S}_{21}$
Base emitter resistance	r_{be}	$\dfrac{1}{r_{be}} = \dfrac{dI_b}{dV_{be}} = \dfrac{i_b}{v_{be}}$ extrapolated from large signal characteristics, represents conductive base emitter diode together with C_{be}. Low frequencies: r_{be} dominant, high frequencies: C_{be} dominant	Real part of $\underline{S}_{11}$
2nd priority			
Base emitter capacitance	C_{be}	Capacitance of forward biased diode, largest capacitance, $C_{be} > C_{bc} > C_{ce}$	Imaginary part of $\underline{S}_{11}$
Base channel resistance	R_{bb}	High series resistance due to low base doping required for high current gain, forms lowpass together with C_{in}	Real part of $\underline{S}_{11}$ [10]
Output resistance	r_{ce}	$\dfrac{1}{r_{ce}} = \dfrac{dI_c}{dV_{be}} = \dfrac{i_c}{v_{be}}$ extrapolated from large signal characteristics, parasitic resistive channel leakage effects are less significant than in FETs and may be neglected	Real part of $\underline{S}_{22}$
Base collector capacitance	C_{bc}	Input to output feedback, introduces high virtual input capacitance at high voltage gain due to Miller effect. Can be considered in $C_{in} > C_{be}$	Imaginary part of $\underline{S}_{12}$
Collector emitter capacitance	C_{ce}	Relative small capacitance, to be considered for fine tuning of output impedance matching	Imaginary part of $\underline{S}_{22}$
3rd priority			
Base contact resistance	R_b	Similar to FET, refer to Table 5.5	
Collector contact resistance	R_c		
Emitter resistance	R_e		
Base contact inductance	L_b		
Collector contact inductance	L_c		
Emitter inductance	L_e		

---

[10] Values at low and high frequencies have to be used to distinguish between R_{bb} and r_{be}.

Like in Eqs. (5.24) and (5.32), the transit frequency and maximum frequency of oscillation of bipolar transistors can be approximated by

$$f_t = \frac{g_m}{2\pi \cdot C_{be}} \qquad (5.67)$$

and

$$f_{max} = \sqrt{\frac{f_t}{8\pi \cdot R_{bb} \cdot C_{bc}}} . \qquad (5.68)$$

Since several minor RC constants are neglected, these simple equations yield rather optimistic values and represent upper limits with deviations, which may be in the order of 30 %. To improve the accuracy, the Miller effect may be taken into account in Eq. (5.67).

State-of-the-art BJTs provide 50 GHz f_t, 70 GHz f_{max}, an NF_{min} of below 3 dB up to 20 GHz, and a collector emitter voltage of around 3 V [Uga95]. Due to the excellent properties of the HBT, in the future, the efforts spend in the enhancement of conventional BJTs will be small.

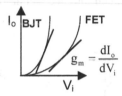

Fig. 5.40. Qualitative comparison of output current vs input voltage of a typical RF BJT and FET, relations are exponential and quadratic, respectively. I_o: output current, V_i: input voltage

5.7.4 Noise Modelling

In Fig. 5.41a, a simplified equivalent noise circuit of the bipolar transistor is shown. The noise of BJTs is particularly determined by shot noise within the pn-junctions [Zie60]. Referring back to Sect 4.3.2, shot noise is generated by random motion of charge carriers. For the common emitter configuration, the shot noise sources in the base and emitter can be characterised by

$$\overline{I_{nb}^2} = 2 \cdot q \cdot I_b \cdot \Delta f \qquad (5.69)$$

and

$$\overline{I_{nc}^2} = 2 \cdot q \cdot I_c \cdot \Delta f , \qquad (5.70)$$

respectively, where q, I_b, I_c and Δf are the elementary charge, the base DC current, the collector DC current and the frequency bandwidth, respectively. The base and collector shot noise source are correlated since they are influenced by the same cause, which is the emitter current. For simplification, we may neglect this correlation, which is more complex than for the FET yielding

$$\overline{I_{nb} \cdot I_{nc}^*} = 0 \, . \tag{5.71}$$

For most applications, this simplification still gives satisfying results. Equations (5.69) and (5.70) reveal that the favourable noise performance is achieved at low DC bias. On the other hand low DC bias decreases the g_m and the gain. Thus, the noise contributions of the load and following stages may become significant indicating a tradeoff. Moreover, we have to consider the thermal noise of the base connection resistance. According to Sect. 4.3.1, this noise contribution can be described by

$$\overline{I_{nbb}^2} = \frac{4 \cdot kT \cdot \Delta f}{R_{bb}} \, . \tag{5.72}$$

According to Eq. (5.66), the value of R_{bb} depends on the doping in the base. To obtain a high current gain, the doping has to be low, resulting in a high R_{bb} and consequently a strong noise contribution. Usually, one aims for a compromise. In addition, flicker noise has to be taken into account at lowest frequencies.

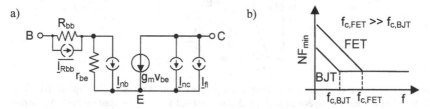

Fig. 5.41a,b. BJT noise: **a** simplified equivalent circuit in emitter configuration; **b** comparison of low frequency noise characteristics of BJT and FET, f_c: corner frequency of flicker noise

As aforementioned in Sect. 4.3.3 and as illustrated in Fig. 4.41b, bipolar transistors have weaker flicker noise than FETs since the flicker corner frequency f_c is lower. With an f_c of around 1–10 kHz, the frequency where the noise increase starts is around 100 times lower for BJTs than for FETs. Since the up-converted flicker noise determines the noise in nonlinear circuits such as oscillators, BJTs are excellent candidates for low noise oscillators.

5.8 Heterojunction Bipolar Transistor

In Sect. 5.7 we have observed that the BJT is subject to tradeoffs limiting the design flexibility. On one hand, low base doping is required to achieve high current. On the other, due to the increased R_{bb}, low base doping degrades the RF performance. In this context, the HBT (Heterojunction Bipolar Transistor) offers remarkable advantages since it relaxes the conflicting goals between the current gain and the RF performance.

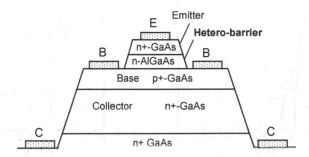

Fig. 5.42. Cross-section of a GaAs/AlGaAs HBT

HBTs can be fabricated on the basis of SiGe, GaAs/AlGaAs or InP/InGaAs. SiGe HBTs provide good performance together with low costs. GaAs/AlGaAs or InP/InGaAs based technologies offer excellent properties, but are more expensive. The SiGe HBT technology is strongly pushed by the industry, whereas GaAs and InP HBTs are favourable candidates for low-volume applications requiring highest performances. The cross-section of a typical GaAs/AlGaAs HBT is plotted in Fig. 5.42.

Similar to HEMTs, HBTs feature a heterojunction barrier, which is implemented in the base emitter junction. In Fig. 5.43, the corresponding band diagram of the HBT is shown and compared with the conventional BJT. The hetero-barrier ΔE_v reduces the parasitic injection of holes from the base in the emitter yielding the following relation:

$$I_{pbe,HBT} \approx I_{pbe,BJT} \cdot \exp\left(\frac{-\Delta E_v}{kT}\right).\tag{5.73}$$

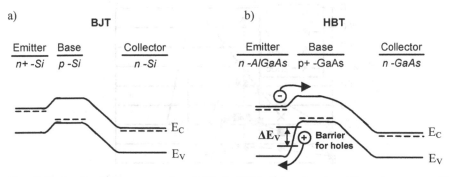

Fig. 5.43a,b. Band diagrams of: **a** BJT; **b** HBT, hetero-barrier ΔE_v reduces parasitic injection of majority carriers from the base into the emitter, thus significantly increasing the current gain

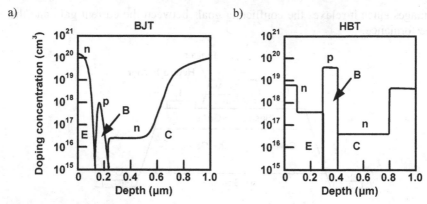

Fig. 5.44a,b. Doping profile: **a** BJT; **b** HBT, high gain associated with hetero-barrier, base can be highly doped to decrease R_{bb}

Thus, the current gain

$$\beta_{HBT} \sim \frac{1}{I_{pbe}} \approx \beta_{BJT} \cdot \exp\left(\frac{-\Delta E_v}{kT}\right) \qquad (5.74)$$

of the HBT is much higher than for the BJT. This high current gain has an important consequence. The base doping can be raised to decrease R_{bb}, while still having sufficient current gain. In Fig. 5.44, the corresponding doping profiles of the BJT and the HBT are shown.

The f_t of IBM's commercial SiGe HBTs is shown in Fig. 5.45. From 1998 to 2006, the f_t has been improved from 30 GHz to over 350 GHz. Within this time frame, the f_{max} has been maturated from 40 GHz to 210 GHz. This progress is remarkable.

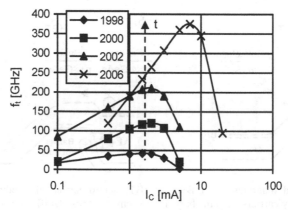

Fig. 5.45. Transit frequency of leading-edge industrial SiGe HBTs of IBM from 1998 to 2006; compare [Rie03]

5.9 BiCMOS

Due to the low costs in mass-fabrication, the superior properties for digital circuits in terms of complexity and static power consumption, CMOS technology plays the key role for commercial applications. The speed and RF performance can be enhanced by including SiGe HBTs into the CMOS technology. Thus, the designer has more flexibility concerning circuit optimisation. For example, CMOS can be used for logic circuits, whereas the SiGe devices can be employed for the speed- and power-critical analogue and RF circuits. Such a BiCMOS technology requires a few more processing steps and is therefore more expensive. However, an aggressively scaled CMOS technology can have higher costs than a BiCMOS process employing moderate lithography. In typical mass fabrication, a 0.13 μm BiMCOS technology may have similar costs than a 90 nm CMOS technology.

In the following, we compare the key small signal properties of these two technologies on basis of IBM technologies [Jag04]. The presented n-channel MOSFET features an l_g of 90 nm. Realised as npn device, the HBT has an emitter width (corresponds to w_b for manufacturing reasons) of 0.13 μm and an emitter length of 2.5 μm.

In Fig. 5.46a, the output current is plotted versus input voltage. As expected for typical devices, the slope of the HBT is much steeper than that for the MOSFET. Recall that the transistors represent exponential and quadratic characteristics, respectively. Consequently, the HBT can provide much higher g_m. As shown in Fig. 5.46b, the peak g_m of the HBT is approximately 50 times higher. To achieve the peak g_m, higher supply current has to be fed for the HBT. The input voltages for maximum g_m are relatively close together. Towards low currents, the g_m gap decreases. However, at a low supply current of 0.1 mA, the g_m of the HBT is still 8 times larger than that of the MOSFET.

Figure 5.46c illustrates the characteristics of the f_t versus the supply current. Although the g_m of the HBT is so much higher, the peak f_t of around 225 GHz is only moderately higher than that of the MOSFET yielding 150 GHz. This is attributed to the fact that the HBT exhibits significantly higher parasitic capacitances. Let us recapitulate that the HBT input features a forward biased pn-diode capacitance, whereas the input of the MOSFET is based on an MOS depletion capacitance. We can draw the conclusion that the advantages of the HBT in terms of g_m and gain become smaller towards highest frequencies.

The noise performances of the devices versus the supply current are plotted in Fig. 5.46d. At moderate to high supply current, the HBT exhibits lower NF_{min}, whereas the FET exhibits superior performance towards very low supply current.

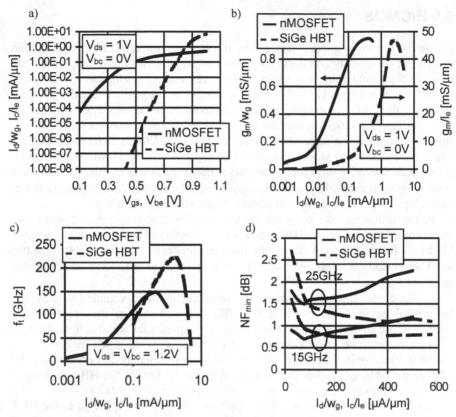

Fig. 5.46a–d. Comparison of 90 nm CMOS and 130 nm BiCMOS SiGe HBT technology with emitter length of l_e=2.5 μm [Jag04]: **a** drain and collector current vs gate source and base emitter voltage; **b** transconductance vs gate source and base emitter current; **c** transit frequency vs gate source and base emitter current; **d** minimum noise figure vs current at optimum drain source/collector emitter voltage

5.10 Overview of Noise Performances

The typical noise performances of different kinds of transistors are illustrated in Fig. 4.47 vs frequency. Due to the low substrate parasitics and high f_t, transistors fabricated in III/V technologies exhibit lower noise figure than silicon devices. The best performance is achieved with InP HEMTs.

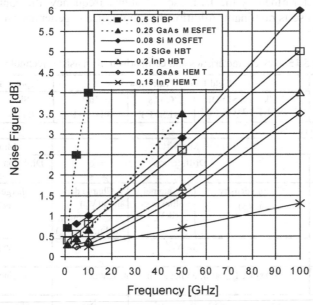

Fig. 5.47. Comparison of typical RF transistor noise performances

5.11 Conclusions

In my career, I have lost a countless number of games
– that is the reason for my success.
Michael Jordan, basketball player

IC transistor technologies determine important circuit properties such as operation frequency, gain, noise, large signal performance, DC power consumption, circuit complexity and costs. Key characteristics of the different transistor types are compared in Table 5.15. Generally, we can distinguish between bipolar and FET transistors realised on silicon and III/V based substrates. Due to the high substrate resistivity, low parasitics and high electron mobilities the latter approach yields favourable prerequisites in terms of speed. Based on charge control techniques such as drift or diffusion, transistors are capable of controlling a strong output signal by means of a weak input signal. In ideal case, they provide the function of a voltage controlled current source with the transconductance as major parameter. Applications of transistors comprise signal amplification, switching and frequency conversion. The performance of transistors is strongly impacted by the bias. Depending on the specific application it has to be set carefully. Due to parasitics and RC time constants, transistors exhibit low-pass filter characteristics. Most of the performance parameters of transistors worsen towards high frequencies. Useful measures for the speed properties are the f_t and the f_{max}. At very low frequencies, the noise is determined by the $1/f$ noise. At higher frequency, the contribution of the thermal noise is dominant. Additionally, shot noise is generated in bipolar devices.

Table 5.15. Key characteristics and comparison of transistor technologies

Type	Field Effect Transistor				Bipolar Transistor	
	Si bulk MOSFET	Si SOI MOS-FET	GaAs MESFET	InP PHEMT	Si BJT	SiGe HBT
Major characteristics						
Charge type	Majority				Minority	
Charge transport	Drift, lateral				Diffusion, drift, vertical	
Charge control	Charge density $n(V_{gs})$		Channel thickness $d(V_{gs})$	1. Charge density $n(V_{gs})$ 2. Channel thickness $d(V_{gs})$	Charge gradient $n(V_{be})$	
Current source	$I_d \sim V_{gs}^2$				$I_c \sim exp(V_{be})$	
DC control power	≈ 0 since $I_{g,dc} \approx 0$ when neglecting gate leakage				>0 since $I_{b,dc} > 0$	
Complementary transistor	Yes	Yes	No	No	Yes	Yes
Costs prototype/mass fabrication	High/ low	High/ low	Low/ high	High/ very high	High/ moderate	High/ moderate
State-of-the-art 2007						
Maximum number of transistors on IC	>1 billion Itanium processor	>1 Mio	<10000	<500	<10000	<10000
Technology	l_g=90 nm	l_g=90 nm	l_g=0.1 µm	l_g=0.1 µm	w_e=0.3 µm	w_e= 0.12 µm
Max. f_t	140 GHz	243 GHz	168 GHz	562 GHz	50 GHz	375 GHz
Max. f_{max}	150 GHz	208 GHz	n.a.	300 GHz	70 GHz	210 GHz
$I_{dc,\,max}$	0.7 A/mm	0.7 A/mm	0.47 A/mm	0.8 A/mm	J_c=3.4 mA/ µm^2	J_c=20 mA /µm^2
$g_{m,\,max}/W_g$	1.3 S/mm	1.3 S/mm	0.45 S/mm	1.23 S/mm	n.a.	n.a.
$V_{supply,\,max}$	$\approx$1.2 V	$\approx$1.2 V	>2 V	$\approx$1 V	$\approx$3 V	$\approx$1.5 V
NF_{min} up 20GHz	<1.5 dB	<1 dB	<1.2 dB	<0.4 dB	<3 dB	<0.8 dB
References	[Pek04, Jag06]	[Plo05]	[Nis96]	[Yam02]	[Uga95]	[Rie04]

Today, the highest f_t and f_{max} amounting to 562 GHz and 300 GHz, respectively are achieved with InP PHEMTs. One major reason for this superior performance is the high InP electron mobility. InP HEMTs transistors provide an excellent NF_{min} of below 0.4 dB up to 20 GHz. Unfortunately, due to the limited transistor yield, III/V based circuits have high costs in mass-fabrication and enable only low circuit complexities.

In the last years, the speed and noise gap between III/V and silicon based technologies has been significantly decreased. SiGe HBTs with f_t of 350 GHz, f_{max} of 210 GHz and an NF_{min} below 1 dB up to 20 GHz are available. Moreover, 90-nm SOI CMOS technologies with f_t of 243 GHz, f_{max} of 208 GHz and an NF_{min} below 1 dB up to 20 GHz have been demonstrated. An f_t up to 210 GHz is expected for 65-nm CMOS. CMOS technologies exhibit a very low static power consumption allowing a very high transistor density without suffering from thermal problems. Hence, CMOS plays a dominant role within the whole IC market. An Intel Itanium processor for example, has over 1.7 billion transistors on a single CMOS chip. Due to the high static power consumption and the thermal loading this would definitely not be possible in III/V technologies. To make single chip solutions feasible, RF designers also have to use CMOS technologies making the design of analogue circuits a demanding task. One of the major challenges is the low output voltage of the transistors. SiGe HBTs offer a relative high maximum supply voltage of above 2 V. Only half of this supply voltage is provided by aggressively scaled CMOS and HEMT technologies.

5.12 Tutorials

1. What are the drawbacks of the vacuum triode?
2. Briefly review the invention of the transistor. Who have been the main inventors? Did they commercially exploit their ideas? What do we learn concerning social aspects?
3. Who invented the integrated circuit? What was the key motivation concerned with the development of the integrated circuit?
4. What are the key tasks of transistors?
5. How can we model an ideal transistor?
6. What are the main effects used for the controlled transport of channel current? Which transistor types are based on these principles?
7. Compare the drift velocity vs the electrical field of silicon and GaAs. What is the consequence for the carrier transport and the speed of transistors?
8. What does mobility mean and how is it defined? Compare both electron and hole mobilities of silicon and GaAs. What about the electron/hole mobility gap of GaAs with respect to silicon? Draw conclusions concerning the capability for complementary logic.
9. Why does complementary logic has no static power consumption? Explain. Draw the schematics of a simple CMOS gate.
10. What kind of FETs and bipolar transistors are you aware of?
11. Review the main differences regarding costs for silicon and III/V based technologies. You are requested to design a prototype RFIC circuit with lowest budget. Which technology do you use? What about circuits in mass-fabrication?

12. What about the power capabilities of the transistor materials? Which are the relevant parameters? Compare them.

13. Explain the functional principles of MESFETs, HEMTs and MOSFETs. Draw their cross-sections. Which materials are typically used? Explain the qualitative characteristics of I_d vs V_{gs} and V_{ds} and identify specific regions in the IV curves. What are the similarities and differences of the approaches? How is the gate isolation realised? Why is the isolation necessary? How is the channel control accomplished? Why is there a current saturation at high V_{ds}?

14. What about the collector current saturation in BJTs? Why does it saturate as well?

15. Explain the differences of E- and D-FETs considering both n-channel and p-channel devices. Take into account the IC characteristics, fabrication, power capability and speed.

16. Illustrate the small signal equivalent circuit of FETs. Identify priorities for the parameters. What happens with g_m if the output current is zero? What about the relation between the DC input and output currents in the ideal and the realistic case?

17. Which port of an FET is more high-ohmic – the input or the output? What consequences does this have concerning impedance matching for a low-power consuming low-noise amplifier and a high-power amplifier?

18. Derive the transit frequency of FETs by assuming a very simple equivalent circuit. Is the result too low or rather too high? Why? How could we improve the equation?

19. Generally, how can we improve the speed of FETs?

20. What are the main noise processes and sources in FETs and BJTs? What are the differences? Which noise effects dominate at high frequencies? Does r_{ds} or r_{ce} generate noise?

21. Which parameters impact F_{min}?

22. Why are gain and noise impedance matching not equal?

23. Discuss the ideal gate width scaling rules for FETs. Consider the intrinsic and extrinsic resistances, capacitances, currents, voltages and transconductances.

24. In ideal case, does f_t and f_{max} change with gate width? Why?

25. What are the drawbacks concerning the use of f_t and f_{max} in reality?

26. What is the content of Moore's law?

27. Discuss the effect of the gate length scaling on key FET parameters. Comment on the problems associated with the scaling in reality.

28. Why is it important to scale the gate oxide thickness together with the gate length?

29. What are short channel effects? Outline an example.

30. Does the small gate leakage current really matter, e.g. in an Itanium processor chip?

31. Which techniques can be used to mitigate the gate leakage?

32. What are the ideas behind advanced CMOS technologies such as SOI, strained silicon, high-k, multi-gate FET, cooling, etc.? Discuss the advantages and

disadvantages of each approach. How much f_t improvement has been demonstrated with these technologies?

33. Suggest that all approaches can be implemented independent from each other. Which f_t could we expect in the nearby future?

34. What may be a problem associated with logic circuits in SOI?

35. What effect has scaling on passive devices such as inductors and interconnects? What can we do to mitigate this effect?

36. Is high-k also advantageous for passive devices? Can we implement the optimum dielectric for both active and passive devices?

37. What are the advantages and disadvantages of the HEMT in comparison to the FET?

38. Compare the conventional, the pseudomorphic and the metamorphic HEMT, and discuss the corresponding properties.

39. Explain the functional principle of a BJT including the tasks and the bias of the diodes. Draw the cross-section. Which materials are used? Explain the charge transport. Discuss the qualitative characteristics of I_c versus V_{be} and V_{ce}, and identify the specific regions.

40. Discuss the differences between FETs and BJTs in terms of input impedance, DC input current, g_m, operation ranges in the IV curves, feasibility for complementary logic, etc.

41. Explain the basic idea behind the HBT and the advantages compared to the BJT in terms of gain, f_t and noise. Draw the cross-section of an HBT. Which materials are used for HBTs?

42. Derive the current gains for both the BJT and the HBT.

43. List the approximate performance values of leading-edge transistors.

44. Do you have an own idea for a novel approach to improve transistors? Go for the Nobel Prize!

References

[Akb04]   M. S. Akbar, H.-J. Cho, R. Choi, C. S Kang, C. Y. Kang, C. H. Choi, S. J. Rhee, Y. H. Kim, J. C. Lee, "Optimized NH_3 annealing process for high-quality HfSiON gate oxide", IEEE Electron Device Letters, Vol. 25, No. 7, pp. 465–467, July 2004.

[Bäc02]   W. Bächtold, Mikrowellen-Elektronik, Vieweg Verlag, 2002.

[Ber05]   L. Berlin, "The Man behind the Microchip", Oxford University Press, New York, 2005.

[Ber90]   M. Berroth and R. Bosch, "Broad-band determination of the FET small-signal equivalent circuit", IEEE Transactions on Microwave Theory and Techniques, Vol. 38, pp. 891, 1990.

[Bon98]   P. K. Bondyopadhyay, "Moore's law governs the silicon revolution", Proceedings of the IEEE, Vol. 86, No. 1, pp. 78–81, Jan. 1998.

[Cha03]   R. Chau, S. Datta, M. Doczy, J. Kavalieros, M. Metz, "Gate dielectric scaling for high-performance CMOS: from SiO_2 to high-K gate insulator", IEEE International Workshop on Gate Insulator, pp. 124–126, Nov. 2003.

[Dam88]   G. Dambrine et. al., "A new method for determining the FET small-signal equivalent circuit", IEEE Transactions on Microwave Theory and Techniques, Vol. 36, pp. 1151, 1998.

[Doy03]   B. S. Doyle, S. Datta, M. Doczy, S. Hareland, B. Jin, J. Kavalieros, T. Linton, A. Murthy, R. Rios, R. Chau, "High performance fully-depleted tri-gate CMOS transistors", IEEE Electron Device Letters, Vol. 24, No. 4, pp. 263–265, April 2003.

[Dor04]   A. Van Dormael, "The french transistor", Proc. of IEEE Conference on the History of Electronics, June 2004.

[Dun03]   J. S. Dunn, D. C. Ahlgren, D. D. Coolbaugh, N. B. Feilchenfeld, G. Freeman, D. R. Greenberg, R. A. Groves, F. J. Guarin, Y. Hammand, A. J Joseph, L. D. Lanzerotti, S. A. St. Onge, B. A Orner, J.-S. Rieh, K. J. Stein, S. H. Voldman, P.-C. Wang, M. J. Zierak, S. Subbanna, D. L. Harame, D. A Herman and B. S. Meyerson, "Foundation of RF CMOS and SiGe BiCMOS techchnologies", IBM Journal on Research and Development, Vol. 47, No. 2/3, March/May 2003.

[Eas02]   L. F. Eastman, U. K. Mishra, "The toughest transistor yet", IEEE Spectrum, pp. 28–33, May 2002.

[Edu05]   http://jas2.eng.buffalo.edu/applets/

[Ell1]   F. Ellinger, "RF integrated circuits for wireless receivers at millimeter wave frequencies", Hartung-Gorre Verlag, ISBN 3-86628-007-6, 2005.

[Ell40]   F. Ellinger, M. Kossel, M. Huber, M. Schmatz, C. Kromer, G. Sialm, D. Barras, L. Rodoni, G. von Büren, and H. Jäckel, "High-Q inductors on digital VLSI CMOS substrate for analog RF applications", IEEE/SBMO International Microwave and Optoelectronics Conference, pp. 869–872, Sept. 2003.

[Ell53]   F. Ellinger, J. Kucera and W. Bächtold, "Improvements on a nonlinear GaAs MESFET model", IEEE MTT-S International Microwave Symposium, Baltimore, June 98, pp. 1623–1626.

[End03]   A. Endo, "Progress of a laser plasma EUV light source for lithography", Annual Meeting of the IEEE Lasers and Electro-Optics Society, Vol. 1, pp. 220–221, Oct. 2003.

[Fra92]   D. J. Frank, S. E. Laux and M. V. Fischetti, "Monte Carlo simulation of a 30 nm dual-gate MOSFET: how short can Si go?" International Electron Devices Meeting, pp. 553–556, Dec. 1992.

[Fuk79]   H. Fukui, "Determination of basic device parameters of a GaAs MESFET", Bell Systems Technical Journal, Vol. 58, pp. 771, 1979.

[Gae77]   F. H. Gaensslen, V. L. Rideout, E. J. Walker and J. J. Walker, "Very small MOSFETs for low temperature operation", IEEE Transaction on Electron Devices, Vol. ED-34, p. 218, 1977.

[Har00]   K. Harazaki, Y. Hasegawa, Y. Shichijo, H. Jabuchi, K. Fujii, "High accurate optical proximity correction under the influences of lens aberration in 0.15 µm logic process", IEEE International Conference on Microprocesses and Nanotechnology, pp. 14–15, July 2000.

[Hol86]   R. P. Holstrom, W. L. Bloss and J. Y. Chi, "A gate probe method of determining parasitic resistance in MESFETs", IEEE Electron Device Letters, Vol. ED-7, p. 410, 1986.

[Ibm41]   www-306.ibm.com/chips/services/foundry/technologies/cmos. Html

[Ibm42]   www.research.ibm.com/resources/press/strainedsilicon/

[Ibm43]   http://www.a1-electronics.net/Res_GenInterest/2003/IBM_Low-k_Dielectric.jpg

[Ici05]   www.icinsights.com/news/

[Ick05]   www.icknowledge.com

[Ieo01]   M. Ieong, E. C. Jones, T. Kanarsky, Z. Ren, O. Dokumaci, R. A. Roy, L. Shi, T. Furukawa, Y. Taur, R. J. Miller, H.-S. P. Wong, "Experimental evaluation of carrier transport and device design for planar symmetric/asymmetric double-gate/ground-plane CMOSFETs", International Electron Devices Meeting, pp. 19.6.1–19.6.4, Dec. 2001.

[Ieo02]   M. Ieong, H.-S. P. Wong, E. Nowak, J. Kedzierski and E. C. Jones, "High performance double-gate device technology challenges and opportunities", IEEE International Conference on Quality Electronic Design, pp. 492–495, March 2002.

[Jae88]   R. C. Jaeger and F. H Gaensslen "Low temperature semiconductor electronics", IEEE Conference on Thermal Phenomena in the Fabrication and Operation of Electronic Components, pp. 106–114, May 1998.

[Jag04]   Jagannathan, D. Greenberg, D. I. Sanderson, J.-S. Rieh, J. O. Plouchart and G. Freeman, "Speed and power performance comparison of state-of-the-art CMOS and SiGe RF transistors", Topical Meeting on Silicon Monolithic Integrated Circuits in RF Systems, pp. 115–118, Sept. 2004.

[Jag06]   B. Jagannathan, D. Chidambarrao, J. Pekarik, "300 GHz transistor performance in production CMOS technologies", IEEE Device Research Conference, pp. 199-200, June 2006

[Lan94]   W. Langheinrich, A. Vescan, B. Spangenberg and H. Beneking, "The resolution of the inorganic electron beam resist, LiF(Al3)", Microelectronics engineering, Vol. 23, p. 287, 1994.

[Léc05]   C. Lécuyer, Making Silicon Valley, MIT Press, Cambridge, 2005.

[Lee04]   T. H. Lee, "The Design of CMOS Integrated Circuits", Cambridge University Press, 2004.

[Maa03]   S. A. Maas, Nonlinear Microwave and RF Circuits, Artech House, 2003.

[Mill83]   S. Millman, "A history of engineering and science in the Bell system", physical sciences (1925–1980), AT & T Bell Laboratories Short Hills, N.J., 1983

[Moo65]   G. Moore, "Cramming more components onto integrated circuits", Electronics, Vol. 38, No. 8, April 19, 1965.

[Nis96]   K. Nishimura, K. Onodera, S. Aoyama, M. Tokumitsu and K. Yamasaki,
           "High performance 0.1 μm self-aligned-gate GaAs MESFET
           technology", European Solid State Circuit Conference, pp. 865–868,
           1996.

[Pek04]   J. Pekarik et al., "RFCMOS technology from 0.25 μm to 65 nm: state
           of the art", IEEE Custom Integrated Circuits Conference, 217-224,
           2004.

[Plo03]   J.-O. Plouchart, "SOI nano-technology for high-performance system
           on-chip applications", IEEE SOI Conference, pp. 1–4, Oct. 2003.

[Plo05]   J.O. Plouchart, N. Zamdmer, J. Kim, R. Trzcinski, S. Narasimha,
           M. Khare, L. F. Wagner, S. L. Sweeney, S. Chaloux, "A 243-GHz Ft
           and 208-GHz Fmax, 90 nm SOI CMOS SoC technology with low-
           power mmwave digital and RF circuit capability", IEEE Transaction
           on Electron Devices, Vol. 52, No. 7, pp. 1370–1375, July 2005.

[Rob02]   F. Robin, "Theoretical and technological investigations of the gate
           structure of InP HEMTs for high-frequency and power applications",
           Dissertation, No. 14849, ETH Zürich, 2002.

[Rim01]   K. Rim, S. Köster, M. Hargrove, J. Chu, P. M. Mooney, J. Ott,
           T. Kanarsky, P. Ronsheim, M. Ieong, A. Grill and H.-S. P. Wong,
           "Strained Si NMOSFET for High performance CMOS technology",
           IEEE Conference on VLSI Technology, pp. 59–60, June 2001.

[Rie03]   J.-S. Rieh, B. Jagannathan, H. Chen, K. Schonenberg, S.-J. Jeng,
           M. Khater, D. Ahlgren, G. Freeman, S. Subbanna, "Performance and
           design considerations for high speed SiGe HBTs of f_t/f_{max} =
           375GHz/210GHz", IEEE Conference on Indium Phosphide and
           Related Materials, pp 374-377, May 2003.

[Rie04]   J.-S. Rieh, B. Jagannathan, D. R. Greenberg, M. Meghelli, A. Rylyakov,
           F. Guarin, Z. Yang, D. C. Ahlgren, G. Freeman, P. Cottrell, D. Harame,
           "SiGe Heterojunction bipolar transistors and circuits towards terahertz
           communication applications", IEEE Transactions on Microwave
           Theory and Techniques, Vol. 52, pp. 2390–2408, Oct. 2004.

[Rim02]   K. Rim, E. P. Gusev, C. D. Emic, T. Kanarsky, H. Chen, J. Chu, J. Ott,
           K. Chan, D. Boyd, V. Mazzeo, B. H. Lee, A. Mocuta, J. Welser,
           S. L. Cohen, M. Leong, H.-S. Wong, H.-S, "Mobility enhancement in
           strained Si NMOSFETs with HfO/sub 2/ gate dielectrics",
           Symposium on VLSI Technology, pp. 12–13, June 2002.

[Ror96]   N. Rorsman, M. Garcia and C. Karlsson and H. Zirath, "Accurate
           small signal modeling of HFETs for millimeter wave applications",
           IEEE Transactions on Microwave Theory and Techniques, Vol. 44,
           pp. 432–437, 1996.

[Sch00]   R. Schmidt, "Low temperature CMOS experience at IBM", IEEE
           Symposium on Semiconductor Thermal Measurement and
           Management, pp. 112–113, March 2000.

[Sch03]   A. J. Scholten, "Noise modeling for RF CMOS circuit simulation",
           IEEE Transaction on Electron Devices, V. 50, No. 3, pp. 618–632,
           March 2003.

[Sho72]   W. Shockley, "How we built the transistor, revolution in miniature: the history and impact of semiconductor electronics", New Scientist Dec. 1972.

[Sun87]   J. Y.-C. Sun, Y. Taur, R. H. Dennard and S. P. Klepner, "Submicrometer-channel CMOS for low-temperature operation", IEEE Transactions on Electron Devices, Vol. ED-34, p. 19, 1987.

[Sze88]   S. M. Sze, High-speed Semiconductor Devices, Wiley, New York, 1988.

[Tau97]   Y. Taur, D. A. Buchanan, W. Chen, D. J. Frank, K. E. Ismael, S.-H. Lo, G. A. Sai-Halasz, R. G. Viswanathan, H.-J. C. Wann, S. J. Wind and H.-S. Wong, "CMOS scaling into the nanometer regime", Processings of the IEEE, Vol. 85, No. 4, pp. 486–504, April 1997.

[Tau01]   Y. Taur, "Analytic solutions of charge and capacitance in symmetric and asymmetric double-gate MOSFETs" IEEE Transactions on Electron Devices, Vol. 48, No. 12, pp. 2861–2869, Dec. 2001.

[Tes04]   A. Tessmann, A. Leuther, C. Schworer, H. Massler, W. Reinert, M. Walther, R. Losch, M. Schlechtweg, "Millimeter-wave circuits based on advanced metamorphic HEMT technology" Infrared and Millimeter Waves, International Conference on Terahertz Electronics, pp. 165–166, Oct. 2004.

[Tra05]   Transistorized, www.pbs.org/transistor/album1/

[Tro03]   T. Ytterdal, J. Cheng, T. A. Fjeldly, Device Modelling for Analog and CMOS RF Circuit Design, Wiley, 2003.

[Uga95]   M. Ugajin, J. Kodate, Y. Kobayashi, S. Konaka and T. Skai, "Very-high f_T and f_{max} silicon bipolar transistors using ultra-high-performance super self-aligned process technology for low-energy and ultra-high-speed LSI's" International Electron Devices Meeting, pp. 735–738, Dec. 1995.

[Wal04]   J. Walko, "Scaling is dead, long live innovation", IEE Review, Vol. 50, No. 6, p. 23, June 2004.

[Wei04]   Weize Xiong; G. Gebara, J. Zaman, M. Gostkowski, B. Nguyen, G. Smith, D. Lewis, C. R. Cleavelin, R. Wise, Yu Shaofeng, M. Pas, Tsu-Jae King, J. P. Colinge, "Improvement of FinFET electrical characteristics by hydrogen annealing", IEEE Electron Device Letters, Vol. 25 , No. 8,pp. 541–543, Aug. 2004.

[Won97]   H.-S. P. Wong, K. K. Chan, Y. Taur, "Self-aligned (top and bottom) double-gate MOSFET with a 25 nm thick silicon channel", International Electron Device Meeting, pp. 427–430, Dec. 1997.

[Won98]   H.-S. P. Wong, D. J. Frank and P. M. Solomon, "Device design considerations for double-gate, ground-plane, and single-gated ultra-thin SOI MOSFET's at the 25 nm channel length generation", International Electron Device Meeting, pp. 407–410, Dec. 1998.

[Yam02]   Y. Yamashita, A. Endoh, K. Shinohara, K. Hikosaka, T. Matsui, S. Hiyamizu and T. Mimura, "Pseudomorphic $In_{0.52}Al_{0.48}As$/ $In_{0.7}Ga_{0.3}As$ HEMTs with ultrahigh f_t of 562 GHz", IEEE Electron Device Letters, Vol. 23, No. 10, pp. 573–575, Oct. 2002.

[Yan86]   L. Yang and S. Long, "New method to measure the source and drain resistance of a GaAs MESFET", IEEE Electron Device Letters, Vol. ED-7, p.75, 1986.

[Zam02]   N. Zamdmer, J. O. Plouchart, J. Kim, L.-H. Lu, S. Narasimha, P. A. O'Neil, A. Ray, M. Sherony and L. Wagner, "Suitability of scaled SOI CMOS for high-frequency analog circuits", European Solid State Device Research Conference, Sept. 2002.

[Zam04]   N. Zamdmer, J. Kim, R. Trzcinski, J.-O. Plochart, S. Narasimha, M. Khare, L. Wagner, S. Chaloux, "A 243-GHz Ft and 208-GHz Fmax, 90-nm SOI CMOS SoC technology with low-power millimeter-wave digital and RF circuit capability", pp. 98–99, IEEE Symposium on VLSI Technology, 2004.

[Zie60]   A. Van der Ziel, "Shot noise in transistors", Proc. IRE, Vol. 48, pp. 114–115, Jan. 1960.

[Zie86]   A. Van der Ziel, Noise in Solid-state Devices and Circuits, Wiley, 1986.

6 Passive Devices and Networks

Reduce to the maximum.
Advertisement slogan

Passive devices play a key role for the design of RFICs. They are required for functions such as biasing, filtering and impedance matching. Due to the resistive and reactive parasitics passive devices exhibit losses. We can distinguish between distributed elements such as transmission lines, and lumped elements including inductors, capacitors and resistors [Bah05].

6.1 Transmission Lines

Microstrip and CPW (Coplanar Wave) transmission lines are frequently used for integrated circuits. They are easy to fabricate, provide a high yield and can be scaled in length. In integrated form, transmission lines are mainly suited for frequencies above 10 GHz, where the large size of the lines scales down. At lower frequencies, lumped elements are employed since they are much more compact and less lossy. Transmission lines are not only important to realise specific circuit elements. The connections between different elements are usually performed by means of transmission lines, which have to be taken into account in simulations. For detailed information and precise mathematical descriptions concerning transmission lines, the reader is referred to the subject-specific literature [Wad91, Mei86, Bäc94].

6.1.1 Microstrip

A cross-section of a microstrip line is illustrated in Fig. 6.1. The line can be realised with a simple metal conductor and a ground metal, which may be at the back of the MMIC. The major part of the electric and magnetic field is located within the dielectric material featuring a dielectric constant $\varepsilon_r > 1$, whereas a part of the field propagates in the air with $\varepsilon_r = 1$. An effective dielectric constant $\varepsilon_{r,eff}$ can be calculated considering the fields in both materials. The geometric relation between the conductor width w and the distance between conductor and metal h plays an important role. One of the major measures for transmission lines is the characteristic impedance Z_w determining if the line exhibits inductive or rather capacitive properties. Table 6.1 provides empirical equations for Z_w and $\varepsilon_{r,eff}$ yielding good results within $0 < \dfrac{w}{h} < 10$ representing the range for most engineering applications.

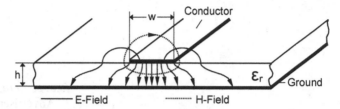

Fig. 6.1. Cross-section of microstrip line, top metal used as conductor line, lower metal represents ground plane

In Fig. 6.2, the corresponding characteristics vs w/h and ε_r are shown. The impedance increases with lowered ε_r and w, and raised h. There are process parameters defined by technology, and design parameters. The dielectric constant is a process parameter. Typical isolators exhibit an ε_r in the range of 2–5. Silicon, GaAs an InP substrates have higher values of 11.8, 12.8 and 12.5, respectively. In addition to w, the designer has also an influence on h in the sense that he can chose between several metal layers yielding different h. The top layer with fixed h is frequently employed, since it provides the thickest metal and the largest distance to the lossy substrate yielding the lowest losses.

Table 6.1. Effective dielectric constant and characteristic impedance vs geometric $\dfrac{w}{h} \leq 1$ relation, compare with [Sch68]

$\dfrac{w}{h}$	≤ 1 (thin line)		≥ 1 (thick line)	
$\varepsilon_{r,eff}$	$\dfrac{\varepsilon_r + 1}{2} + \dfrac{(\varepsilon_r - 1)}{2} \cdot F(\dfrac{w}{h})$			(6.1)
	$F(\dfrac{w}{h}) = \dfrac{1}{2\sqrt{1 + 12\,h/w}} + 0.04(1 - \dfrac{w}{h})^2$ (6.2)		$F(\dfrac{w}{h}) = \dfrac{1}{2\sqrt{1 + 12\,h/w}}$ (6.3)	
Z_w	$\dfrac{60\Omega}{\sqrt{\varepsilon_{r,eff}}} \ln\left(8\dfrac{h}{w} + \dfrac{w}{4h}\right)$ (6.4)		$\dfrac{120\pi\Omega}{\sqrt{\varepsilon_{r,eff}}}$ $\dfrac{}{\dfrac{w}{h} + 2.42 - 0.44\dfrac{h}{w} + \left(1 - \dfrac{h}{w}\right)^6}$ (6.5)	

A transmission line has distributed characteristics, which can be modelled by a cascade of lumped elements as illustrated in Fig. 6.3a. Each increment is mainly determined by the series inductance L_s' and the isolation capacitance C_{ox}'. The resistive losses are taken into account by R_s' and R_{sub}'. Moreover, the substrate capacitance C_{sub}' may be added. In a first order approximation, all measures denoted by (') are normalised by l/n=line length/number of increments. Based on the distributed equivalent circuit and the relation between the voltages and currents, Z_w can be calculated. Referring to [Bäc94], neglecting the series resistance of the conductor and assuming that the substrate represents an ideal ground, the characteristic impedance yields

$$Z_w \approx \sqrt{\frac{L_s'}{C_{ox}'}} . \qquad (6.6)$$

Equation (6.6) reveals important insights of Z_w with respect to its reactive characteristics. We can distinguish between three cases related to the impedance of the interfaces. The unit of Z_w is $\sqrt{\frac{H}{F}} = \Omega$.

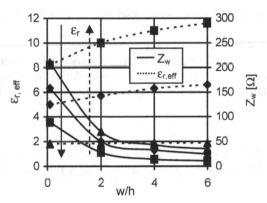

Fig. 6.2. Effective relative dielectric constant and characteristic impedance vs $\frac{w}{h}$ and ε_r=2, 8, 14

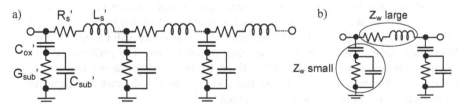

Fig. 6.3a,b. Microstrip line: **a** simplified equivalent circuit with cascaded repetitions, in first order the measures are normalised to l/n=line length/number of increments, L_s': series inductance of conductor, C_{ox}': capacitance across the isolation layer, R_s': series resistance of conductor, C_{ox}': oxide capacitance, G_{sub}'=$1/R_{sub}'$: substrate conductance, C_{sub}': substrate capacitance; **b** illustration of one increment

a. Z_w large
In this case, the value of L_s' is high compared to C_{ox}'. Thus, the line has inductive rather than capacitive characteristics. With line length l and magnetic constant μ_0, the inductance of an ideal thin microstrip line can be approximated by [Yua82]:

$$L_s = \frac{l \cdot \mu_0}{2\pi} \ln\left(\frac{8h}{w} + \frac{w}{4h}\right). \tag{6.7}$$

From the latter equation we can deduce some important insights regarding the inductive line:

- L_s is proportional to l being a key design parameter.
- We observe that L_s can be increased if w is made smaller. Unfortunately, decreased w raises the losses associated with the series resistance demanding for a tradeoff.
- L_s increases also with raised h. However, as mentioned before, h is more or less fixed given that we use for example the top metal.

Equation (6.7) is related to Eq. (6.4) representing Z_w of a thin microstrip line.

b. Z_w=impedance of interface/reference
The line has no reactive effect but introduces a phase shift and time delay. No reflections or signal losses are observed at the boundaries between the line and the connected interface. This is the typical type of transmission lines used for long interconnects.

c. Z_w small
Compared to L_s', the impact of C_{ox}' dominates. The line has capacitive characteristics associated with a large w and a small h (s for CPW line) forming a kind of plate capacitor.

To achieve a high model accuracy within a wide bandwidth, the number of distributed Π elements has to be maximised to obtain a broad range of resonances. By doing so, the calculation time required in CAD software is increased. It has been proved that it is reasonable to approximate a transmission line by a lumped element equivalent circuit with only one increment as depicted in Fig. 6.3b. In this case one has to be aware that the element values are only valid within a limited frequency range, which may be in the range of one octave. The performance of inductive transmission lines may be defined by the quality factors and the self-resonance frequencies. We can map the insights about inductors and capacitors, which will be discussed in Sects. 6.2 and 6.4.

6.1.2 Coplanar Wave

Compared to microstrip lines, CPW lines exhibit many similarities. As shown in Fig. 6.4, the major difference is that the ground plane of CPWs is alongside instead of underneath the strip conductor. This has several consequences. First, the fields extend more significantly into the air leading to higher losses and stronger undesired radiation. A further disadvantage of CPWs is the large circuit area required to realise the ground planes. However, the CPW features also some significant advantages. Most importantly, the ground connections are located close to the plane where other devices such as transistors are implemented. Hence, the parasitics inherent in the ground connections can be decreased.

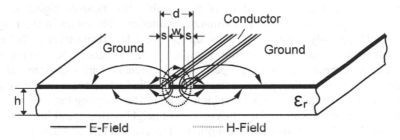

Fig. 6.4. Cross-section of a coplanar line, top metal is used as conductor line and as ground

The calculation of the characteristic impedance is more complicated than for microstrip lines [Wad91, Mei86,]. However, the general properties are very similar. Design parameters are the conductor width w and the distance to the ground s. As expected from the theory of the microstrip line, the impedance of the CPW line and the inductive characteristics raise with decreased ε_r and w, and increased h and s.

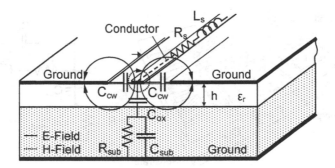

Fig. 6.5. Combination of microstrip and CPW transmission line as found in reality

For s→∞, the coplanar line converges into a microstrip line with ground located at the back of the chip. Because of size and cost constraints, the distances between the conductor line and the ground pads or connections are limited for microstrip lines. Thus, in reality, the microstrip line can also exhibit CPW transmission line characteristics, where an additional coplanar capacitance C_{cw} has to be considered as illustrated in Fig. 6.5.

6.2 Inductors

Progress is the realisation of utopias.
Oscar Wilde

Inductors are employed for tasks such as bias feeding and impedance matching of transistors. Reactively matched transistors enable significantly higher gain/DC power ratios and better noise performance. Inductors with values in the range of 10–0.1 nH typically cover frequencies up to 5–50 GHz, respectively. The higher the inductance value the lower the maximum self-resonance and operation frequencies. Concerning electrical properties, spiral inductors are similar to inductive transmission lines with high Z_w. However, spiral inductors are much more compact since the conductor line is folded into spirals. The coupling between the spiral lines has to be considered. Since the currents of neighbouring spiral lines flow into the same direction, a constructive coupling effect is observed increasing the inductance. Consequently, the inductance per length is high for spiral inductors. We will see in circuit examples that the inductors consume a major part of the total circuit size. Hence, inductors significantly contribute to the chip costs. Consequently, there is a general aim to decrease the size of inductors. Spiral inductors are well suited for the realisation of inductive values of below 0.2 nH. In this case, transmission lines are superior for layout and yield reasons.

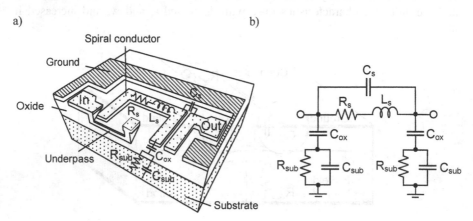

Fig. 6.6a,b. Inductor: **a** cross-section; **b** equivalent circuit

6.2.1 Equivalent Circuit

Figure 6.6a,b illustrates the cross-section and the equivalent circuit of a spiral inductor. The transmission path of the equivalent circuit consists of the desired inductance L_s, the resistance of the spiral lines R_s, and C_s representing the capacitances between the spiral lines and between the spiral lines and the underpath. By means of the oxide capacitance C_{ox} the fields are coupled into the lossy substrate characterised by the resistance R_{sub} and capacitance C_{sub}. The values of the equivalent elements determine the Q (Quality Factor) and the f_{SR} (Self Resonance Frequency). We survey now an inductor with quadratic dimensions and number of turns n, outer spiral dimension d_{out}, inner spiral dimension d_{in}, arithmetic mean value $d_{mean} = \dfrac{d_{out} + d_{in}}{2}$, conductor width w, conductor length l, spiral thickness t, wire conductivity σ, skin depth δ, oxide thickness or in other words distance between conductor and substrate d_{ms}, distance between conductor spirals d_{cc}, permeability in vacuum μ_0, substrate resistivity ρ_{sub}, and substrate capacitance per unit area C_{suba}. With these parameters and according to Fig. 6.7, we can approximate the elements of a square inductor with $n \geq 1$ as follows [Whe28, Won05, Lee04]:

$$L_s \approx \frac{9\mu_0 n^2 d_{mean}^2}{11 d_{out} - 7 d_{mean}} \tag{6.8}$$

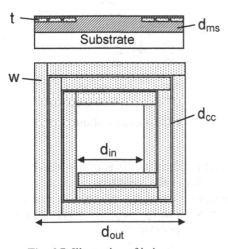

$$R_s \approx \frac{1}{w\sigma\delta(1 - e^{-t/\delta})} \tag{6.9}$$

$$\delta \approx \sqrt{\frac{2}{\omega\mu_0\sigma}} \tag{6.10}$$

$$C_s \approx \frac{n-1}{n} \cdot \frac{l \cdot t \cdot \varepsilon_{ox}}{d_{cc}} + \frac{n w^2 \varepsilon_{ox}}{d_{cc}} \tag{6.11}$$

$$C_{ox} \approx \frac{w \cdot l \cdot \varepsilon_{ox}}{2 \cdot d_{ms}} \tag{6.12}$$

Fig. 6.7. Illustration of inductor dimensions

$$R_{sub} \approx \frac{2\rho_{sub}}{w \cdot l} \tag{6.13}$$

$$C_{sub} \approx \frac{w \cdot l \cdot C_{suba}}{2}. \tag{6.14}$$

The left term in Eq. (6.11) describes the capacitance between the spiral lines, whereas the right term describes the capacitance between spiral lines and underpath.

Equation (6.9) reveals that R_s comprises a constant low frequency component R_{LF}, and a frequency dependant RF component R_{RF}. Hence, with a skin effect corner frequency ω_{skin}, we may also write

$$R_s = R_{LF} + R_{RF} \cdot \sqrt{\frac{\omega}{\omega_{skin}}}. \qquad (6.15)$$

In practice, Eqs. (6.8)–(6.15) provide reasonable accuracy, e.g. better than ±25% for the prediction of L_s, making them well suited for first approximations. More precise determination of the element values should be performed on the basis of CAD field simulations or measurements.

6.2.2 Quality Factor and Self Resonance Frequency

The quality factor describes the relation between the inductive impedance, and the parasitic resistive and capacitive impedances. Many different definitions can be found in literature. For a meaningful comparison it is always important to check which definition has been employed and if it is suitable for specific conditions and inductor technologies.

a)   b)

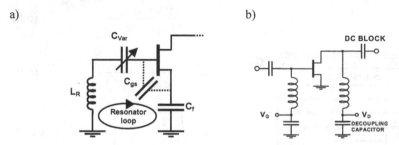

Fig. 6.8a,b. Shorting of one inductor port in practice: **a** LC resonator of oscillator; **b** RF shunted matching inductors or bias chokes

A reasonable equation for the inductor Q is derived on the basis of the individual lossy elements of the inductor [Yue98]. This allows the identification of the relevant loss origins, and consequently, the optimisation of the inductor. We define the inductor Q as the effective reactive energy given by magnetic energy minus electric energy divided by the resistive loss energy per cycle[1]:

$$Q(\omega) = \frac{1}{2\pi} \cdot \frac{E_{magnetic} - E_{electric}}{E_{loss-per-cycle}}. \qquad (6.16)$$

---

[1] This definition is distinct from that of resonators, where the energy is oscillating between the inductance and the capacitance. The resonator Q definition would not make sense since the desired impedance of the inductor is compensated by the parasitic impedance of the capacitor.

With respect to the relative complex equivalent circuit depicted in Fig. 6.6b, the following assumptions and simplifications are made to simplify the calculations:

a) Thin conductors as typically employed for integrated inductors exhibit relatively high ω_{skin} with values in the range of the self-resonance frequency. Given that the inductors are used below ω_{skin}, we assume that R_s is constant.

b) Moreover, we short-circuit one port. In reality, this is valid for many applications, e.g. for LC resonators in oscillators, RF shunted matching inductors and bias chokes as shown in Fig. 6.8.

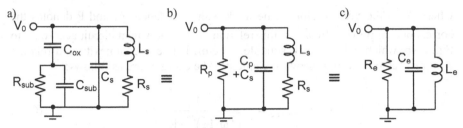

Fig. 6.9a–c. Simplifications for calculation of Q-factor: **a** shorting of one port; **b** R_p, C_p=f(C_{ox}, C_{sub}, R_{sub}); **c** L_e, R_e, C_e=f(L_s, R_s, R_p, C_p)

c) The parallel/series paths are converted into a purely parallel path with R_p, C_p=f(C_{ox}, C_{sub}, R_{sub}). By comparison of Fig. 6.9a,b we obtain

$$\frac{1}{j\omega C_{ox}}+\frac{\dfrac{1}{j\omega C_{sub}}\cdot R_{sub}}{\dfrac{1}{j\omega C_{sub}}+R_{sub}}=\frac{\dfrac{1}{j\omega C_p}\cdot R_p}{\dfrac{1}{j\omega C_p}+R_p}. \tag{6.17}$$

Separation into real and imaginary part yields

$$R_p = R_{sub}+\frac{C_{sub}}{C_{ox}}+\frac{C_p\cdot R_p}{C_{ox}} \tag{6.18}$$

and

$$\omega\cdot C_{sub}\cdot R_{sub}\cdot R_p = \omega\cdot C_p\cdot R_{sub}\cdot R_p+\omega\cdot C_p\cdot R_p\cdot\frac{C_{sub}}{C_{ox}}-\frac{1}{\omega\cdot C_{ox}}, \tag{6.19}$$

respectively, resulting in

$$R_p = \frac{1}{\omega^2\cdot C_{ox}^2\cdot R_{sub}}+\frac{R_{sub}\cdot\left(C_{ox}+C_{sub}\right)^2}{C_{ox}^2} \tag{6.20}$$

and

$$C_p = C_{ox} \frac{1 + \omega^2 \cdot (C_{ox} + C_{sub}) \cdot C_{si} \cdot R_{sub}^2}{1 + \omega^2 \cdot (C_{ox} + C_{sub})^2 \cdot R_{sub}^2} . \tag{6.21}$$

Obviously, the latter two elements are frequency dependent. The energies determining Q are given by

$$E = \left| \frac{P}{\omega} \right| = \left| \frac{\underline{V}^2}{2 \cdot \omega \cdot \underline{Z}} \right|, \tag{6.22}$$

where $\underline{V}$ is the voltage along a network with impedance $\underline{Z}$, and P denotes the corresponding power. We apply this relation on the equivalent circuit according to Fig. 6.9b, which exhibits a reasonable trade-off between complexity and relation with respect to the origin network elements. The energies are as follows:

$$E_{magnetic} = \frac{V_0^2 \cdot L_s}{2 \cdot \left[(\omega \cdot L_s)^2 + R_s^2 \right]}, \tag{6.23}$$

$$E_{electric} = \frac{V_0^2 \cdot (C_s + C_p)}{2} \tag{6.24}$$

and

$$E_{loss} = \frac{V_0^2}{2\omega} \left[\frac{1}{R_p} + \frac{R_s}{(\omega \cdot L_s)^2 + R_s^2} \right]. \tag{6.25}$$

Now, we are ready to determine Q according to Eq. (6.16):

$$Q = \underbrace{\frac{\omega \cdot L_s}{R_s}}_{Q_{LF}} \cdot \underbrace{\frac{R_p}{R_p + \left[(\omega \cdot L_s / R_s)^2 + 1 \right] \cdot R_s}}_{k_{SL}} \cdot \underbrace{\left[1 - \frac{R_s^2 \cdot (C_s + C_p)}{L_s} - \omega^2 \cdot L_s \cdot (C_s + C_p) \right]}_{k_{SR}} \tag{6.26}$$

We can identify three factors:

I. Low frequency Q-factor Q_{LF}
Describes the Q in case of ideal substrate. Only the resistive losses of spiral lines are considered.

II. Substrate loss factor $k_{SL} \leq 1$
Represents the substrate losses. According to Table 5.1, the substrate resistivities for III/V semiconductors (e.g. GaAs and InP) are around 100 times higher than the ones of silicon-based technologies. Consequently, the R_{sub} in III/V technologies can be neglected resulting in a k_{SL} close to unity. For silicon inductors, the

substrate loss is the mandatory loss mechanism leading to a significant degradation of Q.

III. Self resonance loss factor $k_{SR} \leq 1$

Considers the compensation of L_s by means of the parasitic capacitances. At the f_{SR}, k_{SR} and consequently Q vanishes to zero. The f_{SR} limits the maximum operation frequency in circuits. Inductors can only be used for operation frequencies of up to approximately 80% of the f_{SR}. The higher the inductance value, the lower the f_{SR}.

d) We make a further rearrangement to decrease the complexity of the equivalent circuit to the lowest possible. It is a purely parallel connection of one inductor L_e, one resistor R_e and one capacitor C_e as depicted in Fig. 6.9c. The determination of the equivalent capacitor is simple since $C_e = C_p + C_s$. By comparison of Fig. 6.9b,c, we can find identities for L_e, $R_e = f(L_s, R_s, R_p)$. At one fixed frequency, we may transform the series element R_s into a parallel element R_e resulting in the following equation:

$$\frac{1}{R_{se}} - j\frac{1}{\omega L_e} = \frac{1}{R_s + j\omega L_s} = \frac{R_s}{R_s^2 + (\omega L_s)^2} - j\frac{\omega L_s}{R_s^2 + (\omega L_s)^2}. \qquad (6.27)$$

In turn, we get

$$L_e = L_s + \frac{R_s^2}{\omega^2 L_s}, \qquad (6.28)$$

and

$$R_{se} = R_s + \frac{(\omega L_s)^2}{R_s}. \qquad (6.29)$$

If the Q-factor is well above unity corresponding to $\omega L_s > R_s$, we can simplify the latter two identities to $L_e \approx L_s$ and $R_{se} \approx \frac{(\omega L_s)^2}{R_s}$. Since R_e is composed by a parallel connection of R_p and R_{se} we get

$$R_e = \frac{R_p \cdot R_{se}}{R_p + R_{se}}. \qquad (6.30)$$

Because of the low complexity, the equivalent circuit model according to Fig. 6.9c is well suited for simple circuit investigations, e.g. for the analytical description of oscillators which will be presented in Chap. 11. However, due to the transformations made in d), the individual loss mechanisms associated with factors Q_{LF}, k_{SL} and f_{SR} can't be monitored any more. Thus, for analysis and optimisation of inductors, the previous model according to Fig. 6.9b may be superior since it meets the optimum tradeoff between complexity on one hand and fruitful insights concerning specific parasitics on the other hand.

6.2.3 Performance of CMOS Inductors

Now, the properties of typical inductors with inductive values of around 0.31 nH, 0.9 nH and 1.6 nH are reviewed and compared. They have been fabricated on a VLSI silicon substrate with a moderate substrate resistivity ρ_{sub} of around 10 Ω cm [Ell40]. The distance between the lossy substrate and the bottom side of the conductor amounts to d_{ms}=3.5 μm and the conductor thickness is t=1.7 μm. These values are typical for low-cost VLSI CMOS technology. It is clear that better performance would be possible with higher ρ_{sub}, larger d_{ms} and thicker t. However, the improvement of these parameters would results in increased costs. In Table 6.2, details of the equivalent circuit extracted by S-parameter measurements are listed.

Table 6.2. Parameters of inductor

Ind.	Size [μm^2]	n	w [μm]	d_{cc} [μm]	L_s [nH]	C_s [fF]	R_s [Ω]	C_{ox} [fF]	C_{sub} [fF]	R_{sub} [Ω]	$Q_{max}@f$ [GHz]	f_{SR} [GHz]
0L31	120 × 120	1.5	12.5	5	0.31	3	1	50	10	200	13@ 9.5	45
0L9	160× 160	2.5	12.5	5	0.9	6	2	90	20	120	10@ 5	29
1L6	200× 200	3.5	12.5	5	1.6	12	4	140	35	80	6.3@ 4	15

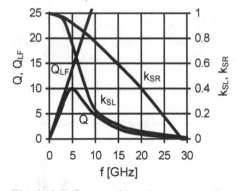

Fig. 6.10. Influence of low frequency quality factor Q_{LF}, substrate loss factor k_{SL} and self-resonance factor f_{SR} on overall quality factor Q of inductor 0L9

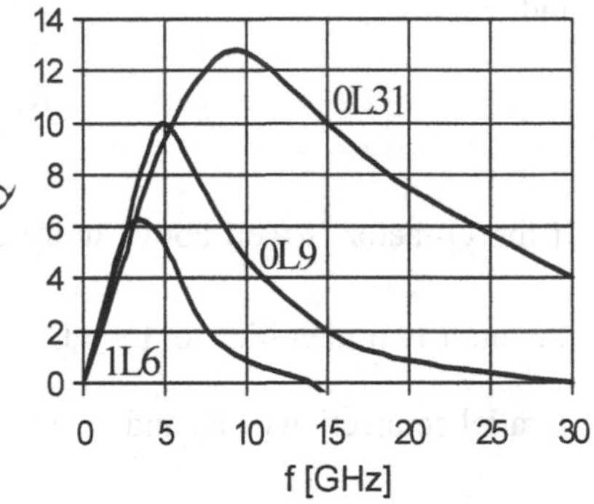

Fig. 6.11. Quality factor vs frequency of different inductors

The impacts of the loss factors with respect to frequency are illustrated in Fig. 6.10 for the 0.9-nH inductor. At low frequencies, Q is determined by Q_{LF}, whereas towards high frequencies, Q is determined by the substrate losses. As expected by definition, Q equals zero at the f_{SR} of the inductor, where k_{SR}=0.

In Fig. 6.11, the characteristics of the three inductors are shown vs frequency. Maximum Q-factors of 6.3 at 4 GHz, 10 at 5 GHz, and 13 at 9.5 GHz were measured for the 1.6-nH, 0.9-nH, and 0.31-nH devices, respectively. The corresponding f_{SR} amounts to 15 GHz, 29 GHz, and 45 GHz, respectively.

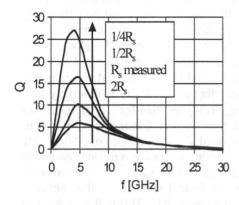

Fig. 6.12. Quality factor Q of inductor 0L9 extracted from measurements with $R_{s,\,measured} = 2\,\Omega$ including variation of spiral series resistance R_s

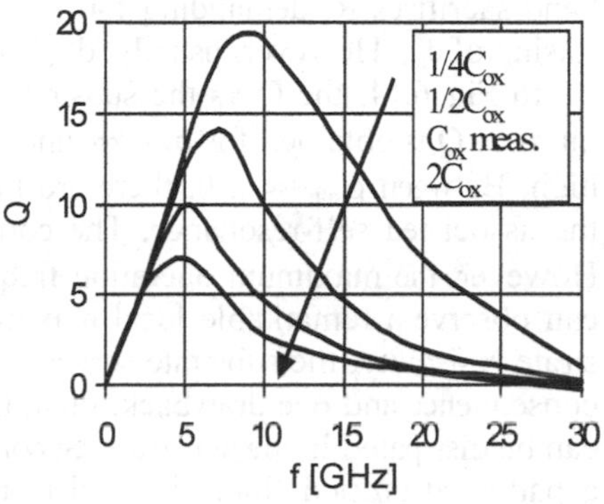

Fig. 6.13. Quality factor Q of inductor 0L9 extracted from measurements with $C_{ox,\,measured}=90$ fF including variation of C_{ox}

The loss inherent in Q_{LF} is determined by R_s, which can be minimised by increasing w and or t. Usually, the later parameter is defined by the technology and can't be modified by the designer. The influence of R_s on Q is shown in Fig. 6.12. A decreased R_s leads to a significant improvement of Q at low frequencies, whereas the improvement towards high frequencies is very small. According to the next two figures, at highest frequencies, Q is determined by the coupling to the substrate by means of C_{ox}, and by the value of the substrate resistivity.

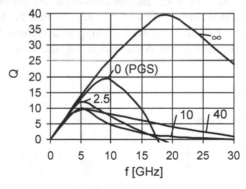

Fig. 6.14. Quality factor Q of inductor 0L9 extracted from measurements with $\rho_{sub}=10\,\Omega$ cm versus frequency and variation of substrate resistivity ρ_{sub} with unit [Ω cm], PGS: patterned ground shield

In Fig. 6.13, the influence of C_{ox} on Q is shown. A decreased C_{ox} significantly improves the Q towards high frequencies, whereas the impact towards low frequencies is small. To reduce this capacitance, w must be lowered, which on the other hand sacrifices R_s demanding for a tradeoff. Moreover, C_{ox} could be reduced by raising of d_{ms}. However, usually, d_{ms} is determined by the used technology.

In Fig. 6.14, the Q vs the substrate resistivity is illustrated. The global maximum of Q is obtained for $\rho_{sub}=\infty$ since the substrate losses are zero and the f_{SR} is high. Between $\rho_{sub}=\infty$ .. 0, there are losses caused by the nonideal substrate and the associated self-resonance. The corresponding maximum Q is relatively low. However, the maximum operation frequency is higher than for $\rho_{sub}=0$, where we can observe a remarkable local maximum for Q at low frequency. At zero substrate resistivity, the substrate serves as a ground shield leading to one favourable consequence and one drawback. First, the substrate losses are zero since no power can be dissipated in the substrate. Second, the f_{SR} and the k_{SR} decrease. In practice, a patterned ground shield is used to mitigate mutual coupling in the substrate, which lowers the effective inductance and the resulting Q. This will be subject to discussions in Sect. 6.2.5.2.

6.2.4 Optimisation

Many parameters and careful tradeoffs must be considered for the inductor design. Optimisation of inductors with respect to individual design constraints is important. Field simulator software such as HFSS (High Frequency Structure Simulator) can be employed to optimise the performance. The properties of currents can be simulated as illustrated in Fig. 6.15. After conversion of the voltage and currents into S-parameter, an equivalent model according to Fig. 6.6b can be fitted in a CAD tool like ADS (Advanced Design System) or cadence. Error functions similar to that described in Sect. 5.4.5 can be employed. By means of the obtained model elements we can determine Q vs frequency. Within the frequency range of interest, it is aimed to maximise Q, while taking into account further constraints such as size and maximum current. A corresponding optimisation routine is shown in Fig. 6.16.

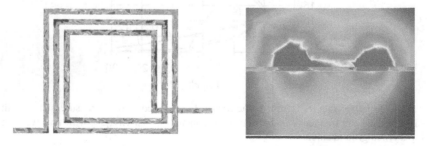

Fig. 6.15. Simulation of current distributions applying HFSS field simulator

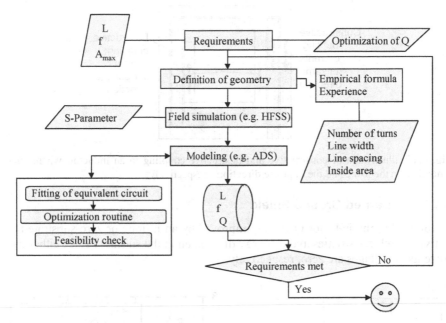

Fig. 6.16. Optimisation procedure for inductors

6.2.5 Improvement Techniques

Low-cost silicon based technologies such as VLSI CMOS suffer from Q-factor degradation due to the considerable substrate and conductor losses. Smart techniques will be treated in the next sections, which reduce the impact of the loss mechanisms.

6.2.5.1 Hollow Spiral Inductors

Referring to Fig. 6.17, we can identify two kinds of current directions in the conductor lines. First, currents flowing in the same direction leading to constructive coupling and increased inductance, which results in higher inductance per area and enhanced Q. Second, we observe contra-directional currents flowing in the conductors located at opposite sides of the inductors. According to the law of Lenz, these currents disturb the common fields yielding lowered inductance, which is not desired. The coupling of these currents can be reduced by leaving out the innermost spirals. However, such hollow inductors require more area. Thus, a trade-off has to be found. For low-Q inductors, e.g. as required for peaking loads, the Q may be sacrificed for maximum inductance per area. Transmission lines do not suffer from contra-directional currents, since they have no turns, and consequently no conductors, where the current flows in opposite directions. On the other hand, transmission lines do not benefit from the positive effect of adjacent conductors, where the currents flow in the same direction.

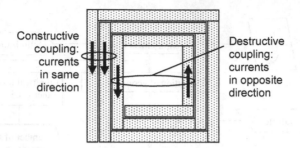

Fig. 6.17. Illustration of constructive and destructive coupling in an inductor, where currents flow in the same and the opposite directions, respectively

6.2.5.2 Patterned Ground Shield

Theoretically, the inductor Q can be improved by an infinite or zero substrate resistivity. Both resistivities avoid energy dissipation in the substrate since either the currents or voltages are zero, respectively.

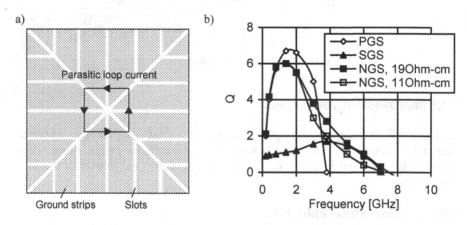

Fig. 6.18. Patterned ground shield a) illustration b) results of 8.5 nH inductor, PGS: patterned ground shield, SGS: solid ground shield, NGS: no ground shield, data from [Yue98]

A conductive ground shield with very low resistivity can be implemented to prevent the field from penetrating into the lossy substrate. Unfortunately, we have not yet considered the law of Lenz. Image current, also known as loop current, is induced into the solid ground shield by the magnetic field of the inductor. This image current flows in the opposite direction compared to the current in the spiral inductor. Similar to the inner-spiral problem addressed in the preceding section, the associated mutual coupling between the currents reduces the magnetic field, the overall inductance, and subsequently the Q. One efficient solution to circumvent this effect is to use a patterned ground shield, which significantly increases the resistance for the image current. The ground shield is patterned with slots orthogonal to the spiral as illustrated in Fig. 6.18a. Attributed to the fact that d_{ms} is lowered by the ground shield, C_{ox} is raised leading to a decrease of k_{SL} and k_{SR}. Usually, at low to moderate frequencies, the degradation due to the decreased k_{SR} is smaller than the improvement caused by the increase of k_{SL}. Thus, the pattern ground shield pays off as demonstrated in Fig. 6.18b.

6.2.5.3 Silicon on Insulator

Section 5.4.10.1 has revealed that SOI technologies yield improved transistor performance since the substrate losses are reduced. A thin DC isolator is implemented between the active devices and the main substrate. Consequently, for the main substrate, a high ρ_{sub} can be chosen without impacting the DC properties of the active devices. This high ρ_{sub} helps to improve the performance of the passive elements. A thick isolation layer would further decrease the coupling to the lossy substrate. However, because of the constraints regarding low-cost and high-yield processing, most of today's commercial technologies feature only a very thin SOI isolator layer. With a thickness below 0.1 μm, they provide DC isolation rather than additional RF isolation. To ensure mechanical stability, the thickness of the main substrate is much larger and typically above 0.3 mm.

Table 6.3. Results of inductors in 90-nm SOI technology

L	Turns	Size	Q_{max}	f_{SR}
0.25 nH	1.25	50×55 μm²	21@41 GHz	>100 GHz
2 nH	3.25	95×95 μm²	12@8 GHz	25 GHz
3 nH	5.25	110×110 μm²	10@5 GHz	18 GHz

a)   b)

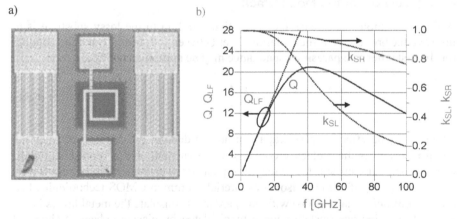

Fig. 6.19a,b. SOI inductor with 0.25 nH: **a** photo of test structure including measurement pads which are deembedded; **b** Q factor including the loss factors

The results of the optimised inductors realised in 90-nm SOI technology are summarised in Table 6.3 [Ell1]. A chip photo of the 0.25-nH inductor incorporated into a test structure, the Q factor and the associated loss factors are shown in

Fig. 6.19a,b. An excellent Q factor of 21 was measured at 40 GHz. The f_{SR} is well above 100 GHz.

6.2.5.4 MEMS

The microelectrical mechanical system (MEMS) technology is one of the most efficient methods to reduce the parasitic coupling to the lossy substrate [Reb03, Jin05]. The idea is to etch a part of the substrate under the devices away leading to a much larger d_{ms}. In Fig. 6.20, a MEMS inductor is illustrated. Due to the low dielectric constant in the air and the relatively large distance to the lossy substrate, the losses caused by the substrate impairments are significantly decreased. Typically, the achievable Q factors are similar to that in III/V substrates. Disadvantages of the MEMS technology are higher costs and limited mechanical stability.

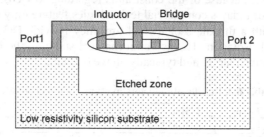

Fig. 6.20. Illustration of MEMS inductors

6.2.5.5 Low Dielectric Metal Stack

A further method to decrease the capacitive coupling to the lossy substrate is to improve the properties of the isolation layer between the substrate and the conductor. In first order, the parasitic capacitance may be approximated by

$$C = \frac{\varepsilon_0 A_o}{d} \cdot \varepsilon_r , \qquad (6.31)$$

where A_o is the effective plate capacitor area, d denotes the distance between the metal plates, ε_0 is the dielectric constant in the vacuum, and ε_r represents the dielectric material constant. Obviously, the parasitic capacitance can be reduced by decreasing the ε_r of the used isolator material. Common CMOS technologies feature fluorosilicate glass (SiO_2) with an ε_r around 4 to isolate the metal layers in the stack. The dielectric constant can be manipulated by plasma enhanced chemical vapour deposition of materials such as hydrogen (H) and carbon (C) into SiO_2. With this SiCOH material, the dielectric constant can be reduced to values ranging from 3.3 to 2.7 thereby lowering the parasitic capacitance in the order of 30% [Gri04]. The reduced coupling within the isolation layer is also advantageous for the interconnections realised by one of the stacked metal layers. Corresponding coupling between signals can have a severe impact on a circuit. VLSI technologies provide 4–12 metal layers. A cross-section through a common CMOS metal stack is shown in Fig. 6.21. Usually, the thickness of the metals increase towards the top of the wafer, e.g. by a factor of 2–4 per layer. The top metal is used for the realisation of the contact pads.

Please do not mix the ε_r requirements for low leakage transistor oxides and interconnect isolation layers. While the first one demands for high-k (ε_r), the latter requires low-k. Fortunately, these constraints are not conflicting, since they appear in different layers.

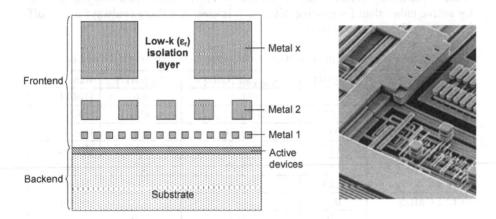

Fig. 6.21. Illustration of cross-section through metal stack of CMOS VLSI technologies

Fig. 6.22. Photograph of cross-section through metal stack of typical VLSI CMOS technology using copper metal [Ibm44]

6.2.5.6 Copper Metal

For over 40 years, the semiconductor industry has applied aluminium for the wiring in CMOS since it is relatively easy to process together with silicon. Due to the trend of CMOS scaling and the associated decrease of the width and thickness of the metal lines, the series resistance has become more and more problematic as mentioned in Sect. 5.4.9.5 and shown in Eq. (5.49).

Typically, copper allows a sheet resistance reduction of approximately 40 % [Ibm41], which helps to mitigate the impact of the scaled dimensions. A cross-section of a metal stack fabricated with copper metal is depicted in Fig. 6.22. The oxides are etched for visibility. The realisation of copper in integrated circuits sounds easy. However, copper as well as silver or gold interact with silicon. The most significant effect is the inter-diffusion and the generation of traps. Even a few stray atoms are enough to generate short-circuits in the thin isolation layers. One method to overcome this problem is to implement diffusion barriers at the copper surfaces. There is evidence that such technologies will be available soon.

6.2.5.7 Comparison of Inductor Technologies

Results of different inductors with state-of-the-art performances are listed in Table 6.4. As expected, the performance of MEMS inductors is superior because of the good substrate isolation. III/V technologies such as GaAs provide also good properties due their high substrate resistivity. The performance of RF CMOS technology can be improved by increasing of the distance between the conductors and the lossy substrate and by employing thick conductor lines. SOI technology benefits from the increased substrate resistivity compared to conventional substrates. The lowest performance is yielded by VLSI CMOS technology optimised for active rather than for passive devices. It is clear that there is always a tradeoff between performance and costs.

Table 6.4. Comparison of state-of-the art inductors using different IC technologies

Technology	L_s [nH]	Q_{max}@f [GHz]	f_{SR} [GHz]	Reference
MEMS	2.7	36@5.2	6.6	[Jia00]
GaAs	0.53	31.6@32.5	82.6	[Ell2]
RF CMOS	0.6	19@10	n.a.	[Ibm45]
	0.75	22@30	64	[Dub02]
SOI CMOS	0.86	27@2	27	[Zam02]
	0.25	21@41	>100	[Ell1]
VLSI CMOS	0.9	10@5	29	[Ell40]
	0.31	12.7@9.5	>40	

6.3 Transformers

Transformers are successfully employed for various applications including impedance transformation, power combiners/splitters, power amplifiers, balanced mixers, single ended to differential converters and phase shifters. Since transformers are composed of coupled inductor coils, lots of the insights gained in the inductor section can be mapped to transformers.

A signal fed into the primary conductor coil L_p induces current into the secondary coil L_s. This is caused by magnetic coupling via the mutual inductance M. Referring to the ideal transformer schematic sketched in Fig. 6.23a, the voltage and currents are related as follows:

$$\underline{V}_p = \underline{Z}_{11} \cdot \underline{I}_p + \underline{Z}_{21} \cdot \underline{I}_s = j\omega L_p \cdot \underline{I}_p + j\omega M \cdot \underline{I}_s \,, \qquad (6.32)$$

and

$$\underline{V}_s = \underline{Z}_{12} \cdot \underline{I}_p + \underline{Z}_{22} \cdot \underline{I}_s = j\omega M \cdot \underline{I}_p + j\omega L_s \cdot \underline{I}_s \,. \qquad (6.33)$$

The two latter equations can be modelled by a T-equivalent circuit with $\underline{Z}_1 = \underline{Z}_{11} - \underline{Z}_{21}$, $\underline{Z}_2 = \underline{Z}_{22} - \underline{Z}_{21}$ and $\underline{Z}_3 = \underline{Z}_{21}$. According to Fig. 6.23b this leads to the network depicted in Fig. 6.23c. Both n:1 and 1:n transformers can be realised, which step the voltage down and up, respectively. Energy conservation demands for inverse current relations. Consequently, impedances are transformed with a ratio of $\frac{1}{n^2}$ and n^2, respectively, revealing that these transformers are well suited for impedance transformation. In the following, we focus on the n:1 transformer. With

$$\underline{V}_s = \frac{1}{n} \cdot \underline{V}_p, \tag{6.34}$$

$$\underline{I}_s = n \cdot \underline{I}_p, \tag{6.35}$$

and Eqs. (6.32)–(6.33), we get

$$n = \sqrt{\frac{L_p}{L_s}}. \tag{6.36}$$

The input coupling coefficient k_p is determined by the voltage relation $\frac{\underline{V}_p}{\underline{V}_s}$ given that $\underline{I}_p = 0$. According to Eqs. (6.32)–(6.33), this yields $k_p = \frac{M}{L_s}$. Similarly, we get $k_s = \frac{M}{L_p}$. The overall coupling is defined by the mean value of the individual coupling factors

$$k = \sqrt{k_p k_s} = \frac{M}{\sqrt{L_p L_s}}. \tag{6.37}$$

For an ideal transformer with k=1 we obtain $M = \sqrt{L_p L_s}$. Typical values of k achieved in IC technologies are in the range 0.55–0.9.

The derived T-circuit does not yet feature two important characteristics of a transformer, which are the DC isolation and the transformation ratio. An ideal transformer with $\underline{V}_s = \frac{1}{n} \cdot \underline{V}_{se}$ and $\underline{I}_s = n \cdot \underline{I}_{se}$ can be incorporated. Referring to Eqs. (6.32)–(6.33) we get

$$\underline{V}_p = j\omega L_p \cdot \underline{I}_p + j\omega M \cdot n \cdot \underline{I}_{se} \tag{6.38}$$

and

$$\underline{V}_{se} = j\omega M \cdot n \cdot \underline{I}_p + j\omega L_s \cdot n^2 \cdot \underline{I}_{se} \tag{6.39}$$

leading to Fig. 6.23d. With Eqs. (6.36) and (6.37) we can modify the network into Fig. 6.23e. High coupling factors are required to obtained minimum losses and maximum bandwidth. Assuming ideal coupling with k=1 leads to the network of Fig. 6.23f.

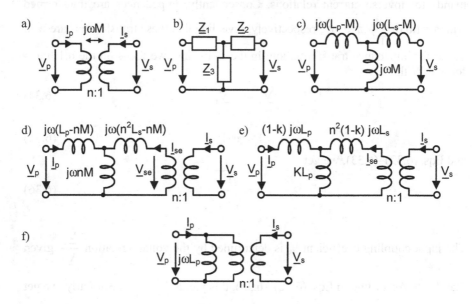

Fig. 6.23a–f. Ideal transformer: **a** circuit schematics; **b** T-type equivalent representation; **c** equivalent inductances; **d** incorporation of transformer core; **e** modification including the coupling coefficient; **f** k=1

Until now we have neglected the resistive and capacitive parasitics, which are important for the prediction of the transfer functions up to high frequencies. They are included in Fig. 6.24a. The impact of the resistances, and the ground and inter-winding capacitances with values of $C_{pe} = C_p + C_{ps}(1 - \frac{1}{n})$, $C_{pse} = \frac{C_{ps}}{n}$ and

$C_{se} = \frac{C_s}{n^2} + C_{ps}(\frac{1}{n} - 1)$ can be derived by extension of the previous equations [Tra01]. Note that for simplicity, the substrate resistance has been assumed to be infinity. The network may be further expanded by C_{ox}, C_{si} and R_{si}. In this context, we can compare the equivalent circuits of the inductors discussed in Sect. 6.2.

We are now able to derive some design strategies. For optimum low frequency response, the inductor value L_p has to be as high as possible to prevent short-circuiting of the input. On the other hand, increasing of L_p raises the parasitic resistances and thus the losses. The upper frequency limit is given by the self-resonance frequency of the coils, which may be approximated by

$\omega_{SRF} \approx \dfrac{1}{\sqrt{L_p C_{eff}}}$ with C_{eff} as the effective total parasitic capacitance. Evidently, ω_{SRF} decreases if L_p becomes larger corresponding to raised C_{eff} as well.

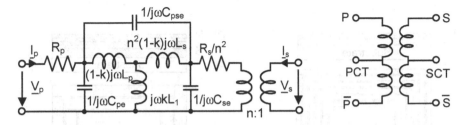

Fig. 6.24. Equivalent circuit of transformer including the parasitic series resistances and the parasitic ground and inter-winding capacitances

Fig. 6.25. Ideal transformer featuring centre taps for differential or balun operation

Due to the complexity of the equivalent transformer circuits, precise analytical calculations can be tedious. Convenient modelling with high accuracy is possible with CAD field simulator tools such as HFSS, FastHenry, etc.

Possible realisations of lumped transformers are treated now. Parallel inductors located in the same metal plane can be applied (as depicted in Fig. 6.26a) [Shi81, Frl89]. Optionally, two vertically stacked metal layers may be used (as shown in Fig. 6.26b). Similar techniques can also be applied for inductors to decrease the chip size [Fin85]. The parasitic capacitance of one coil is shielded by the other coil thereby reducing substrate losses as long as the lower winding does not mitigate this benefit due to the smaller distance to the substrate. Wide metal widths may be chosen to enhance the low frequency properties due to the reduced parasitic series resistances.

A further approach employs concentric coils, where the inner coil is embedded into the hollow area of the outer coil [Moh98]. Drawback is the reduced Q of the inner coil. According to Fig. 6.26c, the ratio between L_P and L_S and consequently the transformation ratio can be defined by sectioning of one of the windings into several individual turns, which are connected in parallel.

Differential circuits are favourable for many applications. Centre taps connected to real or virtual ground are required for this tasks as sketched in Fig. 6.25b. Such taps can be accomplished by corresponding connection of a point in the coil splitting the inductances into two parts. However, with common inductors it is challenging to achieve full symmetry required for differential operation. Referring to Fig. 6.26d, full symmetry can be obtained by the Rabjohn transformer [Rab91]. High quality transformers operating up to 30 GHz have been reported using this approach in standard silicon technology [Keh01, Keh03]. Typical losses of integrated transformers are around 3 dB. Compared to the hybrid-based couplers presented in the preceding section, transformer coupled divider have more bandwidth and smaller dimensions at the expense of higher losses due to the magnetic coupling.

For detailed studies and further information about transformers the reader is referred to the specific literature [Lon00, Bah05].

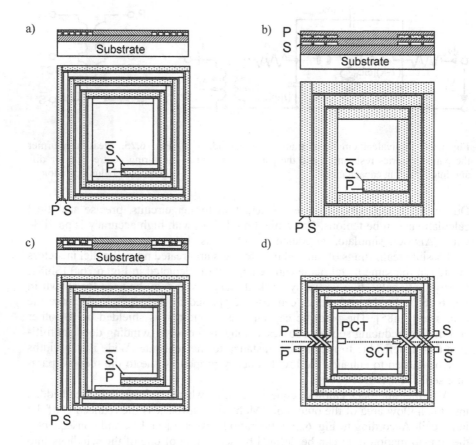

Fig. 6.26a–d. Layout of lumped transformers: **a** parallel conductors in same plane; **b** stacked; **c** 1:3 transformer with parallel secondary turns; **d** square symmetric with centre taps allowing for differential or balun operation according to Fig. 6.25b

6.4 Capacitors

Integrated circuits can't be imagined without capacitors employed for impedance matching, DC blocking, and RF shunting. Generally, the Q-factor of capacitors can be computed by

$$Q(\omega) = \frac{1}{2\pi} \cdot \frac{E_{electric} - E_{magnetic}}{E_{loss-per-cycle}} , \qquad (6.40)$$

where $E_{electric}$, $E_{magnetic}$ and $E_{loss-per-cycle}$ describe the desired electric energy, the parasitic magnetic energy, and the parasitic resistive energy appearing in the capacitor, respectively.

For the sake of simplicity, we neglect any parasitic inductances, put one port to ground and assume that the resistive losses are incorporated into the series resistor R_s. In this case, the Q-factor can be estimated with reasonable accuracy by

$$Q(\omega) = \frac{1}{\omega \cdot R_s \cdot C_s} , \qquad (6.41)$$

where C_s denotes the desired capacitance. With typical Q-factors of beyond 20 at 5 GHz, the Q of capacitors is typically higher than for inductors. That is the reason why usually the Q-factors in resonators and matching networks are determined by the inductor rather than the capacitor performance. However, Eq. (6.41) reveals that the Q of capacitances degrades with frequencies, which is not necessarily the case for inductors. Thus, at very high frequencies, the capacitor may be the limiting element in a reactive LC network.

6.4.1 MOS Capacitors

Every CMOS technology features MOS capacitors since they are basically equal to a MOSFET where the drain and the source are connected together. Since one of the capacitor terminals is located in the lossy substrate, MOS capacitors exhibit a high parasitic substrate capacitance resulting in a high insertion loss when used for series connections yielding Q-factors below 20. Thus, they are not well suited for series DC blocking or impedance matching capacitors. However, they provide a very high capacitance per area of around 5–320 fF/μm^2 making them well suited for compact RF shunts. In Fig. 6.27a,b, a cross-view and an equivalent circuit are illustrated. For a typical capacitor with C_s=3.3 pF, the values for the parasitic elements are as follows: L_s=15 pH, R_s=1 Ω, C_{sub}=50 fF, and R_{sub}=1000 Ω. In an analogous manner, MESFET or HEMT based capacitances can be realised providing similar capacitances per area. However, due to their higher substrate resistivity, the substrate parasitics in IIII/V technologies are much smaller.

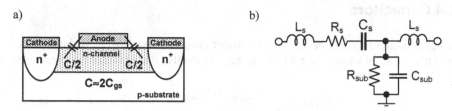

Fig. 6.27a,b. MOS capacitor: a cross-view; b RF equivalent circuit

6.4.2 MIM Capacitors

Metal insulator metal (MIM) type capacitors according to Fig. 6.28a have lower parasitics than MOS capacitors. High Q-factors of well above 20 have been demonstrated. Since both capacitor plates are implemented in the metal front-end better isolation to the lossy substrate is achieved. The capacitance is given by

$$C_s = \frac{\varepsilon_0 \cdot A}{d} \cdot \varepsilon_r,$$ (6.42)

where A is the plate area, ε_0 represents the dielectric constant in the vacuum, ε_r denotes the relative dielectric material constant and d is the distance between the metal plates. Generally, compared to MOS devices, MIM capacitors have a lower capacitance per area. The capacitance per area can be improved by increasing of ε_r. However, this option is not featured in low-cost technologies. A fruitful technique to increase the capacitance per area is the stacking of the metal layers as illustrated in Fig. 6.28b [Apa02]. Capacitances per area of 0.1–3 fF/μm^2 can be achieved for MIM caps. Typical equivalent element values of a stacked device with C_s=1.5 pF are L_s=0.03 nH, R_s=1.5 Ω, C_{ox}=30 fF, R_{sub}=500 Ω, and C_{sub}=20 fF. The equivalent circuit of the MIM capacitor is illustrated in Fig. 6.28c.

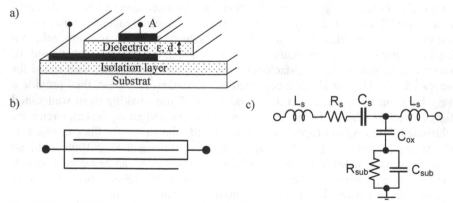

Fig. 6.28a–c. MIM capacitor: a cross-view; b illustration of capacitor structure employing five top metals; c equivalent circuit

6.5 Varactors

Varactors are tuneable capacitors. They are required for key circuits such as volt-age-controlled oscillators, phase shifters and adjustable filters. With the maximum and minimum values of C_v denoted by C_{vmax} and C_{vmin}, respectively, the tuning ratio can be defined by

$$t_v = \frac{C_{v\,max}}{C_{v\,min}}.$$ (6.43)

Varactors can, e.g. be realised by MOS and MES structures. In integrated form, these diodes offer a capacitive tuning range around 1.5–3. In VLSI technology, we frequently do not have special varactor diodes. Referring back to Sect. 6.4.1, where we have discussed MOS capacitors, common FETs with drain and source connected together forming one port and the gate representing the second port, can be employed. To minimise the DC current the device is reverse biased. Hence, usually, negative control voltages are applied. The maximum negative amplitude of the bias voltage is limited by the breakdown voltage of the device. The capacitance of a Schottky (MES) diode may be approximated by

$$C_v - \frac{1}{\sqrt{1 - \dfrac{V_c}{V_p}}} \cdot C_0,$$ (6.44)

where C_0 denotes the capacitance at $V_c=0$ V, $V_c<V_p$ is the control voltage, and V_p is the junction potential.

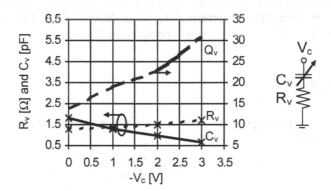

Fig. 6.29. Capacitance, parasitic resistance and Q-factor of a typical FET varacter diode at 5 GHz, MESFET in 0.6-µm technology with w_g=1000 µm varactor, source and drain connected together

The capacitive characteristics of a typical MESFET Schottky varactor diode is illustrated in Fig. 6.29 including the resistive parasitics modelled by a series resistance R_v. Assuming short connections, the inductive parasitics can be neglected. The drain/source port is grounded and V_c is applied at the gate. Within a voltage varied from 0 V to −3 V, the capacitance exhibits $t_v = 3$. The capacitance is decreasing with lowered control voltage as typically observed for MES or MOS diodes. The latter one changes from accumulation to depletion mode. For further information concerning the realisation and optimisation of MOS based varactors, the reader is referred to literature [Fob02, Mag03, Zam04].

Slightly higher t_v of around 3–5 can be obtained using pn-diodes available in BJT or BiCMOS technologies. Hyper-abrupt doping profiles yield further improvement of the t_v up to 10 at the expense of a larger control voltage range of around 10 V. For cost and fabrication reasons, such varactors are not available in common IC technologies. High performance off-chip varactors are available. However, fully integrated solutions are usually preferred. The typical performances of varactor diodes are listed in Table 6.5.

Table 6.5. Typical properties of varactor diodes

Technology	90-nm CMOS	90-nm SOI	0.6-μm MESFET	Si pn-diode	Off-chip pn-diode
t_v	1.5–2	2–2.5	2.5–3	3–5	5–10
ΔV	~1 V	~1 V	~3 V	~6 V	~10 V
Q@5GHz	10–16	12–20	14–30	>30	>30

As for capacitors, the Q-factor of varactors can be defined by Eq. (6.41). The corresponding results of the MESFET varactor have been included in Fig. 6.29. Up to moderate frequencies, FET diodes provide Q-factors of around 20. However, since the Q falls with increased frequency, the Q can be considerably lower at high frequencies. In Sect. 11.6.1, corresponding investigation for a LC resonator employed in a 60-GHz CMOS VCO will be performed.

6.6 Resistors

With resistivity ρ, length l, thickness d and width w, the resistance of a device can be determined by

$$R_s = \frac{\rho}{d} \cdot \frac{l}{w}. \tag{6.45}$$

Frequently represented by the unit of Ω/square, the first factor is defined by the technology. The second factor can be set by the designer. Example: if l=5w, the resistance amounts five times the Ω/square value. To minimise the capacitive parasitics, w is chosen as small as possible with respect to the current to be handled.

Finally, the free design parameter is l. Two different resistor implementations are discussed now.

6.6.1 Thin Film Resistors

A thin metal layer with high resistance can be employed to realise resistors with high precision and low sensitivity on voltage variations. Thin film resistors can handle high currents in the range of 0.5–5 mA/µm with respect to w. However, attributed to the high conductivity of the metals, the resistances are relatively low. Due to the low layer thickness, gate metals (poly-material) are frequently used providing a relatively high resistance of around 150–500 Ω/square. Resistors employing NiCr (nickel chromium) metals exhibit a lower resistance of around 20–100 Ω/square. A cross-view and an equivalent circuit are depicted in Fig. 6.30a,b.

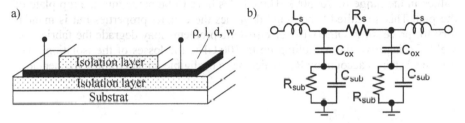

Fig. 6.30a,b. Thin film resistor: **a** cross-view; **b** equivalent circuit

6.6.2 Implanted Resistors

As illustrated in Fig. 6.31a, a semiconductor layer can be doped to define the resistance. These resistors have large geometric resistances of around 100–2000 Ω/square. Hence very compact devices can be realised. Unfortunately, only poor precision is achieved due to the high sensitivity with respect to process tolerances and bias conditions. Similar to the MOS capacitor, the implanted resistor exhibits strong capacitive parasitics since the resistive channel is located in the lossy substrate. Moreover, implanted resistors can only handle a limited current. They are well suited for biasing of FET gates, since the gate current is very low. In Fig. 6.31b, an appropriate equivalent circuit is shown.

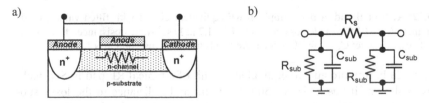

Fig. 6.31a,b. Implanted resistor: **a** cross-view; **b** equivalent circuit

6.7 Signal Pads

Signal pads are mandated to connect the chip inputs and outputs with components located outside the chip such as antennas, etc. Generally, RF pads have to be made as small as possible to reduce the parasitics. On the other hand, for reliable connections, the pad has to be large enough to allow proper contacts. These connections can be performed by wire or ball grid bonding demanding for minimum pad sizes of around 50×50 μm^2 and 150×150 μm^2, respectively. Furthermore, pads are required for on-wafer measurements. Due to the large pad size, a strong capacitive coupling to the lossy substrate is observed impacting the performance and the design. The top metal is used for the realisation of the pad since it has the maximum distance to the substrate. Corresponding parasitics have to be taken into account for the RF impedance matching and optimisation process. Because of design rule and density constraints, the metal density of large metal plates is limited to certain values in the range of 70–50%. Thus, holes have to be made into the top plate of the pad. This so called "cheesing" degrades the contact properties but is mandatory for processing. Otherwise, electrostatic problems may degrade the fabrication yield. To improve the modelling up to 100 GHz, the losses of the isolation layer can be taken into account by R_{ox}, which has a negligible impact for low frequencies.

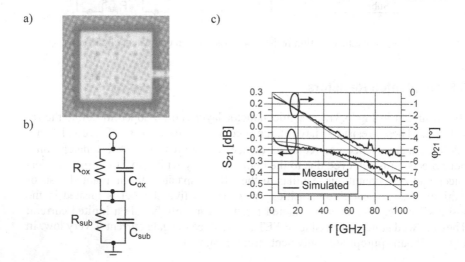

Fig. 6.32a–c. Signal pad: **a** photograph; **b** RF equivalent circuit with fitted values, oxide capacitance: $C_{ox}=10$ fF, oxide loss resistor: $R_{ox}=1.7$ kΩ, substrate resistance: $R_{sub}=60$ Ω, substrate capacitance: $C_{sub}=500$ fF; **c** measured and simulated insertion loss

In Fig. 6.32a,b, a photo and an equivalent circuit of a typical RF pad in SOI technology is shown. It has a size of 50 $\mu m \times 50$ μm and a distance to the lossy substrate of approximately 5 μm. The values of the equivalent circuit are $C_{ox}=10$ fF, $R_{ox}=1.7$ kΩ, $C_{sub}=20$ fF and $R_{sub}=60$ Ω. Being in the sub-Ω region due to the large width of the plates, the series resistances may be neglected. In Fig. 6.32c, the measured and simulated insertion loss of two pads connected in series is plotted. Due to the low parasitics of the employed SOI technology, up to 60 GHz and at the 50 Ω reference impedance, a low insertion loss of less than 0.3 dB was measured. In low-cost bulk technologies with small distance between pad metal and substrate, we have to consider much larger C_{ox} of up to 100 fF, which can have a considerable impact on the RF performance.

Usually, IC devices such as transistors are measured on-wafer. In those measurements, the impact of the input and output pads are included. Since we want to know the properties of the naked transistors as used in an IC, the parasitics of the pads have to be deembedded. One simple method is to subtract the pad admittance from the FET admittance thereby getting rid of the major pad parasitic being the capacitance.

6.8 Wiring

At RF frequencies, it is mandatory to take into account the non-ideal signal and ground connections. They may have significant inductive and/or capacitive parasitics, which can significantly impact or degrade the performance of a circuit. We can distinguish between on-chip and off-chip connections. An Itanium processor has well above 1 billion on-chip connections. Last but not least, we have to connect the IC with the system. Usually, ICs have up to 100 off-chip connections. Most of them are ground connections.

For test proposes and low number of connections, on wafer-measurements can be performed with micro-tips. For measurements up to millimetre-wave frequencies, the maximum number of RF contacts to be measured with those tips is usually around six. Two of them are for the signals (one in, one out) and four of them are for the grounds.

6.8.1 On-Chip Interconnects

Figure 6.33a sketches the simplified cross-view of a common RFIC process featuring three metal layers for on-chip connections and inductor design. The drain of a FET is connected with a MIM capacitor and a NiCr resistor. In Fig. 6.33b, the corresponding circuit schematic is depicted. For long connections and small distances between the wires, RF coupling between the lines has to be considered. The lengths have to be minimised and the lines have to be placed as far away from each other as possible. The latter requirement is traded off with the aim for increased circuit density.

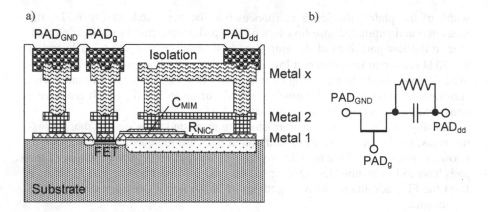

Fig. 6.33a,b. Typical RFIC process: **a** cross-view; **b** schematic of implemented circuit

6.8.2 Off-Chip Wiring

Off-chip connections are required to connect the IC with a system board. There are two main methods:

6.8.2.1 Wire Bonding

According to Fig. 6.34a, thin metal bond wires made of gold, copper or aluminium can be used to connect the IC chip with low-loss transmission lines printed on an assembly board. The required IC bond pads have edge lengths of 30–150 μm. To accomplish a stable connection between the bond wire and the bond pad, ultra-sonic waves, high temperature and pressure are employed in combination. Usually, those bond wires have lengths of 0.3–2 mm with an inductance of approximately 1 nH/mm. According to Eq. (6.7), the inductance increases with length. In many cases, this parasitic inductance is not desired and has to be kept as low as possible by minimising the length. One possibility is to mill a notch into the assembling board with thickness equal to that of the IC. Thus, the distance between the chip edge and the bond pad can be reduced. In specific cases, the inductive bond wire connections may be used as part of the impedance matching. Compared to integrated inductors, those bond wires provide a higher Q. The variation of the wire length during assembly must be considered to prevent from yield degradations.

6.8.2.2 Flip-Chip Bonding

The RFIC is turned and the pad grid of the IC is connected by solder balls on precisely located transmission lines as illustrated in Fig. 6.34b. The inductance is very low since the distance between IC pads and board contacts is very small. This method is used for automated mass-fabrication with high yield and for high frequency applications. A disadvantage of this method is that relative large IC pads with edge sizes of 100–350 μm are required increasing the capacitive parasitics and the chip size.

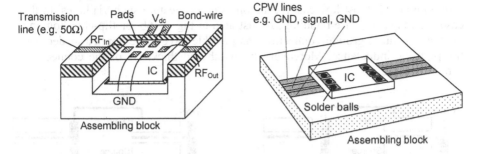

Fig. 6.34a,b. Off-chip connections: **a** wire bonding; **b** flip-chip bonding

We can conclude that wire bonding has high inductive and low capacitive parasitics, whereas ball grid bonding exhibits high capacitive but low inductive parasitics. The corresponding impact depends on the application, the operation frequency and the interface impedance. However, in mass fabrication, ball grid bonding is cheaper making it to the superior technique for most commercial applications.

6.8.3 Ground Connections

It is a well-known mistake of newcomers in the area to optimise the signal path rather than the ground path. However, most signals are referenced to ground. If the ground is not defined, neither are the signals. Hence, good RF grounding is an important issue for RFICs. The nodes to be connected to ground can be very low-ohmic. One example is the ground of a common source stage. These nodes are much more sensitive compared to signal connections having higher impedances. Reviewing the geometric dimensions of ground connections reveals that the impedance of these wires or transmission lines is mainly inductive. Since $Z_w = j\omega L$, the impact increases with frequency. We will discuss in Sect. 7.1.2 that series feedback has a significant effect on the input impedance and the effective g_m of transistors.

Severe problems may appear for cascaded stages if common ground connections are used as shown in Fig. 6.35a. A part of the amplified signal is fed back to the first stage, amplified, fed back, etc., which may lead to instability and oscillation as treated in Sect. 4.1. It is emphasised that such instability may not only degrade the performance. Being a killer criterion, instability can make a circuit totally useless since external control is impossible. An efficient way to decrease the probability for this kind of instability is to use two separate grounds as shown in Fig. 6.35b. In this case, there is no direct feedback path.

We can identify at least two types of ground connections. First, the one from the individual device to the pad ground ring of the IC. Second, the IC ground has to be connected with the overall ground of the system board. Referring to Fig. 6.36a, via holes filled with metal can be used to connect the upper and lower side of the chip. In turn, the IC can be directly soldered on the ground of the assembling board. Most low-cost technologies do not feature such via holes. In this case we can apply wire bonding as illustrated in Fig. 6.36b, or flip-chip bonding. As many parallel ground connections as possible have to be applied to decrease the parasitic inductance. Moreover, the width and thickness of the connections should be large.

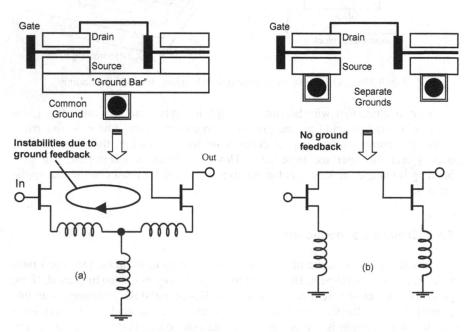

Fig. 6.35a,b. Example of grounding problem for cascaded transistor circuits: **a** common ground may lead to instabilities; **b** separate grounds help to circumvent problem

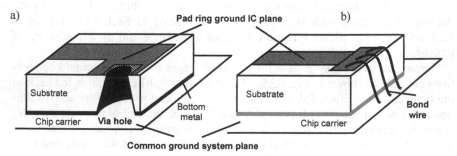

Fig. 6.36a,b. Ground connection between IC and system: **a** via hole filled with metal; **b** wire bonding

6.9 Simple Filters

Filters are required to extract a desired frequency spectrum from a broad range of signals. If a circuit passes all signals from DC to the filter cut-off frequency ω_c and rejects the rest of the spectrum, it is a low-pass filter. Conversely, a high-pass filter passes all signals above ω_c and suppresses those at lower frequencies. If a circuit passes only a finite frequency band that does not include DC or infinite frequency, it is a bandpass filter. A bandstop filter attenuates the signal at a defined frequency band. It is straightforward that bandpass and bandstop filters are specified by two cut-off frequencies. A notch filter blocks a single frequency. Further applications of filters involve impedance matching and phase shifting. Moreover, they can serve as sub-blocks in couplers.

Table 6.6. Simple low- and high-pass filters

Type	Circuit schematics	Calculation of elements	
Lowpass, φ negative			
Π-type		$L = -\dfrac{Z_0 \cdot \sin\varphi}{\omega}$	(6.46)
		$C = -\dfrac{1}{Z_0 \cdot \omega} \cdot \tan\left(\dfrac{\varphi}{2}\right)$	(6.47)
T-type		$L = -\dfrac{Z_0}{\omega} \cdot \tan\left(\dfrac{\varphi}{2}\right)$	(6.48)
		$C = -\dfrac{\sin\varphi}{Z_0 \cdot \omega}$	(6.49)
Highpass, φ positive			
Il-type		$L = \dfrac{Z_0}{\omega} \cdot \tan\left(\dfrac{\varphi}{2}\right)$	(6.50)
		$C = \dfrac{1}{Z_0 \cdot \omega \cdot \sin\varphi}$	(6.51)
T-type		$L = \dfrac{Z_0}{\omega \cdot \sin\varphi}$	(6.52)
		$C = \dfrac{1}{Z_0 \cdot \omega} \cdot \tan\left(\dfrac{\varphi}{2}\right)$	(6.53)

In the last decades, filter theory on the basis of distributed and lumped elements has been extensively treated in literature [Poz05, Mis04]. Major design parameters for distributed elements are the characteristic length φ and the characteristics line impedance Z_w. For circuits such as couplers and power combiners/splitters the functional principle of distributed circuits proves to be very descriptive. Conversion into lumped element equivalents and vice versa is possible. The equivalent element values of basic lumped element filters are summarised in Table 6.6. These simple filters often provide the basis for more complex circuits.

As an example, we study the characteristics of a Π-lowpass element illustrated at the top of Table 6.6. The ABCD parameter representation is well suited for the analysis since it allows convenient cascading of shunt and series element. The ABCD matrix is given by

$$
\begin{vmatrix} \underline{A} & \underline{B} \\ \underline{C} & \underline{D} \end{vmatrix} = \begin{vmatrix} 1 & 0 \\ jY_C & 1 \end{vmatrix} \begin{vmatrix} 1 & jX_L \\ 0 & 1 \end{vmatrix} \begin{vmatrix} 1 & 0 \\ jY_C & 1 \end{vmatrix} = \begin{vmatrix} 1 - Y_C X_L & jX_L \\ jY_C(2 - Y_C X_L) & 1 - Y_C X_L \end{vmatrix}
\tag{6.54}
$$

where $Y_C = \omega C$ and $X_L = \omega L$. The transmission term S_{21} of the scattering matrix yields

$$
\underline{S}_{21} = \frac{2}{\underline{A} + \dfrac{\underline{B}}{Z_0} + \underline{C} \cdot Z_0 + \underline{D}} = \frac{2}{2(1 - y_C x_L) + j(x_L + 2y_C - y_C^2 x_L)}.
\tag{6.55}
$$

where $y_C = Y_C Z_0$ and $x_L = \dfrac{X_L}{Z_0}$ are the normalised impedances of the capacitance and the inductance, respectively. The corresponding transmission phase, which is equal to the characteristic length of an equivalent transmission line can be computed by

$$
\varphi = \tan^{-1} \left[\frac{y_C^2 x_L - 2y_C - x_L}{2(1 - y_C x_L)} \right].
\tag{6.56}
$$

If we neglect resistive parasitics and assume lossless transmission with ideal impedance match, the following relations have to hold at the centre frequency:

$$
\left| \underline{S}_{11} \right| = \frac{\left| \underline{A} + \dfrac{\underline{B}}{Z_0} - \underline{C} \cdot Z_0 - \underline{D} \right|}{\left| \underline{A} + \dfrac{\underline{B}}{Z_0} + \underline{C} \cdot Z_0 + \underline{D} \right|} = 0
\tag{6.57}
$$

$$
\left| \underline{S}_{22} \right| = \frac{\left| -\underline{A} + \dfrac{\underline{B}}{Z_0} - \underline{C} \cdot Z_0 + \underline{D} \right|}{\left| \underline{A} + \dfrac{\underline{B}}{Z_0} + \underline{C} \cdot Z_0 + \underline{D} \right|} = 0.
\tag{6.58}
$$

The previous two identities can be found in S-parameter to ABCD parameter conversion tables. In this context, the reader is referred to Table 3.5. Solving for Eqs. (6.57) and (6.58) results in

$$\frac{B}{Z_0} = C \cdot Z_0 , \tag{6.59}$$

since $\underline{A} = \underline{D}$ as shown in Eq. (6.54). Hence, we get

$$x_L = \frac{2y_C}{1 + y_C^2} . \tag{6.60}$$

Using Eqs. (6.56) and (6.60) and rearranging the terms according to trigonometric identities results in

$$C = -\frac{\tan\left(\dfrac{\varphi}{2}\right)}{\omega \cdot Z_0} . \tag{6.61}$$

Referring to Eqs. (6.60) and (6.61), we obtain

$$L = -\frac{Z_0 \sin(\varphi)}{\omega} . \tag{6.62}$$

The periodic characteristics of the involved trigonometric functions have to be taken into account. Maximum filtering is achieved at $\varphi = -90°$ leading to $C = \dfrac{1}{\omega \cdot Z_0}$ and $L = \dfrac{Z_0}{\omega}$. In this case, $\omega_c = \dfrac{1}{\sqrt{LC}}$. The design equations for the other filter configurations are listed in Table 6.6.

6.10 Combiners and Dividers

> *There is only one thing which is more expensive than education*
> *and this is no education.*
> B. Franklin

Passive power combiners and couplers are required for various circuits including power amplifiers, phase shifters and frequency converters. They are partly based on the filters discussed in the preceding section.

6.10.1 N-Way Wilkinson Divider

In Fig. 6.37a, the circuit schematic of an N-way Wilkinson divider is sketched. All terminals have the same real reference impedance Z_0. By virtue of symmetry, the input signal is split into N equi-phase and equi-amplitude parts, where N can be even or odd. The output nodes are connected by means of the resistors R_x. Since all outputs have the same potential at node x, no voltage difference exists across the resistors R_x. Consequently, no power is dissipated in these resistors. Up to now we have assumed ideally matched terminations. Given non-ideal terminations, reflections can occur at one of the output terminals. The reflected signal is split into two parts. One part travels into the remaining outputs via the resistors. The second part is transferred to the input, reflected, divided again and fed into the output terminals. The reflected signals cover two-times the length of a path line. Given path lengths of 90°, the two parts of the reflected wave cancel out. It has been shown that complete cancellation can be obtained if the impedance of the line R_x and the line impedance Z_w are chosen as follows [Wil60]:

$$R_x = Z_0 , \tag{6.63}$$

$$Z_w = \sqrt{N} \cdot Z_0 . \tag{6.64}$$

The input impedance seen at each line is determined by the $\lambda/4$ length transformer yielding $Z_{ine} = \dfrac{Z_w^2}{Z_0}$. Since all lines are in parallel we get $Z_{in} = \dfrac{Z_{ine}}{N}$. Together with Eq. (6.64), we obtain $Z_{in} = Z_0$ verifying the input match.

a)   b)

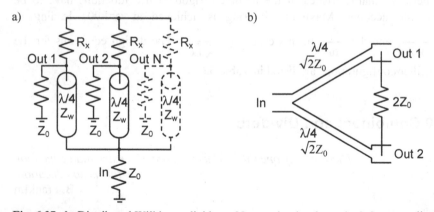

Fig. 6.37a,b. Distributed Wilkinson divider: **a** N-way circuit schematic; **b** 2-way realisation

Impedance matching and good isolation between the output terminals can be obtained within a bandwidth of around 20%. Higher bandwidth can be achieved by multistage designs. Unequal power divider ratios are possible [Mic05]. Variations of the topology have been reported trading off bandwidth, size, etc. [Sca02], [Gys75], [Sta89]. A possible realisation of a distributed 2-way Wilkinson splitter is shown in Fig. 6.37b.

Similar to other distributed circuits, the general drawback of the Wilkinson divider is the size of the transmission lines. At slightly lower bandwidth, the bulky

$\lambda/4$ lines can be substituted by lumped element equivalents based on the simple Π-type lowpass. The schematic of a lumped 2-way divider is plotted in Fig. 6.38a. Referring to Eqs. (6.46) and (6.47), setting $\varphi = -90°$ and $Z_w = Z_0 \cdot \sqrt{2}$ yields

$$L = \frac{Z_0 \cdot \sqrt{2}}{\omega} \quad \text{and} \quad C = \frac{1}{Z_0 \cdot \sqrt{2} \cdot \omega}. \tag{6.65}$$

Figure 6.38b shows the results of a device optimised for a centre frequency of 5.2 GHz. Dividers can also be used as combiners. However, there is one remarkable difference. Lossless combining is only possible if all N input signals have the same phase otherwise power losses occur across the resistances R_x.

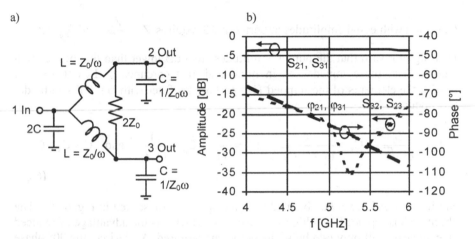

Fig. 6.38a,b. LC lumped equivalent Wilkinson divider: **a** circuit schematics; **b** transmission losses, phases and isolation using typical high-Q MMIC elements at centre frequency of 5.2 GHz, $Z_0 = 50\ \Omega$, $L = 2.16$ nH and $C = 0.43$ pF

6.10.2 90° Coupler

Various 90° couplers have been reported in literature. One of the simplest is the so-called branch-line coupler. In Fig. 6.39a,b, the distributed and equivalent lumped element types are sketched. The $\lambda/4$ lines divide the input signal at port 1 into two output signals available at port 2 and 3. Due to the 180° phase offset of the signal components adding at port 4, no signal appears at port 4 yielding $S_{41} = 0$. With respect to the input signal, the phase offsets at port 2 and 3 are -90 and $-180°$. Lossless operation demands

$$\left| \underline{S}_{21} \right|^2 + \left| \underline{S}_{31} \right|^2 = 1. \tag{6.66}$$

The latter two parameters can be evaluated on basis of the wave parameter analysis [Bäc94] yielding

$$\underline{S}_{21} = -j\frac{\underline{Z}_s}{Z_0} \quad \text{and} \quad \underline{S}_{31} = -\frac{\underline{Z}_s}{\underline{Z}_p}. \tag{6.67}$$

with shunt impedance $\underline{Z}_p$ and series impedance $\underline{Z}_s$. The relations in Eq. (6.67) verify the 90° phase offset between the outputs. Together with Eq. (6.66) we obtain

$$\left|\frac{\underline{Z}_p}{Z_0}\right|^2 + \left|\frac{\underline{Z}_s}{\underline{Z}_p}\right|^2 = 1. \tag{6.68}$$

Coupling with equal amplitudes at port 2 and 3 requires $\underline{Z}_s = \frac{Z_0}{\sqrt{2}}$ and $\underline{Z}_p = Z_0$. It is straightforward that the output amplitudes are 3 dB lower than that of the input signal. With these impedances in mind and in accordance to Eqs. (6.46) and (6.47), the elements of the lumped Π-type lowpass filter with φ=−90° can be determined:

$$L_s = \frac{Z_0}{\omega \cdot \sqrt{2}}, \quad C_s = \frac{1}{\omega \cdot Z_0 \cdot \sqrt{2}}, \quad L_p = \frac{Z_0}{\omega}, \quad C_p = \frac{1}{\omega \cdot Z_0} \quad \text{and} \quad C = C_s + C_p. \tag{6.69}$$

Similar properties can be obtained with the approach illustrated in Fig. 6.39c. The shunt path is replaced by a highpass structure, which has the advantage of reduced size since overall only two bulky inductors are required. As desired, the 90° phase difference between the outputs and the isolation at port 4 are preserved. Referring to Table 6.5, the elements can be calculated as follows:

$$L = \frac{Z_0}{\omega \cdot \sqrt{2}}, \qquad C_1 = \frac{1}{\omega \cdot Z_0} \quad \text{and} \quad C_2 = \frac{1}{\omega^2 \cdot L} - C_1. \tag{6.70}$$

In Fig. 6.38d, the characteristics of a 5.2-GHz circuit are shown. Due to the combined lowpass/highpass topology, the bandwidth is limited. Branch-line couplers can also be used as power combiner.

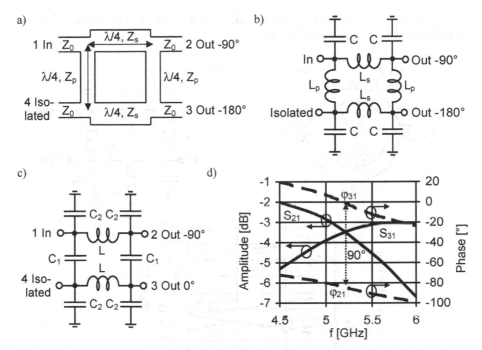

Fig. 6.39a–d. 90° hybrid: **a** schematics for distributed approach; **b** schematics based on lumped elements; **c** modified circuit requiring only two inductors; **d** simulations employing typical high-Q MMIC elements for centre frequency of 5.2 GHz and Z_0=50 Ω, L=1.31 nH, C_1=0.51 pF, C_2=0.2 pF

6.10.3 180° Rat-Race Coupler

For many applications such as single-ended to differential converters, power division with phase difference of 180° is needed. With respect to the 90° branch-line coupler, this phase shift can be obtained by increasing the phase offset between the output ports by means of an additional line with length of $\lambda/2$. One realisation is the so-called rat-race coupler as sketched in Fig. 6.40a, where the total circumference is 3/2 λ. Given that the impedances of all lines equal $Z_0 \cdot \sqrt{2}$, power fed at port 1 divides equally between ports 2 and 4 with phase differences of 180°. The signal components approaching port 3 are destructively combined. A lumped element network is shown in Fig. 6.40b comprising three −90° lowpass sections and one +90°=−270° highpass section. According to Eqs. (6.46), (6.47), 6.52) and (6.53), the elements for these two filter types can be determined by

$$L = \frac{Z_0 \cdot \sqrt{2}}{\omega} \qquad \text{and} \qquad C = \frac{1}{Z_0 \cdot \sqrt{2} \cdot \omega}. \qquad (6.71)$$

In Fig. 6.40c, the results of a 5.2-GHz circuit are depicted. The rat-race coupler can also be used as power combiner.

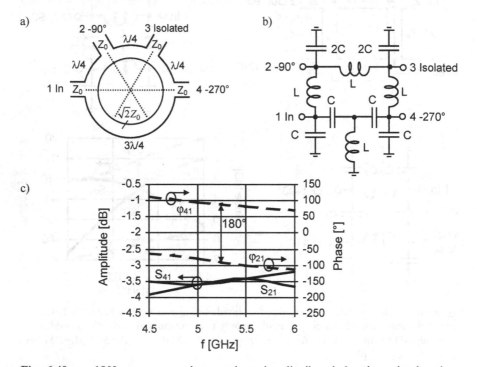

Fig. 6.40a–c. 180° rat-race coupler: **a** schematics distributed; **b** schematics based on lumped elements; **c** simulations of **b** employing typical high-Q MMIC elements for centre frequency of 5.2 GHz and Z_0=50 Ω, L=1.88 nH, C=0.5 pF

6.11 Tutorials

1. Discuss the typical frequency range of active-, lumped element- and transmission line-impedance matching approaches. What limits the operation of the elements?
2. Draw a cross-section of a microstrip line and include the E- and H-fields. Discuss the proportionalities of the parameters regarding the line impedance. With respect to geometry and material properties, how can we influence the inductive or capacitive characteristics?
3. Discuss the advantages and disadvantages of a coplanar line.
4. Why do inductance values required for circuits inversely scale with frequency?
5. Why does it not make sense to use spiral inductors for very high frequencies? Which element may be better suited?
6. Describe the equivalent circuit of an inductor in silicon.

7. Explain the self resonance. Why can inductors only be used below the self resonance?

8. Generally, how is the Q of an inductor and a capacitor defined?

9. Which reasonable simplification can we make for the calculation of the inductor Q? Which parasitics and effects decrease the Q?

10. Compare the substrate resistivities and the substrate loss factors in silicon and III-V technologies.

11. How can we improve silicon-based inductors? Explain at least three methods.

12. What is the impact of currents flowing in the same and into opposite direction with respect to the inductor Q?

13. You want to optimise one inductor for high frequencies and one for low frequencies. How do you choose the line width? Why?

14. Describe the impact of the distance between the substrate and the conductor of an inductor regarding the Q vs frequency.

15. For which applications do we employ transformers? Derive the impedance transformer ratio by means of the current/voltage relations.

16. Deduce an equivalent T-circuit for the transformer including the coupling factor. What happens if the coupling is ideal?

17. Which effects limit the bandwidth of transformers? Derive corresponding design strategies.

18. What kind of transformer realisations do you know? What are the pros and cons? How can we implement a differential transformer? What are the typical losses of integrated transformers?

19. Describe the two major capacitor types, illustrate the cross-views and the equivalent circuits. What are their advantages and disadvantages?

20. For which applications do we need varactors? How can we realise them? What are the typical capacitive tuning ranges? What are the main parasitics? What is the Q compared to inductors? Comment on the frequency dependence.

21. Describe the two major resistor types and illustrate the cross-views and equivalent circuits. What are the advantages and disadvantages?

22. How can we connect ICs with the system or measurement equipment? Do we have to consider the connections?

23. Has a bonding wire or a long and thin transmission line inductive or capacitive characteristics? How can we approximately model the characteristics of a bond wire with length of 2 mm?

24. What are the main differences between wire and flip-chip bonding?

25. Discuss the potential problems of non-ideal grounding. How can we mitigate the problem?

26. Derive the design equations for Π and T-type highpass-filters on the basis of ABCD parameters. The results are listed in Table 6.6. What is the major advantage of ABCD parameters?

27. How does the Wilkinson-divider work? How are the port reflections decreased? Are there any divider losses? Why is there no loss associated with the implemented resistors? Can we use the Wilkinson divider also as combiner? What about the corresponding losses? Consider the phases of the paths.

28. Explain the functionality of both distributed and lumped element 90° and 180° couplers. What limits the bandwidth and determines the losses?

References

[Apa02] R. Aparicio and A. Hajimiri, "Capacity limits and matching properties of integrated capacitors", IEEE Journal of Solid-State Circuits, Vol. 37, No. 3, pp. 384–393, March 2002.

[Bah05] Inder Bahl, Lumped Elements for RF and Microwave Circuits, Artech House, 2005.

[Bäc94] W. Bächtold, Lineare Elemente der Höchstfrequenztechnik, VDF Verlag, Zürich 1994.

[Dub02] Dubuc, E. Tournier T, Parra I, et al., "High quality factor and high self-resonant frequency monolithic inductor for millimeter-wave Si-based IC's", IEEE Microwave Theory and Techniques Symposium, pp. 193–196, 2002.

[Ell1]   F. Ellinger, RF Integrated Circuits in VLSI SOI CMOS Technology for Wireless Receivers at Millimetre Wave Frequencies, ISBN: 3-86628-007-6, Hartung Gorre Verlag, Konstanz, 2005.

[Ell2]   F. Ellinger, Monolithic Integrated Circuits for Smart Antenna Receivers at C-Band, ISBN 3-89649-663-8, Hartung-Gorre Verlag, Konstanz, 2001.

[Ell40] F. Ellinger, M. Kossel, M. Huber, M. Schmatz, C. Kromer, G. Sialm, D. Barras, L. Rodoni, G. von Büren, and H. Jäckel, "High-Q inductors on digital VLSI CMOS substrate for analog RF applications", IEEE/SBMO International Microwave and Optoelectronics Conference, pp. 869–872, Sept. 2003.

[Fin85] H. J. Finlay, UK patent application, 8800115.

[Fob02] N. Fong, G. Tarr, N. Zamdmer, J.-O. Plouchard, C. Plett, "Accumulation MOS varactors for 4 to 40 GHz VCOs in SOI CMOS", IEEE International SOI Conference, pp. 158-160, Oct. 2002.

[Frl89] E. Frlan, S. Meszaros, M. Cuhaci, J. Wight, "Computer-aided design of square spiral transformers and inductors", IEEE Microwave Theory and Techniques Symposium, pp. 661–664, June 1989.

[Gri04] A. Grill, "From DLC to DLC-SIO$_2$ hybrid low-K dielectrics for ULSI interconnects", www.eng.auburn.edu/department/ee/ADC-FCT2001/ADCFCTabstract/067.htm

[Gys75] U. H. Gysel, "A new N-way power divider/combiner suitable for high-power applications", IEEE International Microwave Symposium Digest, Vol. 75, No. 1, pp. 116–118, May 1975.

[Ibm41]   www-306.ibm.com/chips/services/foundry/technologies/cmos. html
[Ibm44]   www.a1-electronics.net/General_Interest/2003/IBM_Low-
          k_Copper_Cu-11.shtml
[Ibm45]   IBM 6HP technology, IBM web page
[Jia00]   H. Jiang, Y. Wang, J-L. A. Yeh and N. C. Tien, "On-Chip spiral
          inductors suspended over deep copper-lined cavities", IEEE
          Transactions on Microwave Theory and Techniques, Vol. 48, No. 12,
          pp. 2415–2423, Dec. 2000.
[Jin05]   Y. Jin and C. Nguyen, "A 0.25 μm CMOS T/R switch for UWB
          wireless communication," IEEE Microwave and Wireless Component
          Letters, 2005.
[Keh01]   D. Kehrer, W. Simburger, H.-D. Wohlmuth, A. L. Scholtz, "Modeling
          of monolithic lumped planar transformers", IEEE Custom Integrated
          Circuits Conference, pp. 401–404, May 2001.
[Keh03]   D. Kehrer, H.-D. Wohlmuth, C. Kienmayer, A. L. Scholtz, "A 1V
          monolithic transformer-coupled 30 Gb/s 2:1 multiplexer in 120nm
          CMOS" IEEE Microwave Symposium Digest, Vol. 3, pp. 2261–2264
          June 2003.
[Lee04]   T. H. Lee, The design of CMOS radio-frequency integrated circuits,
          Cambridge, 2004.
[Lon00]   J.R. Long, "Monolithic transformers for silicon RF IC design", IEEE
          Journal of Solid-State Circuits, Vol. 35, No. 9, pp. 1368–1382, Sept.
          2000.
[Mag03]   J. Maget, M. Tiebout, R. Kraus, "MOS varactors with n- and p-type
          gates and their influence on an LC-VCO in digital CMOS", IEEE
          Journal of Solid-State Circuits, Vol. 38, No. 7, pp. 1139-1147, July
          2003
[Mei86]   Meinke, Gundlach, "Taschenbuch der Hochfrequenztechnik", Springer,
          1986.
[Mic05]   www.microwaves101.com/encyclopedia/Wilkinson_splitters.cfm
[Mis04]   D. K. Misra, Radio-frequency and Microwave Communication Circuits,
          Online, Wiley, 2004.
[Moh98]   S. S. Mohan, C. P Yue, M. del Mar Hershenson, S. S. Wong, T. H. lee,
          "Modeling and characterisation of on-chip transformers", International
          Electronic Device Meeting, pp. 531–534, Dec. 1998.
[Poz05]   D. M. Pozar, Microwave Engineering, Wiley, 2005.
[Rab91]   G. G. Rabjohn, "Monolithic microwave transformers", Master thesis,
          Carleton University, Ottawa, Canada, Apr. 1991.
[Reb03]   Gabriel M. Rebeiz, "RF MEMS: theory, design, and technology",
          Wiley-Interscience ; 2003.
[Sca02]   M. C. Scardelletti, G. E. Ponchak, T. M. Weller, "Miniaturized
          wilkinson power dividers utilizing capacitive loading", IEEE
          Microwave and Wireless Components Letters, Vol. 12, No 1, pp. 6–8,
          Jan. 2002.
[Sch68]   V.M. Schneider, "Microstrip lines for integrated circuits", The Bell
          System Technical Journal, May–June 1968.

[Shi81] K. Shibata, K. Hatori, Y. Tokumitsu, H. Komizo, "Microstrip spiral directional couplers", IEEE Transactions on Microwave Theory and Techniques, Vol. 29, pp. 680–689, July 1981.

[Sta89] J. Staudinger, "An ultra wide bandwidth power divider on MMIC operating 4 to 10 GHz", Microwave and Millimeter-Wave Monlithic Circuits Symposium, 127–131, 1989.

[Tra01] C. Trask, "Wideband transformers, an intuitive approach to models, characterisation and design", Applied Microwave Wireless, pp. 30.41, Nov. 01.

[Wad91] B. C. Wadell, "Transmission line design handbook", Artech House, Norwood, 1991.

[Whe28] H. A. Wheeler, "Simple inductance formulas for radio coils", Proc. IRE, Vol. 16, No. 10, pp. 1398–1400, Oct. 1928.

[Wil60] E. J. Wilkinson, "An N-way hybrid power divider", IEEE Transaction Microwave Theory and Techniques, pp. 116–118, Jan. 1960.

[Won05] B. P. Wong, A. Mittal, Y. Cao, G. Starr, Nano-CMOS circuit and physical design, Wiley, 2005.

[Yua82] H.-T. Yuan, Y.-T. Lin and S.-Y. Chiang, "Properties of interconnection on silicon, sapphire, and semi-insulating gallium arsenide substrates", IEEE Transactions on Electron Devices, ED-29, No. 4, p. 639–644, 1982.

[Yue98] C. P. Yue, S. S. Wong, "On-chip spiral inductors with patterned ground shield for Si-based RF IC's", IEEE Journal of Solid-State Circuits, Vol. 33, No. 5, pp. 743–752, May 1998.

[Zam02] N. Zamdmer, J. O. Plouchart, et al., "Suitability of scaled SOI CMOS for high-frequency analog circuits", European Solid State Device Research Conference, September 2002.

[Zam04] N. Zamdmer, J. Kim, R. Trzcinski, J.-O. Plochart, S. Narasimha, M. Khare, L. Wagner, S. Chaloux, "A 243-GHz Ft and 208-GHz Fmax, 90-nm SOI CMOS SoC technology with low-power millimeter-wave digital and RF circuit capability", pp. 98-99, IEEE Symposium on VLSI Technology, 2004.

7 Basic Amplifier Circuits

To enhance the comprehension you have to change the perspectives.
Antoine de Saint-Exupéry

7.1 Topologies

The design of complicated circuits can be ascribed to simple basic topologies as treated in this section. We will derive key small signal properties such as the complex input and output impedance, and voltage and current gain in the frequency domain.

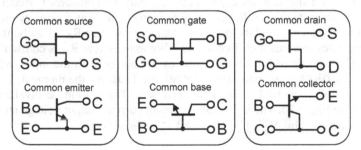

Fig. 7.1. Basic circuit topologies using FETs and BJTs

Reasonable simplifications are made to enable plain-vanilla calculations, meaningful circuit insights, feasibility checks and optimisations. More accurate circuit calculations can be performed using complex device models and CAD tools. The applied simplifications hold for typical transistor parameters, applications and frequency ranges. However, it is always fruitful to check the limits of the assumptions. One example: up to which frequency is the supposition correct that the input impedance of a common source circuit is mainly capacitive? As boundary we may assume that the reactive impedance should be at least three times larger than the resistive impedance. With typical values of C_{gs}=100 fF and R_{gs}=10 Ω, we can estimate a frequency of $1/(3 \cdot 2\pi \cdot R_{gs} \cdot C_{gs})$=53 GHz.

Usually, transistors are 3-port devices. One exception is the bulk MOSFET, which has four ports. Suppose that the bulk port is connected to ground. Consequently, this type of transistor can also be treated as 3-port device. Depending on the port serving as the common node for both the input and the output, we can distinguish between three configurations per transistor type as illustrated in Fig. 7.1. The common node is connected directly or indirectly (e.g. by means of a further device) to a reference potential, which frequently is the ground. Common-source, -gate and -drain circuits can be realised using FETs. A detailed comparison of these circuits has been presented in [Jia05]. Analogously, the BJT topologies are the common -emitter, -base and -collector circuit, respectively. The basic configurations of the FET and the BJT exhibit many similarities. To reduce redundancies, we will focus on the FET and limit the BJT discussions to specific issues.

7.1.1 Common Source Circuit

As per definition and illustrated in Fig. 7.2a, the source of the transistor is connected to ground and serves as reference for the gate and drain representing the input and output node, respectively. For the sake of simplicity, the gate and drain DC supply networks are not drawn in the circuit schematic and are assumed to be lossless. More information about practical implementations of DC bias networks will be given in Sect. 7.3 and Table 7.2. The transistor acts as a voltage controlled current source driving a specific output load R_L. According to Sect. 5.4, the g_m and the C_{gs} are the most relevant device model parameters. In a second order analysis, the device feedback capacitance C_{gd}, the output impedance r_o, the gate source resistance R_{gs}, and the channel capacitance C_{ds} may be considered. According to Sect. 5, r_o represents the parallel connection between the low frequency small signal output resistance r_{ds} and the parasitic channel resistance R_{ds}. Usually, the value of R_{ds} decreases towards high frequencies. We assume that R_{ds} is constant vs frequency. The gate drain capacitance couples the output with the input, which increases the difficulty for analytical calculations. Thus, for the moment, C_{gd} is neglected. We will come back to this parameter later. The resulting equivalent circuit is depicted in Fig. 7.2b. The input voltage, input current, output voltage and output current are denoted by $\underline{V}_{in}$, $\underline{I}_{in}$, $\underline{V}_{out}$ and $\underline{I}_{out}$, respectively.

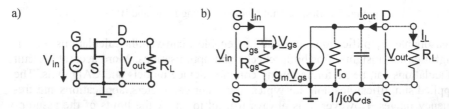

Fig. 7.2a,b. Common source circuit: **a** simplified circuit schematic, where the DC supply networks, which are assumed to be lossless, are not included; **b** simplified equivalent small signal circuit

7.1.1.1 Input Impedance

The input impedance of the device yields

$$\underline{Z}_{in} = \frac{\underline{V}_{in}}{\underline{I}_{in}} = \frac{1}{j\omega C_{gs}} + R_{gs}. \tag{7.1}$$

For typical FETs operating at frequencies well below f_t, the capacitive part of the input network dominates. The demand to reduce the current consumption of the devices requires small transistor widths resulting in high impedances, typically in the $k\Omega$ range at RF frequencies. In this case, the impedance matching to low and resistive impedances such as the 50 Ω reference impedance may be challenging.

7.1.1.2 Output Impedance

As per definition of an ideal input voltage source, in the small signal case, the input is short-circuited. Thus, the output impedance of the device can be calculated by

$$\underline{Z}_{out} = \frac{\underline{V}_{out}}{\underline{I}_{out}}\bigg|_{\underline{V}_{in}=0} = r_o \left\| \frac{1}{j\omega C_{ds}} \right. . \tag{7.2}$$

Since $C_{ds} < C_{gd} < C_{gs}$, the impact of the capacitive part of the channel impedance is usually weak up to frequencies below f_t. Hence, the output impedance is mainly resistive making the output impedance matching less difficult compared to the impedance matching at the input. Recall that transistor scaling leads to a decreased r_o and C_{ds}. Thus, C_{ds} must only be taken into account for high frequencies.

7.1.1.3 Voltage Gain

At frequencies well below f_t we find $\dfrac{1}{j\omega C_{gs}} \gg R_{gs}$. Consequently, we may assume that $\underline{V}_{gs} \approx \underline{V}_{in}$. With a voltage at the drain node of

$$\underline{V}_{out} = -g_m \cdot \underline{V}_{in} \cdot \underline{Z}_{out} \| R_L \tag{7.3}$$

we obtain a voltage gain of

$$\underline{a}_v = \frac{\underline{V}_{out}}{\underline{V}_{in}} = -g_m \cdot \underline{Z}_{out} \| R_L . \tag{7.4}$$

In the event that $R_L < |\underline{Z}_{out}|$ and/or at frequencies well below f_t, the real part of $\underline{a}_v$ is dominant[1]. Hence, indicated by the negative sign in Eq. (7.4), the input and output voltages exhibit a phase offset of $180°$. Given that $R_L \| |\underline{Z}_{out}| > \dfrac{1}{g_m}$, voltage

---

[1] This assumption will be made throughout this chapter.

gain can be achieved. High voltage gain demands for a large R_L. On the other hand, R_L may be defined by the input impedance of the following stage, which for bandwidth constraints may require a low value. Furthermore, we have to keep in mind that an increase in R_L is only efficient as long as $R_L < |\underline{Z}_{out}|$. In any case, the maximum possible voltage gain of the circuit is bounded by

$$\underline{a}_v = \frac{\underline{V}_{out}}{\underline{V}_{in}} = -g_m \cdot \underline{Z}_{out} , \tag{7.5}$$

which is a technology specific constant and is not dependent on the FET width given that the current density through the device is kept constant.

7.1.1.4 Current Gain

We neglect R_{gs}. With an input current of

$$\underline{I}_{in} = j\omega C_{gs} \cdot \underline{V}_{in} \tag{7.6}$$

and an output current of

$$\underline{I}_{out} = g_m \cdot \underline{V}_{in} \cdot \frac{\underline{Z}_{out}}{\underline{Z}_{out} + R_L} \tag{7.7}$$

we obtain a current gain of

$$\underline{a}_i = \frac{\underline{I}_L}{\underline{I}_{in}} = -\frac{\underline{I}_{out}}{\underline{I}_{in}} = -\frac{g_m}{j\omega C_{gs}} \cdot \frac{\underline{Z}_{out}}{\underline{Z}_{out} + R_L} . \tag{7.8}$$

Maximum current gain of $\underline{a}_i = -\dfrac{g_m}{j\omega C_{gs}}$ is achieved at $R_L \to 0$. In this case, we refer to the short-circuit current gain. A $|\underline{a}_i|$ greater than unity can be achieved depending on the operation frequency and the f_t being relevant for the g_m/C_{gs} factor.

7.1.1.5 Power Gain

We have observed that the common source circuit is capable of yielding both current and voltage gain. Consequently, high power gain can be achieved making the common source circuit well suited for amplifiers.

7.1.1.6 Miller Effect

In the previous discussions we have neglected the effect of the device feedback capacitance C_{gd} to ease the calculations. We investigate now the impact of this capacitance on the input by deriving an equivalent input capacitance. To achieve equal input admittance conditions with respect to the two networks shown in Fig. 7.3a,b, the following relation has to hold:

$$j\omega C_{in} = \frac{\underline{I}_{gs} + \underline{I}_{gd}}{\underline{V}_{gs}} . \tag{7.9}$$

With

$$\underline{I}_{gs} = j\omega C_{gs} \cdot \underline{V}_{gs} \tag{7.10}$$

we get

$$\underline{I}_{gd} = j\omega C_{gd} \cdot \underline{V}_{gd} = j\omega C_{gd} \left(\underline{V}_{gs} - \underline{V}_{ds} \right)$$

$$= j\omega C_{gd} \cdot \left(1 - \frac{\underline{V}_{ds}}{\underline{V}_{gs}} \right) \cdot \underline{V}_{gs} = j\omega C_{gd} \cdot (1 - \underline{a}_v) \cdot \underline{V}_{gs} . \tag{7.11}$$

Suppose that the imaginary part of $\underline{a}_v$ can be neglected. This is a reasonable approximation for frequencies well below f_t. Hence, the equivalent capacitance seen at the input becomes

$$C_{in} = C_{gs} + C_{gd} \cdot (1 - \underline{a}_v) = C_{gs} + C_m , \tag{7.12}$$

where C_m represents the so-called Miller capacitance. We can draw the following conclusions:

- At large $\underline{a}_v$, C_m and consequently C_{in} may be much higher than C_{gs}. Usually, C_{gd} is around 2–10 times smaller than C_{gs}. However, the term $(1 - \underline{a}_v) \approx (1 + g_m \cdot R_L)$ can be well above this ratio.

- Since $f_t \sim \dfrac{1}{C_{in}}$ and as a first approximation $f_{max} \sim \sqrt{f_t}$, the speed decreases

  due to the Miller effect. A compromise between the gain and the bandwidth has to be made. The high power gain of the common source circuit cannot be exploited towards highest frequencies.
- The high input capacitance makes the input matching to resistive source impedances difficult.

The Miller capacitance can be reduced by the cascode topology or the transimpedance/transadmittance stage (TIS/TAS), which will be discussed in Sects. 7.1.6 and 7.4.4, respectively.

Is there a considerable Miller capacitance at the output of a common source amplifier? The answer is no, since the magnitude of the reverse voltage gain $\dfrac{\underline{V}_{gs}}{\underline{V}_{ds}}$ is below unity.

a)   b)

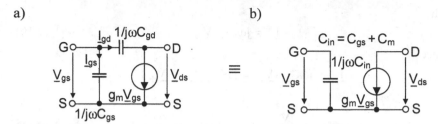

Fig. 7.3a,b. Equivalent small signal network of the common source circuit for investigation of the Miller effect and derivation of an equivalent input capacitance: **a** with feedback capacitance C_{gd}; **b** C_{gd} absorbed into C_{in}

7.1.2 Common Source Circuit with Source Feedback

Let's review the impact of an impedance $\underline{Z}_s$ located between the source and the ground according to Fig. 7.4a. This impedance may be implemented on purpose to impact the circuit properties by means of a series feedback, or may be the result of a parasitic ground connection. In the latter case, $\underline{Z}_s$ has to be minimised by proper layout. Recall that no ideal RF grounds exist in practice.

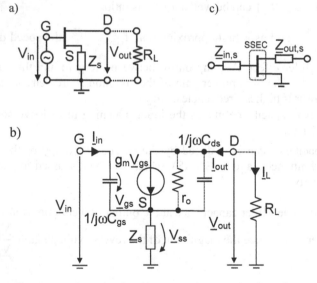

Fig. 7.4a–c. Common source circuit with source degeneration impedance: **a** simplified circuit schematic, where the DC supply networks, which are assumed to be lossless, are not included, Z_s represents a resistive or reactive network; **b** small signal equivalent circuit including the key device elements; **c** small signal equivalent circuit (SSEC) for impedance considerations only

7.1.2.1 Input Impedance

Again, for the calculation of the input impedance, we do not consider $\underline{Z}_{out}$. Referring to Fig. 7.4b, we obtain a voltage across $\underline{Z}_s$ of

$$\underline{V}_{ss} = \underline{Z}_s \cdot \underline{V}_{gs} \cdot \left(g_m + j\omega C_{gs} \right) \tag{7.13}$$

and

$$\underline{I}_{in} = \underline{V}_{gs} \cdot j\omega C_{gs}. \tag{7.14}$$

Now, we can write for the input impedance

$$\underline{Z}_{in} = \frac{\underline{V}_{in}}{\underline{I}_{in}} = \frac{\underline{V}_{gs} + \underline{V}_{ss}}{\underline{I}_{in}} = \frac{1 + \underline{Z}_s \left(g_m + j\omega C_{gs} \right)}{j\omega C_{gs}}$$

$$= \frac{1}{j\omega C_{gs}} \left(1 + \underline{Z}_s \cdot g_m \right) + \underline{Z}_s, \tag{7.15}$$

where an additional contribution due to $\underline{Z}_s$ is given by

$$\underline{Z}_{in,s} = \frac{\underline{Z}_s \cdot g_m}{j\omega C_{gs}} + \underline{Z}_s \tag{7.16}$$

as illustrated in Fig. 7.4c. Depending on the type of $\underline{Z}_s$ we can distinguish between the following cases and properties:

A. Resistive $\underline{Z}_s = R_s$
The capacitive part of $\underline{Z}_{in}$ is increased by the factor $R_s \cdot g_m$, making the matching to a resistive impedance more difficult.

B. Inductive $\underline{Z}_s = j\omega L_s$
A real part with a value of $\dfrac{L_s \cdot g_m}{C_{gs}}$ is generated simplifying broadband matching to low-ohmic and resistive impedances. Due to the increased resistive content of the network, the stability is improved. Consequently, an inductive $\underline{Z}_s$ may be used for circuit stabilisation. Moreover, it has been shown that inductive source degeneration can be applied to move the optimum impedances for minimum noise figure and maximum gain closer together, thereby improving the overall performance of the LNAs [Leh85]. However, as a drawback, L_s decreases the gain, which may not

be desired. In practice, such an inductive impedance is generated by the layout-specific connection between the device and the ground.

C. Capacitive $\underline{Z}_s = \dfrac{1}{j\omega C_s}$

In this case, a negative real part with value of $-\dfrac{g_m}{\omega^2 \cdot C_s \cdot C_{gs}}$ is generated, which can make the circuit unstable. As elaborated in Section 11.5.1, this effect is exploited for oscillators. Due to the proportionality associated with $\dfrac{1}{\omega^2}$, the impact increases towards low frequencies.

7.1.2.2 Output Impedance

To simplify the derivation of the output characteristics, we neglect the small current through the input. The following relations hold:

$$\underline{Z}_{out} = \frac{\underline{V}_{out}}{\underline{I}_{out}}, \tag{7.17}$$

$$\underline{V}_{ss} = \underline{Z}_s \cdot \underline{I}_{out}, \tag{7.18}$$

and

$$\underline{V}_{out} = \underline{I}_{out} \cdot \underline{Z}_s + \underline{Z}_{out,CS} \cdot \left(\underline{I}_{out} - g_m \cdot \underline{V}_{gs}\right), \tag{7.19}$$

where $\underline{Z}_{out,CS}$ represents the $\underline{Z}_{out}$ of the common source circuit without any feedback. As per definition $\underline{V}_{in} = 0$. Consequently $\underline{V}_{gs} = -\underline{V}_{ss}$. Now, we obtain a simple expression for the output impedance of

$$\underline{Z}_{out} = \underline{Z}_{out,CS} \cdot \left(1 + g_m \cdot \underline{Z}_s\right) + \underline{Z}_s \approx r_o \cdot \left(1 + g_m \cdot \underline{Z}_s\right), \tag{7.20}$$

where the additional contribution due to $\underline{Z}_s$ can be estimated by

$$\underline{Z}_{out,s} = r_o \cdot g_m \cdot \underline{Z}_s. \tag{7.21}$$

We can conclude that $\underline{Z}_s$ can significantly increase the $|\underline{Z}_{out}|$. Depending on the type of $\underline{Z}_s$, the contribution can be resistive, inductive or capacitive. A large $|\underline{Z}_{out}|$ lowers the losses associated with the current transfer to the load being in parallel with $\underline{Z}_{out}$. Moreover, the load current is less dependent on the device impedance. These properties are favourable for current sources.

7.1.2.3 Transconductance

Due to the high $|\underline{Z}_{out}|$, the circuit exhibits current source properties. Thus, it makes sense to specify the gain by means of the transconductance. The source impedance $\underline{Z}_s$ exhibits a voltage drop changing the effective $\underline{V}_{gs}$ at a given $\underline{V}_{in}$ thereby impacting the current source properties. By assuming an infinite output impedance, we can easily calculate the extrinsic transconductance $\underline{g}_{m,ex}$:

$$\underline{g}_{m,ex} = \frac{\underline{I}_{out}}{\underline{V}_{in}} = g_m \cdot \frac{\underline{V}_{gs}}{\underline{V}_{gs} + \underline{V}_{ss}} . \tag{7.22}$$

The relations

$$\underline{V}_{ss} = \underline{Z}_s \left(\underline{I}_{out} + j\omega C_{gs} \cdot \underline{V}_{gs} \right) \tag{7.23}$$

and

$$\underline{I}_{out} = g_m \cdot \underline{V}_{gs} \tag{7.24}$$

yield

$$\underline{g}_{m,ex} = g_m \cdot \frac{1}{1 + \underline{Z}_s \left(g_m + j\omega C_{gs} \right)} . \tag{7.25}$$

Now, we review the impact for different types of $\underline{Z}_s$.

A. Resistive $\underline{Z}_s = R_s$

$$\underline{g}_{m,ex} = g_m \cdot \frac{1}{1 + R_s \left(g_m + j\omega C_{gs} \right)} . \tag{7.26}$$

Obviously, R_s decreases the magnitude of $\underline{g}_{m,ex}$ resulting in a lower voltage and current gain but improved stability. Consider that for frequencies well below f_t the relation $\frac{g_m}{\omega C_{gs}} \gg 1$ holds.

B. Inductive $\underline{Z}_s = j\omega L_s$

$$\underline{g}_{m,ex} = g_m \cdot \frac{1}{1 + j\omega L_s \left(g_m + j\omega C_{gs} \right)} = g_m \cdot \frac{1}{1 + \omega L_s \left(j g_m - \omega C_{gs} \right)} \tag{7.27}$$

with similar consequences as for case A. in terms of gain and stability. However, the frequency dependence is more significant.

C. Capacitive $\underline{Z}_s = \dfrac{1}{j\omega C_s}$

$$\underline{g}_{m,ex} = g_m \cdot \frac{1}{1 + \dfrac{C_{gs}}{C_s} - j\,\dfrac{g_m}{\omega C_s}} . \tag{7.28}$$

Since a negative $\underline{Z}_{in}$ is generated, the circuit may become unstable.

7.1.3 Common Source Circuit with Drain Gate Feedback

The circuit schematics for the common source circuit with impedance $\underline{Z}_{gd}$ connected between the gate and drain are illustrated in Fig. 7.5a,b. In practice, a DC decoupling capacitor is usually implemented in series with $\underline{Z}_{gd}$ to separate the gate source and drain source bias voltages.

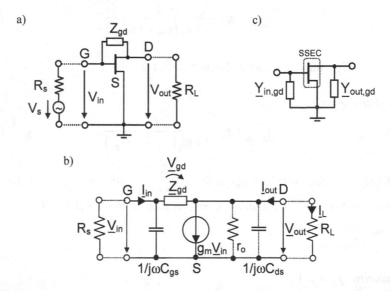

Fig. 7.5a–c. Common source circuit with drain gate feedback impedance: **a** simplified circuit schematic, where the DC supply networks, which are assumed to be lossless, are not included, Z_{gd} represents a resistive or reactive network; **b** small signal equivalent circuit including device elements; **c** small signal equivalent circuit (SSEC) for impedance considerations only

$\underline{Z}_{gd}$ generates a significant coupling between the input and the output. Thus, compared to previous circuit investigations, we consider also the source impedance R_s. The latter element has a significant impact on the output impedance.

7.1.3.1 Input Impedance

By substituting $j\omega C_m$ with $\underline{Y}_{gd}$ we can perform the calculations according to the Miller relation derived in Sect. 7.1.1.6 yielding

$$\underline{Y}_{in} = \frac{\underline{I}_{in}}{\underline{V}_{in}} = j\omega C_{gs} + \underline{Y}_{gd}(1-\underline{a}_v), \qquad (7.29)$$

where the additional $\underline{Y}_{in}$ due to the drain gate feedback according to Fig. 7.5c is given by

$$\underline{Y}_{in,gd} = \underline{Y}_{gd} \cdot (1-\underline{a}_v). \qquad (7.30)$$

A. Resistive $\underline{Z}_{gd} = R_{gd}$

$\underline{Y}_{in,gd} = \dfrac{1}{R_{gd}} \cdot (1-\underline{a}_v)$ reveals that lowering R_{gd} and increasing $|\underline{a}_v|$ decreases $|\underline{Z}_{in}|$ and makes $\underline{Z}_{in}$ more resistive.

B. Inductive $\underline{Z}_{gd} = j\omega L_{gd}$

$\underline{Y}_{in} = j\left[\omega C_{gs} - \dfrac{1}{\omega L_p} \cdot (1-\underline{a}_v)\right]$. Amplified by $\underline{a}_v$, L_{gd} resonates C_{gs} leading to a high $\underline{Z}_{in}$ at resonance, which may cause instability.

C. Capacitive $\underline{Z}_{gd} = \dfrac{1}{j\omega C_{gd}}$

$\underline{Y}_{in} = j\omega C_{gs} + j\omega C_{gd} \cdot (1-\underline{a}_v)$. The Miller effect boosts the impact of C_{gd} and increases the input capacitance corresponding to the aforementioned consequences.

7.1.3.2 Output Impedance

With

$$\underline{Y}_{out} = \frac{\underline{I}_{out}}{\underline{V}_{out}} = \frac{1}{r_o} + \frac{g_m \cdot V_{in} + \underline{V}_{out} \cdot \left(\dfrac{1}{\underline{Y}_{gd}} + \dfrac{1}{j\omega C_{gs} + 1/R_s} \right)^{-1}}{\underline{V}_{out}} \qquad (7.31)$$

and

$$\underline{V}_{in} = \underline{V}_{out} \cdot \frac{\underline{Y}_{gd}}{j\omega C_{gs} + \underline{Y}_{gd} + \dfrac{1}{R_s}} \qquad (7.32)$$

we obtain

$$\underline{Y}_{out} = \frac{1}{r_o} + \underline{Y}_{gd} \cdot \frac{j\omega C_{gs} + g_m + \dfrac{1}{R_s}}{j\omega C_{gs} + \underline{Y}_{gd} + \dfrac{1}{R_s}} . \qquad (7.33)$$

The additional $\underline{Y}_{out}$ due to the drain gate feedback according to Fig. 7.5c is given by

$$\underline{Y}_{out,gd} = \underline{Y}_{gd} \cdot \frac{j\omega C_{gs} + g_m + \dfrac{1}{R_s}}{j\omega C_{gs} + \underline{Y}_{gd} + \dfrac{1}{R_s}} , \qquad (7.34)$$

demonstrating that R_s plays a non-negligible role in this feedback scenario. On the condition that $R_s = 0$ we get $\underline{Y}_{out,gd} = \underline{Y}_{gd}$.

A. Resistive $\underline{Z}_{gd} = R_{gd}$
The impedance of $\underline{Z}_{out}$ is lowered with decreased R_{gd}.

B. Inductive $\underline{Z}_{gd} = j\omega L_{gd}$
L_{gd} resonates C_{gs} softening the decrease of $\underline{Z}_{out}$.

C. Capacitive $\underline{Z}_{gd} = \dfrac{1}{j\omega C_{gd}}$
Increases the capacitive content of $\underline{Z}_{out}$. However, if $R_s = \infty$ (unloaded input), $C_{gs} \gg C_{gd}$ and $\omega_t = \dfrac{g_m}{C_{gs}}$ large, we get $\underline{Y}_{out,gd} = \omega_t \cdot C_{gd}$, which is resistive. This value is used for the calculation of f_{max}; refer to Eq. (5.31).

7.1.3.3 Transimpedance

We have observed that $\underline{Z}_{gd}$ decreases the impedances of the circuit leading to voltage source like characteristics. Hence, it is reasonable to describe the gain properties by means of a transimpedance:

$$\underline{Z}_t = \frac{\underline{V}_{out}}{\underline{I}_{in}}. \tag{7.35}$$

The reader may calculate the $\underline{Z}_t$ as an exercise.

7.1.4 Common Gate Circuit

Now, we study the properties of the common gate circuit. Referring to Fig. 7.6a, the gate acts as a common ground and reference potential. We derive the schematics by rearranging our well-known equivalent circuit of the common source topology depicted in Fig. 7.6b. However, the current source is now located between the input and the output rather than at the output only as desired to model the device as a voltage controlled current source. The latter model is favourable for the understanding of many circuits.

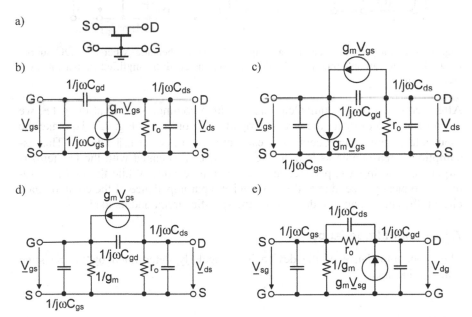

Fig. 7.6a–e. Derivation of the common gate equivalent circuit: **a** schematic; **b** equivalent circuit; **c** modification of the current source into two equal current sources with shared node at the gate; **d** replacement of the current source by means of $1/g_m$; **e** final equivalent circuit with common gate line

We apply a smart trick to obtain a modified equivalent model. According to Fig. 7.6c, the current source can be split into two series current sources with a shared node at the gate. The drain to source current stays the same. Furthermore, the gate to source current is not influenced since no current can flow in or out of the gate node because the current sources are identical. Thus, we can conclude that our rearrangement does not change the circuit properties.

As shown in Fig. 7.6d, one of the current sources is now connected at the same

nodes as $\underline{V}_{gs}$. Since $r = \dfrac{V_{gs}}{I_{gs}} = \dfrac{V_{gs}}{g_m \cdot V_{gs}} = \dfrac{1}{g_m}$, this current source can be re-

placed by a resistor $r = \dfrac{1}{g_m}$.

Finally, according to Fig. 7.6e, we are able to rearrange the circuit with the common gate line as required. To preserve the direction of the voltage arrows with respect to ground, $\underline{V}_{gs}$ and $\underline{V}_{gd}$ are changed into $\underline{V}_{sg}$ and $\underline{V}_{dg}$. Moreover, the direction of the current source is inverted.

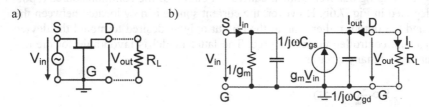

Fig. 7.7a,b. Common gate circuit: **a** simplified circuit schematic, where the DC supply networks, which are assumed to be lossless, are not included; **b** simplified equivalent small signal circuit

After derivation of the equivalent circuit with a common ground at the gate[2], we are now able to evaluate the circuit properties according to Fig. 7.7. To make the input and output independent from each other, we neglect $r_o \parallel 1/j\omega C_{ds}$. This assumption is reasonable as long as the impedance associated with the last term is larger than the input impedance and the load impedance. While the first requirement is typically meet due to the low device input impedance of the common gate circuit, the latter condition depends on the specific circuit and its load.

7.1.4.1 Input Impedance

The input impedance of the device is particularly determined by the resistive transconductance:

---

[2] Equal results are obtained by deriving the small signal parameters by means of the large signal equations. In this case, no network modifications but some calculations would be required.

$$\underline{Z}_{in} = \frac{\underline{V}_{in}}{\underline{I}_{in}} = \frac{1}{g_m} \left\| \frac{1}{j\omega C_{gs}} \right. . \tag{7.36}$$

At frequencies below f_t, $1/g_m \ll 1/\omega C_{gs}$. Thus, $\underline{Z}_{in}$ is mainly resistive and much lower compared to the common source circuit. A common gate circuit is well suited for impedance matching to real source impedances by choosing an appropriate value of g_m. In turn, g_m depends on the gate width and the bias current. The 50 Ω reference impedance would require a g_m of 20 mS – a value, which can easily be achieved with a FET. In this case, no reactive matching is required. As an additional advantage, a large bandwidth can be achieved with the common gate circuit. Hence, common gate circuits are excellent candidates for input stages in wideband amplifiers.

7.1.4.2 Output Impedance

The output impedance of the device may be approximated by

$$\underline{Z}_{out} = \frac{\underline{V}_{out}}{\underline{I}_{out}} = \frac{1}{j\omega C_{gd}} , \tag{7.37}$$

which is capacitive and has a large value. However, in the simplified equivalent circuit, the effect of the device feedback resistor r_o has been neglected. In reality, r_o adds a resistive contribution to $\underline{Z}_{out}$.

7.1.4.3 Voltage Gain

Suppose that $R_L \ll |\underline{Z}_{out}|$. With a voltage at the drain node of

$$\underline{V}_{out} = g_m \cdot \underline{V}_{in} \cdot R_L \tag{7.38}$$

we obtain a maximum possible circuit voltage gain of

$$\underline{a}_v = \frac{\underline{V}_{out}}{\underline{V}_{in}} = g_m \cdot R_L . \tag{7.39}$$

Since $|\underline{a}_v|$ is similar to that of the common source circuit, the equation may be already familiar to the reader. At high R_L, a voltage gain well above unity can be achieved. If the impedance of R_L is not much larger than $|\underline{Z}_{out}|$, the latter impedance has to be taken into account by parallel connection to R_L. One significant difference with respect to the common source circuit has to be mentioned. In good approximation we can assume that $\underline{a}_v$ has a real part only. Hence, the phase of the voltage gain is positive. Consequently, the input and output voltages of the common gate circuit have equal phases. This has a significant impact on the Miller effect. Why?

7.1.4.4 Current Gain

With an output current of

$$\underline{I}_L = \underline{I}_{in} - \underline{I}_g,\tag{7.40}$$

where $\underline{I}_g$ is the gate current, we get a current gain of

$$\underline{a}_i = \frac{\underline{I}_L}{\underline{I}_{in}} = 1 - \frac{\underline{I}_g}{\underline{I}_{in}}.\tag{7.41}$$

The equation reveals that the common gate circuit does not provide current gain beyond unity. In terms of amplifier applications, this may be a disadvantage. Since $\left|\underline{I}_g\right|$ is very low, the current gain is close to unity.

7.1.4.5 Power Gain

The common gate circuit is capable of offering voltage but no current gain. For narrow band applications, the associate power gain is moderate and usually lower than that of the common source circuit. However, due to the ability for broadband input matching, the common gate amplifier may exhibit similar or even better gain bandwidth properties.

7.1.4.6 Miller Effect

The equivalent Miller capacitance seen at the input amounts to

$$C_m = C_{ds} \cdot (1 - \underline{a}_v).\tag{7.42}$$

For the following reasons, the Miller effect is weak:

1. At low to moderate frequencies the phase of $\underline{a}_v$ is positive, thus $(1 - \underline{a}_v) < 1$

2. Input properties mainly determined by $\dfrac{1}{g_m} < \dfrac{1}{\omega C_{in}}$

3. Value and impact of feedback capacitance C_{ds} are small since $C_{ds} < C_{gd} < C_{gs}$

Consequently, the performance degradation towards high frequencies is not as severe as for the common source circuit. Nevertheless, due to its superior prerequisites in terms of low frequency power gain, the common source circuit usually still provides higher RF power gain than the common gate circuit.

7.1.5 Common Drain Circuit

The circuit schematic of a common drain circuit, where the drain is connected to ground is illustrated in Fig. 7.8a. Similar to the common source circuit with drain gate feedback, the common drain circuit exhibits a significant coupling between the input and the output, e.g. by means of R_L acting as feedback. Furthermore, $\underline{V}_{gs}$ introduces a significant coupling between the input and the output, which cannot be disregarded. Thus, R_L and R_S are considered for the calculation of the input and output device impedances, respectively. By the way, since the output potential follows the source potential, the common drain configuration is also referred to as a source follower. To ease the derivations, we assume the simplest equivalent circuit consisting of C_{gs} and the current source as shown in Fig. 7.8b.

a)   b)

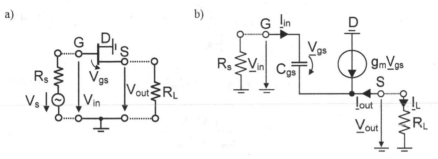

Fig. 7.8a,b. Common drain circuit: **a** simplified circuit schematic, where the DC supply networks, which are assumed to be lossless, are not included; **b** simplified small signal equivalent circuit

7.1.5.1 Input Impedance

As before, the input impedance is given by

$$Z_{in} = \frac{\underline{V}_{in}}{\underline{I}_{in}} . \qquad (7.43)$$

With

$$\underline{V}_{in} = \underline{I}_{in} \cdot \frac{1}{j\omega C_{gs}} + R_L \cdot (\underline{I}_{in} + g_m \underline{V}_{gs}) . \qquad (7.44)$$

and

$$\underline{V}_{gs} = \underline{I}_{in} \cdot \frac{1}{j\omega C_{gs}} \qquad (7.45)$$

we get

$$\underline{Z}_{in} = \frac{1}{j\omega C_{gs}} \cdot (1 + g_m \cdot R_L) + R_L . \qquad (7.46)$$

The factor $g_m \cdot R_L$ can be well above unity. Consequently, $\left| \underline{Z}_{in} \right|$ is large and $\underline{Z}_{in}$ is mainly capacitive. This property is well suited for some applications, where the loading of the previous stage should be minimised. One example is a VCO buffer.

7.1.5.2 Output Impedance

As usual, we calculate the output impedance of the device by means of

$$\underline{Z}_{out} = \frac{\underline{V}_{out}}{\underline{I}_{out}}. \tag{7.47}$$

With

$$\underline{V}_{gs} = -\underline{V}_{out} \frac{\dfrac{1}{j\omega C_{gs}}}{\dfrac{1}{j\omega C_{gs}} + R_s} \tag{7.48}$$

and

$$\underline{I}_{out} = -g_m \underline{V}_{gs} - j\omega C_{gs} \underline{V}_{gs} \tag{7.49}$$

it follows that

$$\underline{Z}_{out} = \frac{1 + j\omega C_{gs} R_s}{g_m + j\omega C_{gs}}. \tag{7.50}$$

Given that $g_m >> \omega C_{gs}$, we get $\underline{Z}_{out} = \dfrac{1}{g_m} \left(1 + j\omega C_{gs} R_s \right)$. Moreover, if $\omega C_{gs} R_s << 1$, we obtain $\underline{Z}_{out} = \dfrac{1}{g_m}$. We can conclude that $\underline{Z}_{out}$ exhibits a low impedance and is mainly resistive. Similarly to the $\underline{Z}_{in}$ of the common gate circuit, the designer can determine the $\underline{Z}_{out}$ of the common drain circuit by setting an appropriate gate width or bias. Obviously, common drain circuits are well suited for output stages of amplifiers with 50 Ω impedances.

7.1.5.3 Voltage Gain

The current at the source node sums as follows:

$$j\omega C_{gs} \cdot \underline{V}_{gs} + g_m \cdot \underline{V}_{gs} - \frac{\underline{V}_{out}}{R_L} = 0. \tag{7.51}$$

Together with

$$\underline{V}_{in} = \underline{V}_{gs} + \underline{V}_{out}, \tag{7.52}$$

we get

$$\underline{a}_v = \frac{\underline{V}_{out}}{\underline{V}_{in}} = \frac{g_m + j\omega C_{gs}}{g_m + j\omega C_{gs} + \dfrac{1}{R_L}} \, . \tag{7.53}$$

The latter equation reveals that $\left|\underline{a}_v\right| \leq 1$. This can easily be verified by

$\underline{a}_v = \dfrac{\underline{V}_{out}}{\underline{V}_{in}} = 1 - \dfrac{\underline{V}_{gs}}{\underline{V}_{in}}$. If $R_L = \infty$, $\left|\underline{a}_v\right| = 1$. $R_L = 0$ yields $\left|\underline{a}_v\right| = 0$. Let us recall that a

voltage gain of $\left|\underline{a}_v\right| > 1$ can be obtained for the common source and gate circuits but not for the common drain topology.

7.1.5.4 Current Gain

The output and input current can be approximated by

$$\underline{I}_{out} = -g_m \cdot \underline{V}_{gs} - j\omega C_{gs} \cdot \underline{V}_{gs} \tag{7.54}$$

and

$$\underline{I}_{in} = j\omega C_{gs} \cdot \underline{V}_{gs} \tag{7.55}$$

leading to

$$\underline{a}_i = \frac{\underline{I}_L}{\underline{I}_{in}} = -\frac{\underline{I}_{out}}{\underline{I}_{in}} = \frac{g_m}{j\omega C_{gs}} + 1 \, . \tag{7.56}$$

Regarding the magnitude, this current gain is similar to that of the common source circuit. Depending on the $\dfrac{\omega_t}{\omega}$ ratio, values well above unity can be achieved.

7.1.5.5 Power Gain

The common drain circuit is able to provide current gain but no voltage gain. Consequently, the power gain is lower than the one of the common source circuit. Despite the relative low power gain, the reactively matched common drain topologies tend to be critical in terms of stability due to the strong input to output feedback. Therefore, resistive elements may have to be implemented, which decrease the possible power gain.

7.1.5.6 Miller Effect

The equivalent Miller capacitance seen at the input is given by

$$C_m = C_{gs} \cdot (1 - \underline{a}_v) \, . \tag{7.57}$$

Since $|\underline{a}_v| < 1$ and the phase of $\underline{a}_v$ is positiv up to frequencies well below f_t, the common drain circuit has a very weak Miller effect. Thus, the decrease in its low frequency power gain towards high frequencies is relatively weak. Nevertheless, due to the significantly higher low frequency gain, the common source circuit still outperforms the common drain topology in terms of high frequency gain.

7.1.6 Cascode Circuit

Because of the Miller effect, the excellent properties of the common source circuit in terms of voltage and current gain cannot be fully exploited towards high frequencies. We will see that the cascode circuit can be successfully applied to mitigate this effect. As depicted in Fig. 7.9a,b, the cascode consists of a common source circuit followed by a common gate circuit. Let's assume equal transistors and bias for both transistors. Since the devices are stacked, the cascode topology requires double the supply voltage compared to the common source topology. This may be a limitation for mobile applications specifying ultra low supply voltages.

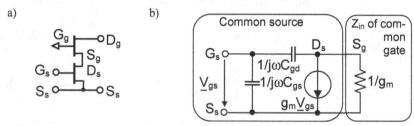

Fig. 7.9a,b. Cascode circuit: **a** schematics; **b** simplified small signal equivalent circuit for the calculation of the Miller capacitance

7.1.6.1 Miller Effect

Loaded with the input impedance $\dfrac{1}{g_m}$ of the common gate circuit, the $\underline{a}_v$ of the common source stage with transconductance g_m exhibits a low value of -1 since $\underline{a}_v \approx -g_m \cdot R_L$. According to Eq. (7.12) we get

$$C_m = (1 - \underline{a}_v) \cdot C_{gd} = 2 \cdot C_{gd}. \qquad (7.58)$$

Obviously, the corresponding value is much smaller than the one for the common source circuit. Consequently, the lowpass characteristic associated with the input capacitance is less pronounced yielding higher cutoff frequencies for the voltage, current and power gain. Moreover, according to Eq. (5.48), due to the improvement of the equivalent ω_t, the reduced input capacitance also lowers the NF_{min}.

However, what about the Miller capacitance of the second stage? Due to the low input impedance of the common gate circuit, the second Miller capacitance is more or less short-circuited resulting in a low RC time constant. Moreover, the voltage gain of the common gate stage is positive. Hence, the Miller effect of the second stage is weak.

7.1.6.2 Input Impedance

The input impedance is similar to that of the common source stage. However, with a reduced Miller and input capacitance:

$$Z_{in} \approx \frac{1}{j\omega C_{gs}} \left\| \frac{1}{2 \cdot j\omega C_{gd}} + R_{gs} \right. .$$

(7.59)

For simplicity, R_{gs} has not been included in Fig. 7.9b.

7.1.6.3 Output Impedance

The input of the cascode stage is represented by the input of the common source circuit. As per definition of an ideal voltage source the input of the common source circuit is short-circuited. Hence, the V_{gs} of the common source stage is zero and the associated current source of the common source circuit is deactivated. Consequently, the drain node of the common source stage can be simply approximated by the output impedance of the common source stage $Z_{out,CS}$. As far as the output impedance is concerned, the equivalent circuit of the cascode is now equal to the common source circuit with source degeneration as explained in Sect. 7.1.2.2. According to Eq. (7.21) and with Z_s replaced by r_o we obtain

$$Z_{out} \approx g_m \cdot Z_{out,CS}^2 \approx g_m \cdot r_o^2 .$$

(7.60)

Since usually $g_m \cdot r_o \gg 1$, the Z_{out} of cascode circuits is much larger than that of the common source approach.

7.1.6.4 Gain

Given that equally sized transistors are used, the voltage gain of the cascode stage is mainly given by the voltage gain of the common gate stage since the common source stage has a voltage gain of unity. Because the common gate stage provides no current gain, the common source stage determines the current gain. The high Z_{out} enhances the power transferred to the parallel load. Consequently, the voltage and current gains of cascode circuits are higher than that of common source configurations. This is why the cascode approach is well suited for circuits fabricated with aggressively scaled technologies. Recall that R_{ds} is reduced by gate length scaling. Thus, the losses associated with R_{ds} play a significant role.

Equations (5.27) and (5.28) indicate that a high output impedance improves the power gain and the f_{max}. The superior performance of the cascode circuit with respect to the common source circuit is demonstrated in Fig. 7.10 for 90-nm MOS transistors.

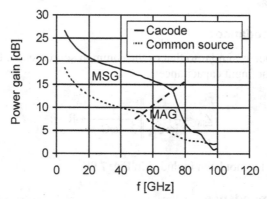

Fig. 7.10. Measured power gain of a common source and cascode amplifier stages using 90-nm SOI N-FET with w_g=64 μm, bias of FETs: V_{gs}=0.5 V, V_{ds}=1 V, I_d=17 mA, MSG: maximum stable gain, MAG: maximum available gain

7.1.7 Conclusions

In Table 7.1, the key properties of the treated circuits are summarised. The common source circuit exhibits both voltage and circuit gain. Consequently, the common source circuit has superior power gain performance. Due to the capacitive and high-ohmic characteristics, impedance matching is mandatory to achieve maximum performances at 50-Ω terminations. The high power gain cannot be exploited towards high frequencies due to the large Miller capacitance seen at the input, which acts as lowpass filter.

The cascode topology provides the best performance towards highest frequencies since it has a much lower Miller effect and a higher output impedance which lowers the output losses of the transistors.

Offering high voltage gain but no current gain, the common gate circuit provides a smaller power gain than the common source topology. Due to the low and resistive input impedance, the input matching is simplified. Thus, the configuration is well suited for input stages of wideband amplifiers. Due to the low impact of the Miller effect, a high bandwidth can be achieved. By proper choice of the bias and gate width of the transistor, 50 Ω input matching can be achieved without inductive elements. This makes the design of compact circuits possible.

The common drain configuration has a high current gain and no voltage gain. It features a low and resistive output impedance simplifying the output matching. Hence, the configuration is a favourable candidate for output stages of wideband amplifiers. Since the voltage gain is smaller than unity, the equivalent Miller capacitance is low minimising the loading of previous stages. Thus, the common drain topology is frequently used as output buffer, e.g. for VCOs.

Proper choice and combination of the different circuit topologies with respect to individual specifications and interfaces allow optimum results.

Table 7.1. Comparison of basic circuit configurations

	Common source	Common gate	Common drain	Cascode
$\underline{Z}_{in}$)[3]	High, mainly capacitive $\approx 1/j\omega C_{gs}$ $+1/j\omega C_{gd}(1-\underline{a}_v)$	Low, mainly resistive $\approx 1/g_m$	High mainly capacitive $\approx 1/j\omega C_{gd}+$ $1/j\omega C_{gs}(1-\underline{a}_v)$	High, mainly capacitive $\approx 1/j\omega C_{gs}+$ $1/j\omega 2C_{gd}$
$\underline{Z}_{out}$	High, $\approx r_o \left\|\dfrac{1}{j\omega C_{ds}}\right.$	High, mainly capacitive $\approx 1/j\omega C_{gd}$	Low, mainly resistive $\approx 1/g_m$	Very high $\approx g_m r_o^2$
$\underline{a}_{vmax}$	$\approx -g_m R_L \|\underline{Z}_{out}$ $\approx -g_m R_L$	$\approx g_m R_L \|\underline{Z}_{out}$ $\approx g_m R_L$	≤ 1	Slightly higher than CS, more current into R_L
$\underline{a}_{imax}$	$\approx -g_m/j\omega C_{in}$	≤ 1	$\approx g_m/j\omega C_{in}$	Higher than CS since lower C_{in}
Miller effect	Strong since $\underline{a}_v$ high and negative	Low since $\underline{a}_v$ positive and input resistive	Low since $\underline{a}_v$ small and positive	Low since $\underline{a}_v$ of CS stage small
MAG	High	Moderate	Low	Very high
BW	Limited by Miller effect	Larger than CS since low input impedance and no Miller effect	Larger than CS since no Miller effect	Larger than CS since reduced Miller effect
F$_{min}$	$\approx 1 + \dfrac{2}{\sqrt{5}}\dfrac{\omega}{\omega_t}\sqrt{\gamma\delta(1-|c|^2)}$ (see Section 5.4.7)	Similar to CS, limitations if gain low	In practice, much higher than for CS	Slightly lower than CS[4] due to reduced Miller effect and higher effective ω_t
Superior application	High power gain, frequency converters	Moderate power gain, input stage, wideband applications	Output buffer, e.g. for VCO	Highest power gain at highest frequencies possible

CS: common source, $\underline{a}_{vmax}$: maximum voltage gain, $\underline{a}_{imax}$: maximum current gain, BW: bandwidth

---

[3] For the calculation of $\underline{Z}_{in}$ we assume that $\underline{a}_v$ exhibits only a real part, which is valid at frequencies well below ω_t.

[4] This holds only for moderate to high frequencies. At low frequencies, the additional noise of the second stage negates the advantages concerning the reduced Miller effect and the higher effective ω_t.

7.1.8 Comparison between FET and BJT Circuits

In principle, the common-source, -gate and -drain FET configurations exhibit many similarities with their BJT based counterparts, which are the common-emitter, -base and -collector circuits, respectively. Let us summarise the major differences, which must be taken into account by the designer:

- Considering typical bias and device dimensions for RF applications, BJTs provide higher g_m, thereby yielding higher low frequency gain. However, due to the high input capacitance of the forward biased base emitter diode, common BJTs do not necessarily offer better RF properties.
- The input and output of FETs can be characterised by a connection of a capacitor and a resistor. At the input, this connection is in series, whereas at the output it is in parallel. Thus, the input impedance is larger and less resistive than the output. At small to moderate transistors sizes, 50 Ω input matching is more difficult.
- As illustrated in Fig. 7.11, in terms of impedance matching, BJTs benefit from lower impedances being more resistive. Both at the output and the input, these impedances are based on a parallel connection of a capacitor and a resistor. Thus, BJTs allow simpler and more broadband impedance matching to low-ohmic and resistive terminations. This must be borne in mind for wideband amplifiers.
- The conditions may differ for power amplifiers employing huge transistors. In this case, the low-ohmic impedances may be a disadvantage with respect to 50-Ω matching.
- The common base circuit as an input matching stage is not as frequently used as the common gate circuit. Proper 50 Ω input matching can already be achieved with the common emitter topology.
- Since the input impedance of the common emitter stage is relatively low, an undesired loading of the previous stage may result. A common collector circuit may be added to increase the impedance seen by a previous stage.

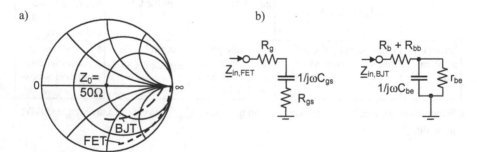

Fig. 7.11a,b. Comparison of the input impedance of a FET and a BJT: **a** typical S_{11} in the Smith chart; **b** simplified equivalent circuit of a FET and a BJT, respectively

7.2 Stabilisation Networks

Based on Sects. 4.1 and 7.1, where stability and basic circuit topologies have been discussed, stabilisation networks will be treated now. If a circuit is unstable, stabilisation is required to make the circuit unconditionally or conditionally stable. Generally, improved stability corresponds to an increased K-factor. The stability improvement sacrifices gain and noise. Thus, an over-stabilisation should be avoided. One must strive for a tradeoff between stability improvement and overall performance. Process variations have to be taken into account. With the support of CAD tools, the elements of the stabilisation circuits can be tuned until the K-factor is just a little bit greater than unity and unconditional stability is reached. Most stabilisation methods are based on the following approaches:

1. Reducing of g_m to lower the gain.
2. Adding of resistive losses to lower the gain and the reflected power.
3. Implementing and manipulating of poles, which optimise the magnitude and phase of the gain and the impedances.
4. Implementation of a destructive feedback, which couples the signal from the output back to the input in a way such that the signal energy is reduced in each signal period. In this context, we distinguish between inverting and non-inverting circuits with an input to output phase relationship of 180° and 0°, respectively, as illustrated in Fig. 7.12.

Exhibiting negative voltage gain, the common source amplifier belongs to the category of inverting amplifiers. It can be stabilised by a feedback with positive phase, most efficiently through a phase of 0°. Due to the positive sign of the voltage transfer function, common gate and drain topologies are non-inverting circuits demanding a feedback with a 180° phase offset. As an important conclusion, the impact of a stabilisation network greatly depends on the circuit topology used. While one element can improve the stability for an inverting amplifier, it can lead to degradations for a non-inverting amplifier.

In the following, we focus on stability networks for common source amplifiers. The insights can be mapped for other topologies.

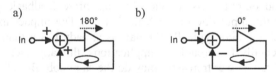

Fig. 7.12a,b. Generic schematics improving stability: **a** inverting amplifier such as common source circuit; **b** non-inverting amplifier such as common gate circuit and common drain circuit

7.2.1 Source Degeneration

We have shown in Sect. 7.1.2 that series feedback by means of a resistive or inductive source impedance as indicated in Fig. 7.13a lowers the effective g_m of a common source stage. Furthermore, the input and output become more resistive thereby introducing additional losses. Thus, the stability is improved. In most cases an inductor is preferred compared to a resistor since it has a much lower noise contribution. Moreover, an inductor simplifies 50-Ω matching through compensation of C_{gs}. For these reasons, the so-called inductive source degeneration is the first candidate for the stabilisation of LNAs. We have to keep in mind that the impedance and the corresponding stabilisation effect of an inductor vanish towards low frequencies, eventually demanding for additional low frequency stabilisation. As long as a narrow band impedance matching or a DC block capacitance reduces the gain at low frequencies, this is not a problem. Due to the opposite transmission phase, a capacitor rather than an inductor is required in the gate ground connection to improve the stability in common gate circuits.

a)   b)   c)   d)

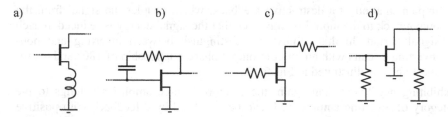

Fig. 7.13a–d. Frequently used stabilisation methods for a common source circuit: **a** source degeneration; **b** drain gate feedback; **c** series resistive loading; **d** parallel resistive loading

7.2.2 Drain Gate Feedback

According to Sect. 7.1.3, a capacitive or resistive drain gate feedback improves the stability of a common source stage since virtual resistive admittances are generated at the input and output ports. Obviously, capacitive feedback generates a large capacitance at the input due to the Miller effect. The impact increases towards high frequencies, where typically the gain is already lower than at low frequencies. Thus, especially for broadband applications, this approach may not be advantageous. Finally, a slower transistor may do the same job. Broadband stabilisation can be achieved by using a DC-decoupled resistor as illustrated in Fig. 7.13b. The latter topology is frequently used for power amplifiers since the output losses are relatively small compared to the level of stabilisation. The disadvantage of both capacitive and resistive drain gate feedback is the increased noise figure. Noise is not an important issue for power amplifiers, whereas it is for LNAs. As it leads to degraded stability, drain gate feedback by means of an inductance is not suited for common source stages. Active feedback using transistor stages may be possible.

7.2.3 Resistive Loading

Transistors can be stabilised within a broad bandwidth by introducing parallel and/or series resistors in the input and output ports as depicted in Fig. 7.13c,d. In this case, we raise the return losses resulting in an increased K-factor. Parallel resistors have the additional benefit that they can be used for bias feeding. Unfortunately, resistances at the input increase the noise figure. Thus, only resistive loading at the output may be an option for LNAs. Resistors in the output of power amplifiers have to be avoided to prevent power losses, whereas at the input, they are well suited for stabilisation and impedance matching. In any case resistors decrease the possible gain.

A design equation for the calculation of the input or output stabilisation resistors has been derived in [Gar99] to achieve unconditional stability:

$$R = \left(\frac{\left| ce_{L,S} + k_{st} \right|^2 - ra_{L,S}^2}{\left(ra_{L,S} + 1 \right)^2 - \left| ce_{L,S} \right|^2} \right)^{k_{st}}, \tag{7.61}$$

where R is the resistor,

$$k_{st} = \begin{cases} 1 & \text{for shunt loading} \\ -1 & \text{for series loading} \end{cases}, \tag{7.62}$$

$ce_{L,S}$ denote the centre of the load or source stabilisation circle, respectively, and $ra_{L,S}$ represent the radius of the load or source stabilisation circle, respectively. For information concerning $ce_{L,S}$ and $ra_{L,S}$, the reader is referred to Section 4.1.1.

7.3 Bias Supply

The properties of transistors and subsequently, the performance of circuits strongly depend on the applied DC supply voltages and the resulting supply currents. Thus, the determination and feeding of the optimum DC bias is very important and the first major step in circuit design after choosing the suitable transistor.

Recall the typical equivalent circuit for a transistor, e.g. that of a FET according to Fig. 5.13. The DC bias impacts the values of the equivalent elements, which are the transconductance, and the reactive and resistive components caused by parasitic effects. As illustrated in Fig. 7.14a,b, the DC bias defines the origin of the RF signal swing[5]. The load impedance given by R_L determines the relation between $I_{D,RF}$ and $V_{ds,RF}$. Frequently, R_L is determined by the reference impedance, e.g. 50 Ω, which can be transformed to provide an optimum impedance for individual

---

[5] In this paragraph, we explicitly distinguish between the DC and RF parts of signals by considering the indicated notation. However, in this book, we may switch between these two parts since the separation should be clear with respect to the individual context.

tasks. In reality, this transformed impedance may be complex to resonate the C_{ds}, which has some impact at high frequencies.

For very small power, the signal swing is weak and small signal analyses, e.g. based on S-parameters can be performed. If the signal swing is significant, then the device operates under large signal conditions. The large signal parameters are related to an array of small signal parameters incrementally approached within the signal swing.

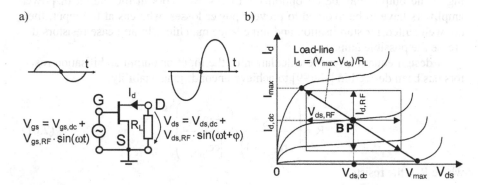

Fig. 7.14a,b. RF and DC components of V_{gs} and V_{ds}: **a** simplified circuit schematics, for simplicity, the bias supply networks are not yet included, $V_{gs,dc}$: DC part of V_{gs}, $V_{ds,dc}$: DC part of V_{ds}, $V_{gs,RF}$: RF part of V_{gs}, $V_{ds,RF}$: RF part of V_{ds}; **b** DC bias point (BP) is origin of RF swing, V_{max}: maximum V_{ds}, I_{max}: maximum I_{ds}

7.3.1 Bias Supply Optimisation

The typical IV characteristics of a FET are illustrated in Fig. 7.15a. We can identify the following bias regions providing optimum performances for specific tasks.

I. Maximum gain
Referring to Chap. 5, increased V_{gs} raises g_m, and subsequently f_t, f_{max}, the voltage gain and the current gain. Referring to Fig. 7.15b, in practice, we can observe this behaviour up to a certain saturation voltage $V_{gs,gmmax}$. A drawback of a high V_{gs} is the significant consumption of DC current. For wireless and battery driven components, a compromise between gain and DC power is aimed for.

II. Minimum noise figure
According to Sect. 5.4.7, the F_{min} of FETs is getting smaller with increased ω_t motivating for a high V_{gs} as well. On the other hand, at high currents, we observe a saturation of ω_t resulting in a degraded noise figure. Consequently, in practice, we typically find an optimum bias point $V_{gs,NFmin}$ at moderate V_{gs} as shown in Fig. 7.15b. In addition, for BJTs, we have to consider the shot noise, which increases with current. Thus, the optimum bias point for noise in BJTs is typically lower than for FETs making a compromise between gain and noise more difficult

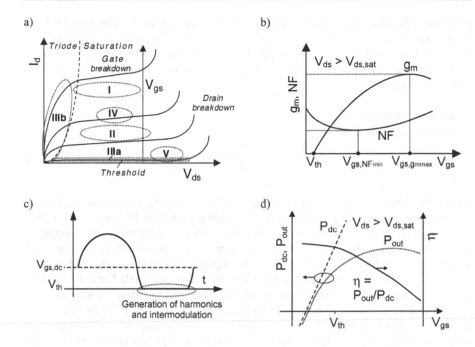

Fig. 7.15a–d. Bias optimisation: **a** regions providing specific performances; **b** typical characteristics of g_m and NF vs V_{gs}; **c** time domain representation showing generation of harmonics and mixing products due to signal clipping in the threshold region; **d** simplified illustration of output power, DC power and efficiency vs V_{gs}

III. High non-linearity

For some applications, high nonlinearities are desired since they allow the generation of harmonics, and subsequently frequency mixing products. Examples of these circuit applications are frequency multipliers and mixers. The level of nonlinearity can be loosely defined as the power ratio between the non-fundamental frequency components with respect to the fundamental frequency signal. Harmonics and intermodulation products are generated if parts of the signal are clipped as illustrated in Fig. 7.15c. If this is the case, a sinusoidal signal with one frequency is transformed to a square-like signal featuring harmonic frequency components. Obviously, by inspection of Fig. 7.15a, the strongest clipping is achieved at bias voltages close to the IV boundaries, which are the threshold region (IIIa), and the border between the resistive and saturation region (IIIb).

IV. Maximum linearity and output power

Obviously, the highest linearity is achieved when the bias point is far away from the IV boundaries being instrumental for the generation of the harmonics. This point is located in the middle of the IV curves providing symmetrical swing across R_L. Upper and lower limitations appear due to the resistive region and the threshold region, respectively. In this bias point high output power can be obtained since the maximum possible voltage and current swings are large. Finally, this bias point provides also good gain and noise performances. Thus, this bias point exhibits a good compromise between various requirements. That is one of the reasons why this bias point is frequently applied for amplifiers, as long as the efficiency and heating are not major concerns. Details about power amplifiers will be presented in Chap. 9.

V. Maximum efficiency
The efficiency is defined as

$$\eta = \frac{P_{out}}{P_{dc}}, \tag{7.63}$$

where the desired RF output power P_{out} and DC power P_{dc} have to be optimised for low-power consuming applications demanding minimum DC power for a specific output power. Needless to say that this property determines the operation time of battery-driven devices. In wireless transceivers, power amplifiers consume a considerable part of the DC power. The highest efficiency is obtained at bias points close to V_{th}, where the average DC power is minimal. Little or no DC power is consumed if no RF power is applied. The corresponding illustrations are plotted in Fig. 7.15d. Drawbacks of this bias point are the low RF gain due to the reduced g_m and f_t, and the high non-linearity due to the clipping of the negative half-wave of the input signal. Filtering is required to suppress undesired harmonics. In Chap. 9, the efficiency issues will be treated in detail.

7.3.2 Bias Supply Networks

For simplicity, the networks required for bias feeding have not yet been included in the previous sections. We have assumed that the bias networks are ideal and lossless. However, bias networks are not ideal. Depending on the bandwidth and the currents to be handled, bias networks exhibit both RF and DC losses. A bias network should yield the following basic characteristics:

1. The DC losses should be as low as possible requiring a low resistive impedance. E.g. if a DC current of 20 mA has to be fed through a bias element with series resistance of 50 Ω, a non-desired voltage drop of 1 V would be generated. We know that the drain current is much higher than the gate current. Thus, for common source and drain circuits, the corresponding restriction is more significant for the output rather than the input bias networks. However, in common gate circuits, a high DC current appears at both the input and output.
2. Within the required RF frequency range, the impedance of the bias network should be as high as possible to avoid losses with respect to the source and load impedances. Such a high RF impedance may be realised by an inductance with a frequency dependent impedance $\underline{Z}(\omega) = j\omega L$. Since high inductance values are usually required, the inductors consume lots of circuit area. Moreover, the inductive impedance is only large enough for high frequencies. At a

relatively low frequency of 1 GHz, we would need an inductance larger than 8 nH to provide an impedance above 50 Ω. At the same time, towards high frequencies, the application is restricted by the self-resonance frequency. Thus, inductive bias networks are mainly suited for narrow-band applications. Higher bandwidth can be obtained with a high-ohmic resistor. If we would have an output bias resistor of 50 Ω and a 50 Ω load, we would lose 3 dB of RF power. As discussed above (see 1.), the value of the bias resistor is limited by the associated voltage drop at high currents.

To make the bias independent of the previous or following stages, DC block capacitors $C_{dc-block}$ are required. RF shunt capacitors $C_{RF-shunt}$ ensure the grounding of the bias elements to define the impedances towards the DC supply. In this case, the impact of arbitrary connections, e.g. due to a bondwire, do not change the RF impedances of circuits. As a second task, the RF shunts block undesired RF signals associated with the DC source. Common realisations of DC bias networks are listed in Table 7.2 together with a summary of their associated pros and cons.

Table 7.2. Frequently used DC bias networks

Schematics	Advantages	Disadvantages
I. Inductive biasing		
	- Low losses, high gain and low noise - Low DC voltage drop across inductances resulting in low power consumption - Inductances can be used for impedance matching - At very high frequencies, inductive transmission lines are used	- Narrowband due to resonance - $\underline{Z} = j\omega L_{bias}$, limited use for frequencies below 1 GHz since high inductance values are required - Good RF shunting required, otherwise non-desired resonances are possible with unknown supply impedances - Inductances require large size
II. Resistive biasing		
	- Broadband, no resonances - Works down to DC - Very compact - Simple - Improved stability	- Significant losses, low gain, high noise figure - High DC voltage drop across R_{bias} at drain node where high current has to be handled - High DC supply voltage required, high DC power consumption
III. Depletion FET current source		
	- Upper transistor is a D-FET biased at $V_{gs}=0V$ and acts as an active load - Bias current depends on the gate width relation n/w - High load resistance at low frequencies allowing high low frequency gain - Works down to DC - Compact - Further information about improved approaches using active loads see [Rob91]	- Voltage drop across D-FET - Additional FET parasitics degrade RF properties - Tradeoff, n small: high resistivity and low RF losses, n large: low DC losses - Sensitive to input power variations, unstable DC operation point leading to limited linearity

Table 7.2. Continued

IV. Depletion FET with DC feedback in source		
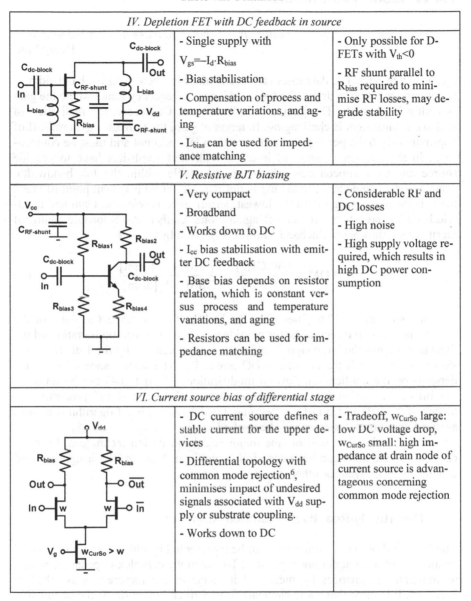	- Single supply with $V_{gs}=-I_d \cdot R_{bias}$ - Bias stabilisation - Compensation of process and temperature variations, and aging - L_{bias} can be used for impedance matching	- Only possible for D-FETs with $V_{th}<0$ - RF shunt parallel to R_{bias} required to minimise RF losses, may degrade stability
V. Resistive BJT biasing		
	- Very compact - Broadband - Works down to DC - I_{cc} bias stabilisation with emitter DC feedback - Base bias depends on resistor relation, which is constant versus process and temperature variations, and aging - Resistors can be used for impedance matching	- Considerable RF and DC losses - High noise - High supply voltage required, which results in high DC power consumption
VI. Current source bias of differential stage		
	- DC current source defines a stable current for the upper devices - Differential topology with common mode rejection[6], minimises impact of undesired signals associated with V_{dd} supply or substrate coupling. - Works down to DC	- Tradeoff, w_{CurSo} large: low DC voltage drop, w_{CurSo} small: high impedance at drain node of current source is advantageous concerning common mode rejection

---

[6] Common mode rejection is the ability to suppress undesired signals, which are not differential.

7.4 Wideband Amplifiers

Failure is the chance to make it better the next time.
Henry Ford

According to Sect. 1.5, data rates can be enhanced by increasing the bandwidth. It is common practice to define the bandwidth as the frequency range where the gain is within 3 dB of the DC gain or mid-band gain. Obviously, the realisation of wideband amplifiers is challenging. In terms of gain, the transistor gain roll-off of approximately 6 dB per octave has to be taken into account and must be compensated in the frequency range of interest. Wideband amplifiers have to provide source and load impedances, which are acceptable within the full bandwidth. Thus, resonant impedance matching is not feasible. From a system point of view, the sub-frequency range with the lowest performance is relevant. Gain and bandwidth (BW) have to be traded-off against one another. A reasonable figure of merit for multipurpose wideband amplifiers is given by

$$\text{FOM} = \frac{\text{BW[GHz]}}{f_t[\text{GHz}]} \cdot \frac{S_{21}[\text{dB}]}{\text{NF[dB]}} \cdot \frac{P_{1\text{dB}}[\text{dBm}]}{P_{\text{dc}}[\text{dBm}]} . \tag{7.64}$$

The first term in Eq. (7.64) describes the bandwidth with respect to the f_t of the used technology, the second factor considers the gain to noise figure ratio and the third term relates the large signal properties represented e.g. by the 1 dB compression point $P_{1\text{dB}}$ with the consumed DC power P_{dc}. The significance of the terms depends on the application. Several modifications of Eq. (7.64) can be found in the literature. Optionally, the representation of the large signal properties by means of the IIP3 instead of the $P_{1\text{dB}}$ may be also reasonable. Logarithmic as well as non-logarithmic measures may be applied.

In this section, we will outline major wideband design techniques. In many cases, these methods can be successfully combined. Many of the insights gained can also be employed for other circuits, e.g. for mixers.

7.4.1 Resistively/Reactively Matched Amplifier

The input and output of transistors can be represented by either a series or parallel connection of a capacitor and a resistor. To match the complex impedances to real terminations, resonances by means of inductances are required as sketched in Fig. 7.16. It is clear that these elements form a filter, which limits the bandwidth. The return losses and bandwidth can be increased by introducing resistors. In this case, we damp the resonance and absorb the reflected power. The amount of gain reduction at low frequencies can be controlled by $R_{b1,2}$, whereas $L_{b1,2}$ reduces the gain decreasing effect towards high frequencies yielding a flat gain characteristic. A positive side effect is enhanced stability. Major disadvantages are the increased noise and the decreased output power at the input and output, limiting the application for LNAs and PAs, respectively.

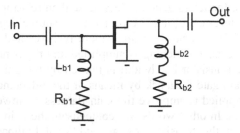

Fig. 7.16. Simplified RF schematic of the resistively/reactively matched amplifier

7.4.2 Active Matching

According to the previous sections, the common gate and the common drain circuit provide low and resistive input and output impedances, respectively. Why not using these properties for matching. Since the impedances are mainly determined by g_m, the transistor size and bias can be optimised to get excellent matching over a wide bandwidth. A design example of such an amplifier will be presented in Sect. 8.3.1. Optionally, as shown in Fig. 7.17, a common source amplifier can be added to maximise the gain of the circuit. Interstage matching networks may be used for gain peaking towards highest frequencies.

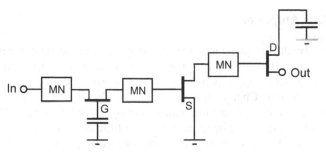

Fig. 7.17. Simplified RF schematic of common gate/common source/common drain amplifier, MN: matching network

7.4.3 Feedback Amplifier

Feedback can improve the bandwidth since the input and output impedances can become more resistive resulting in less pronounced matching resonances. Hence, impedance matching can be performed over a wide frequency range [Che63, Bar04]. Corresponding details have been discussed in Sects. 7.1.2 and 7.1.3.

Moreover, by proper combination of feedback circuits, the frequency-limiting Miller effect can be reduced as discussed in Sect. 7.4.4. A typical schematic of a feedback amplifier is shown in Fig. 7.18. Inductors and resistors can be employed as feedback elements. Inductors generate less noise than resistors. This is the major reason why an inductance L_s may be preferred as source degeneration. Referring back to Sect. 7.1.2, L_s makes the input more resistive and low-ohmic. As a consequence, the impedance matching is simplified and the bandwidth increased. Let us recall that the transistor bandwidth is limited by the gain roll-off towards high frequencies. Drain gate feedback by means of the inductance L_{gd} and the resistance R_{gd} can be applied to improve the gain flatness by lowering the gain towards low frequencies. In other words, we compensate the gain slope of the lowpass filter inherent in the transistor by an additional highpass filter. At low frequencies, L_{gd} can be approximated by a short. The gain level is decreased by means of R_{gd}. Towards high frequency, L_{gd} reduces the feedback and the gain decreasing effect of R_{gd}.

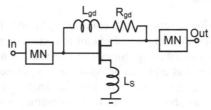

Fig. 7.18. Simplified RF schematic of an amplifier with drain gate feedback and source degeneration

7.4.4 TIS/TAS Amplifier

A further bandwidth enhancement can be achieved by the proper cascading of feedback circuits. This principle has been investigated and proposed by E.M. Cherry and D.E. Hopper [Che63, Che96, Che00]. Accordingly, these amplifiers are also known as Cherry-Hopper amplifiers. Due to the good gain capabilities, common source circuits are frequently employed for RF circuits. Determined by the Miller effect, the large input capacitance of the common source circuit limits the bandwidth. We have already discussed one concept, which overcomes this problem by loading the common source circuit with the low impedance of a common gate stage thereby significantly reducing the Miller capacitance. This is the cascode amplifier treated in Sect. 7.1.6.

A similar idea is exploited in the TIS (Transimpedance)/TAS (Transadmittance) approach illustrated in Fig. 7.19. To minimise the size, achieve high bandwidth and avoid stability problems, resistors are commonly used as feedback elements. A disadvantage of resistive feedback is the noise degradation. However, TIS/TAS circuits are often used for wired applications, where the SNR and subsequently the noise is not a major issue. A common source circuit with source degeneration and consequently high input and output impedances is loaded by a common source circuit with drain gate feedback featuring low impedances. Thus, according to previous sections, the voltage gain of the first stage is low resulting in a low Miller capacitance in the first amplifier. The process variations of transistors requiring complex fabrication steps and highly accurate lithography are much larger than for resistors. In this context, we can identify one advantage of the TIS/TAS concept. Given strong feedback, the process variations of the transistors are significantly reduced yielding stable input and output impedance over a broad range of process variations. Moreover, consider gain variations. The gain decreasing impact of the resistive feedback increases with the gain resulting in an efficient compensation of potential gain variations. Variations of the threshold voltages, which have a significant impact on the DC operating current and the device properties, are partly compensated. In Sect. 8.2.2, a bias stabilisation circuit featuring series DC feedback will be presented.

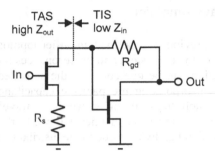

Fig. 7.19. Simplified schematic of TIS/TAS concept

7.4.5 Balanced Amplifier

According to Fig. 7.20, the balanced amplifier consists of two amplifiers and two 90° hybrids. Details about hybrids and couplers have been treated in Sect. 6.10.

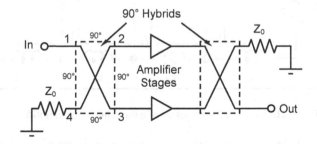

Fig. 7.20. Illustration of balanced amplifier

Irrespective of the individual impedances of the amplifier stages, unwanted reflections at the circuit ports are cancelled or absorbed allowing high return losses. The two amplifiers are fed with a phase offset of 90°. After passing the coupler paths twice, the signals reflected at the amplifier inputs become 180° out-of-phase and chancel out at the coupler input. Signals at port 4 are absorbed in the termination resistor Z_0. Consequently, the return losses of balanced amplifiers are high even if the impedance matching of the employed amplifiers is poor. At the output, the power of the amplifier stages is combined in phase, and undesired reflections are also chancel out or absorbed in Z_0. Compared to a single amplifier, the maximum possible output power is doubled, whereas the gain remains the same. In practice, the losses of the couplers have to be taken into account. Transmission line based hybrids allow bandwidths up to two octaves but consume a large circuit area. The size can be reduced by employing lumped element couplers. The drawback of the latter approach is the reduced bandwidth due to the non-distributed realisation and the loss, which may be up to 2 dB per coupler in integrated circuits.

7.4.6 Travelling Wave Amplifier

The bandwidth of the previously described amplifier topologies is limited by the parasitic capacitances of the transistors and the resonances required for impedance matching. This limitation can be overcome by the so-called travelling wave (or distributed) amplifier incorporating the parasitic capacitances of the amplifiers into artificial transmission lines. In other words, a travelling wave amplifier (TWA) acts like an active transmission line. It is well known that transmission lines can provide very high bandwidths due to the distributed resonances.

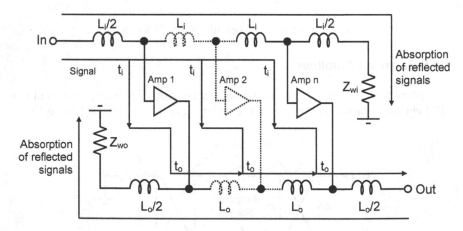

Fig. 7.21. Simplified schematics of a travelling wave amplifier

Figure 7.21 illustrates the schematics of a travelling wave amplifier. The sections of the input and output lines represent lowpass structures with characteristic impedances of

$$Z_{wi} = \sqrt{\frac{L_i}{C_i}} \tag{7.65}$$

and

$$Z_{wo} = \sqrt{\frac{L_o}{C_o}} , \tag{7.66}$$

respectively, with amplifier input capacitance C_i, line input inductance L_i, amplifier output capacitance C_o, and line output inductance L_o. By proper choice of the ratio between the L and C for each line, 50 Ω matching can be achieved up to high frequencies. The input signal travels down the input line and feeds each amplifier. In Fig. 7.22a, an equivalent circuit of each stage is drawn. Maximum signal adding at the output node and subsequently the highest gain and output power can be achieved if the signals are in-phase demanding that the input and output delays of $t_i \approx \sqrt{L_i C_i}$ and $t_o \approx \sqrt{L_o C_o}$ are equal, yielding $L_i C_i = L_o C_o$. In most amplifier configurations, we can find $C_i > C_o$ and thus $t_i > t_o$ at a given L_o. Since the impact of the time misalignment is frequency dependent, we can observe undesired ripples versus frequency if the delays are not equal. According to Fig. 7.22b, there are at least four possibilities for solving the phase problem:

a) Increase in C_o by adding parallel output capacitance C_{oa}
Because of its simple implementation, this approach is frequently used. The associated decrease in the cutoff frequency of the output line has to be considered. As long as the total output capacitance is not larger than that of the input line, there is no significant drawback. More relevant can be the decrease in the output impedance seen by each stage leading to a lower gain. Let us recall that for a common source amplifier stage, the gain is given by $a_v = -Z_{wo} g_m$. However, the losses due to phase misalignments would typically be higher.

a)   b)

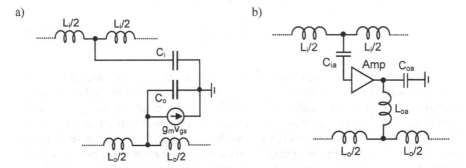

Fig. 7.22a,b. Travelling wave amplifier: **a** common source amplifier stage; **b** phase alignment techniques

b) Input capacitor C_{ia} in series to C_i.

Strategy a) increases C_o by the parallel connection with C_{oa}. Similarly, we can lower C_i by a series connection with C_{ia} [Aga98]. The advantage of this approach is that the cutoff frequency and the bandwidth can be raised. Unfortunately, this decreases the gain since we introduce a voltage drop at the input.

c) Variation of L_o

The phase can be influenced by means of L_o. An increased L_o yields higher gain due to the larger value of Z_{wo}. To obtain appropriate length conditions of the lines, corresponding layout considerations have to be made. Optionally, the output inductance may be raised by implementing an additional inductor L_{oa}. However, compared to methods a) and b) it is not possible to achieve equal input and output impedances, since $Z_{wi}=Z_{wo}$ would demand for $\dfrac{L_i}{C_i}=\dfrac{L_o}{C_o}$. Together with the in-phase condition, this can only be achieved with $L_o=L_i$ and $C_o=C_i$.

d) Variation of L_i

Phase alignment by adaptation of L_i would be possible as well. However, in this case, the input impedance is lower than the output impedance, which may be a problem for many systems.

To absorb potential reflections, the lines are terminated by resistors with impedances of around Z_{wi} and Z_{wo}. This increases the return losses within a broad bandwidth and improves stability, which is typically uncritical for properly designed travelling wave amplifiers. In contrast to cascaded amplifier topologies, the gains of the TWA amplifier stages are added instead of multiplied. Thus, TWAs provide relatively low gain. One half of the output signal travels towards the load and the other half of the output signal travels towards the absorbing resistor. With n sections, the maximum low frequency power gain is given by

$$G = a_v^2 = \frac{1}{4} \cdot n^2 \cdot Z_{wo}^2 \cdot g_m^2 . \tag{7.67}$$

Theoretically, the gain raises with n^2. However, in practice, the increase in n is limited by the parasitics of the lines. After a certain total line length, the signal attenuation associated with additional amplifier line segments is stronger than the individual contribution of these additional amplifiers in terms of amplification. Moreover, the size and costs increase, and the impact of the low-pass sections accumulates with n, thereby lowering the maximum operation frequency and bandwidth. Typically, we can find realisations with n ranging from 3 to 10. The bandwidth of the circuit is particularly determined by the cutoff frequencies of the equivalent transmission line segments. Considering the T-type arrangement of the lumped elements, we get

$$f_{ci} = \frac{1}{\pi\sqrt{L_i C_i}}$$ (7.68)

and

$$f_{co} = \frac{1}{\pi\sqrt{L_o C_o}}$$ (7.69)

for the input and output cutoff frequency, respectively. As for other circuits, the bandwidth and gain has to be traded off. In this context, we can for example identify the transistor width w_g of the amplifier stages as an important design parameter. With large w_g, the gain is high due to the high g_m. On the other hand, the bandwidth is sacrificed due to the large capacitances. What about the choice of the amplifier stages? Common gate and common drain circuits have resistive input and output impedances, respectively. Thus, they are not well suited for the TWA amplifier stages, since capacitive characteristics are required for the incorporation into the artificial lines. Resistive impedances would cause additional line losses. Thus, common source stages are well suited. Further improvements can be achieved with cascode circuits. Attributed to the reduced Miller effect and the increased output resistance, higher bandwidth and lower output line losses can be achieved. Generally, the input and output impedance of FETs are more capacitive than the ones of BJTs. Thus, FETs are slightly better suited for TWAs. However, the high g_m of BJTs may help to compensate the additional line losses. For detailed information about TWAs, the reader is referred to the literature [Won93, Bey84, Aya82] and to the design example presented in Sect. 8.3.1.

7.4.7 Comparison of Wideband Amplifiers

A qualitative comparison of the different wideband amplifier approaches is summarised in Table 7.3. Input and output impedance matching is considered for low-ohmic source and loads with values of around 50 Ω.

Table 7.3. Comparison of wideband amplifier approaches

Topology	Band width	Impedance matching input/output	Stability	Gain	Noise	RF output power	Supply power	IC size /costs
Resistive Matching	Moderate	Tradeoff with gain and noise	Very high	Low	High (determined by input resistances)	Limited by resistor losses at output	Moderate	Very compact
Active matching (common gate and common drain)	High	Good/good	Can be critical for non-differential topologies due to undesired feedbacks in grounds	High	Low to moderate (no noise matching)	Limited due to fixed w	High	Compact
Common source with inductive source degeneration	Moderate to low	Very good/bad	High	Moderate to high	Very low	Limited, increased source impedance reduces current	Moderate to low	Large
Common source with resistive drain gate feedback	Moderate to low	Good/very good	High	Moderate to high	Moderate to high	Moderate	Moderate to low	Very compact
TIS/TAS	High	Good/good	Good	Moderate	Moderate to high	Moderate	Moderate to high	Very compact
Balanced amplifier	Moderate	Good, depends on coupler bandwidth	Good, reflections cancelled, absorbed in coupler	Amplifier gain − 2× coupler loss	Good, amplifier +coupler loss)	Good 2× single amplifier − coupler loss	Moderate	Large to very large
Cascode	Moderate	Needs matching	Can be critical	Very high	Very low	Moderate	Very low	Large
Travelling wave	Very high	Good/good	High, reflected signals absorbed in termination resistors	Low, gain adding instead of multiplication	Good if high n reducing impact of input line termination resistor	Good, current adding but losses in output termination resistor	Very high	Very large

7.5 Tutorials

1. What are the basic FET and bipolar amplifier topologies? What does the common port represent?
2. Draw the simplified equivalent circuit of the common-source, -gate, -drain, -emitter, -base and -collector topology.
3. Derive the equivalent circuit of the common gate circuit from the common source circuit.
4. Discuss and compare the input/output impedances and the voltage/current/power gains of the circuits. What are the pros and cons?
5. Discuss the impact of source degeneration and drain source feedback with respect to gain, bandwidth, port impedances and stability considering inductive, capacitive and resistive elements.
6. How can we realise an oscillator?
7. Derive the impact of a series inductor and capacitor in the source of a common source circuit and in the gate of a common gate circuit. Which conclusions can be drawn?
8. What limits the exploitation of the high low-frequency gain of the common source stage in high frequency applications? What is the Miller effect? Calculate the Miller capacitance for the common source circuit. What about the Miller effect exhibited by the other two basic configurations?
9. Which circuit topology with high RF power gain can mitigate the Miller effect? What is the basic idea? Calculate the corresponding Miller capacitance.
10. Explain why cascode circuits have a higher DC and RF gain than common source stages. What is the benefit for aggressive gate length scaling regarding the output losses?
11. Summarise and explain similarities and differences of FETs and BJTs considering the equivalent input circuit network and impedance, DC input current, g_m and f_t. Regarding the matching issue distinguish between a small transistor required for LNAs and a large transistor applied in power amplifiers.
12. Which parameter is very sensitive to process variations? Explain how DC feedback can be used to compensate this effect. Which device is more sensitive to process variations – the BJT or the FET?
13. Compare depletion and enhancement FETs with respect to the DC properties and the fabrication.
14. Why is the biasing of transistors important? Identify the bias regions in the IC curves, which give optimum performance in terms of gain, NF, linearity, power and efficiency. Explain qualitatively, why the corresponding properties can be expected. Which bias point provides a good tradeoff for all needs?

15. How can we apply a bias to transistors? Why do we need a high RF and low DC impedance for the feeding elements? Why can we use a high-ohmic resistor for the FET gate and not for the drain? Which elements/element values are better suited for the drain bias with high current? What are the drawbacks?

16. At which node would you connect the output bias inductor in an LNA? Directly at the drain or after the matching network? We assume that the matching network passes DC. Design goals are minimum losses and minimum size of the inductor. The low power consuming LNA is matched to 50 Ω.

17. Explain bias topologies, which allow a single supply bias for depletion and enhancement FETs. Draw their schematics.

18. How can we stabilise a circuit? Discuss the advantages and disadvantages of source degeneration, drain gate feedback and resistive loading. If you would stabilise a FET featuring a small gate width, would you prefer series or parallel loading? Why? Which benefits does the inductively degenerated common source stage offer in terms of stability and input matching?

19. Why do we need broadband circuits? Which wideband amplifier concepts do you know? Explain the pros and cons regarding bandwidth, gain, noise, output power and stability. How can we compensate for the frequency dependent gain of transistors?

20. Explain the functional principle of a FET TWA. Why is the achievable frequency bandwidth high? What determines the cutoff frequency? What is the bandwidth-limiting element in the equivalent transistor circuit? Why is the gain relatively low although there are many stages? What determines the line losses? Why are the return losses high? What limits the noise and the output power with respect to the line terminations? Which basic amplifier topology would you use to get the highest gain bandwidth product? Why? Why is there an optimum number of stages? Would you choose bipolar transistors or FETs for a TWA? Discuss size, cost and power consumption issues.

21. To minimise the size and costs you should design a broadband FET amplifier without any matching elements. Explain the relevant concepts. What are the drawbacks?

References

[Aga98] B. Agarwal, A. E. Schmitz, J. J. Brown, M. Matloubian, M. G. Case, M. Le, M. Lui and M. J. W. Rodwell, "112-GHz, 157 GHz, and 180-GHz InP HEMT traveling-wave amplifiers", IEEE Transactions on Microwave Theory and Techniques, Vol. 46, No. 12, pp. 2553–2559, Dec. 1998.

[Aya82] Y. Ayasli, R. L. Mozzi, J. L. Vorhaus, L. D. Reynolds, R. A. Pucel, "A monolithic GaAs 1–13 GHz travelling wave amplifier", IEEE Transactions on Microwave Theory and Techniques, Vol. MTT-30, No. 7, 12, pp. 976–981, July 1982.

[Bar04] D. Barras, F. Ellinger, H. Jäckel and W. Hirt, "Low supply voltage SiGe LNA for ultra-wideband frontends", IEEE Microwave and Wireless Component Letters, Vol. 14, No. 10, pp. 469–471, Oct. 2004.

[Bey84] J. B. Beyer, S. N. Prasad, R. C. Becker, J. E. Nordman, G. K. Hohenwarter, "MESFET distributed amplifier design guidelines", IEEE Transactions on Microwave Theory and Techniques, Vol. MTT-32, No. 3, 12, pp. 268–275, March 1984.

[Che00] E. M. Cherry, "Feedback amplifier configurations", IEE Proceedings on Circuits Devices and Systems, Vol. 147, No. 6, pp. 334–346, Dec. 2000.

[Che96] E. M. Cherry, "Input impedance and output impedance of feedback amplifiers", IEE Proceedings on Circuits Devices and Systems, Vol. 143, No. 4, pp. 334–346, Aug. 1996.

[Che63] E. M. Cherry, D. E. Hooper, "The design of wide-band transistor feedback amplifiers", Processing IEE, Vol. 110, No. 2, Feb. 1963.

[Gar99] M. Garcia, C. Fager, "A design equation for the load resistors that ensure stability for potentially unstable transistors", IEEE/SBMO International Microwave and Optoelectronic Conference, pp. 589-591, 1999

[Jia05] J. Gao, G. Böck, "Relationships between common source, common gate and common drain FETs", IEEE Transactions on Microwave Theory and Techniques, Vol. 53, No. 12, pp. 3825–3831, Dec. 2005.

[Leh85] R. E. Lehmann, D. D. Heston "X-band monolithic series feedback LNA", Transaction on Microwave Theory and Techniques, pp. 1560–1566, 1985.

[Rob91] I. D. Robertson, A. H. Aghvami, "Ultrawideband biasing of MMIC distributed amplifiers using improved active load" Electronic Letters, Vol. 27, No. 21, pp. 1907–1909, 1991.

[Won93] T. T. Y. Wong, Fundamentals of Distributed Amplification, Artech House, Norwood, 1993.

8 Low Noise Amplifiers

The chance knocks on the door more often than expected, however,
in most cases nobody is home.
Will Rogers

Based on the knowledge of the preceding chapters, we are now well prepared to design circuits. Let us begin with LNAs (Low Noise Amplifiers), which are important components in receivers as, e.g. illustrated in Fig. 2.2.

8.1 Design Strategies

The typical design procedure for an LNA is illustrated in Fig. 8.1. First, we have to define reasonable specifications, which may be predefined by standards. The LNA is the first gain stage. According to the equation of Friis, the LNA NF directly adds to the system NF. Thus, minimum noise is one of the most important design requirements for LNAs. Recall that the LNA decreases the noise contribution of the following stages on the condition that the gain of the LNA is high. This necessitates sufficient gain. To provide enough dynamic range, the compression and intermodulation performances must be adequate. These large signal measures can be represented by P_{1dB} and IIP3. For mobile applications, the power consumption has to be as low as possible to maximise the stand-by and operation time. In many systems, the LNA has to be switched on even in stand-by mode, since the received signal must be high enough to activate a wake up circuit, which turns the whole receiver on. It is clear that the need for low power consumption degrades the large signal characteristics since the saturation occurs sooner. In addition, we have to account for reduced bandwidth at given gain and noise figure. These properties are attributed to the strong resonances required for impedance matching of low-power devices featuring thin channel widths and subsequently very high impedances. We can conclude that optimum LNA design involves a compromise between many performance measures, which may cause headache. The following figure of merit is frequently used for circuit optimisations and comparisons:

$$\text{FOM} = \frac{f_c[\text{GHz}]}{f_t[\text{GHz}]} \cdot \frac{S_{21}[\text{dB}]}{\text{NF}[\text{dB}]} \cdot \frac{\text{IIP3}[\text{dBm}]}{P_{dc}[\text{dBm}]}. \qquad (8.1)$$

The first term in Eq. (8.1) describes the centre operation frequency f_c with respect to the f_t of the used technology, the second factor considers the gain to noise figure ratio, and the third term relates the large signal properties characterised by IIP3 with the consumed DC power P_{dc}. Several modifications of Eq. (8.1) can be found in the literature. Optionally, the representation of the large signal properties by means of P_{1dB} may be reasonable as well. For wideband LNAs, f_c may be substituted by the bandwidth.

Based on the technical specifications at one hand and the costs at the other hand we have to choose the optimum technology. In mass fabrication, silicon based technologies may exhibit the lowest costs. However, if we target a prototype design, III/V technologies may be an attractive option. In any case, the properties in terms of speed and large signal performance are usually superior for III/V technologies. On basis of device measures such as F_{min} and MAG extrapolated from f_{max}, and the DC power consumption, we can accomplish first feasibility studies regarding the possible NF, gain and large signal properties, respectively.

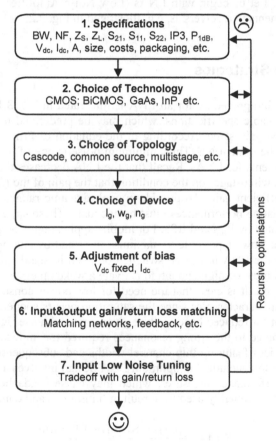

Fig. 8.1. LNA design flow

Then, with given specifications and IC technology we have to choose a suitable circuit topology, the devices and the bias. To reach optimum gain and NF performance, the smallest gate length of the available technology is usually employed. The gate width and the DC current are a tradeoff between current consumption on one hand, and simple impedance matching and bandwidth on the other hand. The supply voltage is defined by the system DC voltage.

Now, we try to get close to the MAG by gain/return loss impedance matching of the input and the output. Hand calculations, or more conveniently, Smith chart evaluations or CAD simulations can be used for this task. To minimise the losses of the passive components, the number of elements has to be kept as low as possible. If feasible, the series DC block capacitors and the shunted bias inductors can be reused for impedance matching. Element values should be chosen, which provide maximum Q at the operation bandwidth. Gain/return loss matching does not generally implicate minimum noise figure. Thus, in a second step, we have to tune the element values for low noise. If the source impedances for maximum gain/return loss and minimum noise are far away, which is rarely the case, the matching topology has to be modified. Luckily, in practice, we see that a good compromise between noise matching and gain matching can be obtained without exorbitant degradation of the other parameter. Unfortunately, the different design steps depend on each other. That means that recursive optimisations are required. In the following sections, we present the design of two narrowband and two wideband LNAs implemented in different technologies.

8.2 Narrow-Band Design

The design of narrow-band LNAs is studied by means of two representative examples.

8.2.1 Simple 30–40 GHz CMOS LNA

We start with the study of an LNA for WLAN applications at 30–40 GHz fabricated in 90 nm SOI CMOS technology [Ell15]. Information about this technology can be found throughout this book, e.g. in Sect. 5.4.6.4. Since the circuit is employed for non-mobile base stations, the DC power consumption is not important. In Fig. 8.2, the simplified circuit schematics of the LNA are depicted. In Chap. 7 we have demonstrated that the cascode circuit is superior for operation at highest frequencies. Advantages are the reduced Miller effect and the increased output resistance. Thus, a cascode circuit is used.

A BSIM (Berkeley Simulation Model) large signal model [Ber04] is applied for the transistor simulations. The passive elements are modelled by LRC equivalent networks as elaborated in Chap. 6 and illustrated in the circuit schematics. For simplicity, these parasitics will not be shown for the circuits presented in further sections. However, the relevant parasitics must always be taken into account for simulations. The tasks and details of each circuit element are summarised in Table 8.1.

To minimise the losses associated with the matching elements, the DC bias elements are reused for impedance matching. This is possible by adjusting the gate width and the bias of the transistors. It is clear that this design strategy is advantageous for low power consumption; however, for this circuit, power consumption was not an important design issue. By means of simple equivalent circuits as shown in Fig. 8.3, we can approximate the input and output impedances of the transformed transistor impedances. For the input impedance we get

$$Z_{in} = \frac{1}{j\omega C_1} + \cfrac{1}{\cfrac{1}{j\omega L_1} + \cfrac{1}{R_{gs} + \cfrac{1}{j\omega C_{in}}}}, \qquad (8.2)$$

where, according to Eq. (7.59), C_{in} can be approximated by $C_{in} = C_{gs} + 2\,C_{gd}$. The values of the free matching parameters C_1 and L_1 have to be chosen in a way that Z_{in} equals the impedance of the termination. The 50 Ω matching demands for $\mathrm{Re}\{Z_{in}\} = 50\ \Omega$ and $\mathrm{Im}\{Z_{in}\} = 0$.

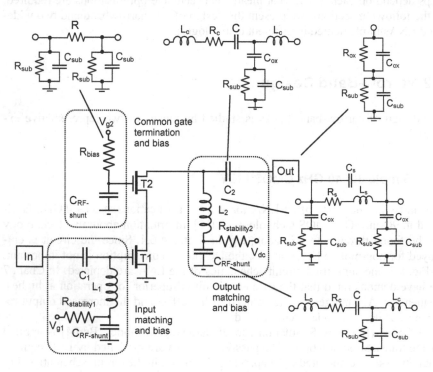

Fig. 8.2. Circuit schematics of 30–40 GHz CMOS cascode LNA, compare [Ell15] © IEEE 2004

Table 8.1. Discussion of circuit elements

Parameter	Discussion
90 nm SOI CMOS	Can be integrated with baseband, system on a chip, low costs in mass-fabrication, high costs for prototyping, SOI allows reduced parasitics because of high substrate isolation, 90 nm allows sufficient gain at target frequency around 40 GHz
Cascode	High power gain (high output resistance) at high RF frequencies (low input capacitance)
T1,2: w=64 μm	Simple and broadband impedance matching possible by reusing bias elements
Bias voltage V_{dc}=2.4 V	0.4 V voltage drop across stability resistors, V_{g1}=0.5 V, V_{g2}=1.5 V, thus each transistor has V_{ds}=1 V, saturation, higher voltage would cause reliability problems
Bias current I_{dc}=17 mA	$I_d{\approx}0.5I_{max}$, saturation, class-A, high gain, good linearity, moderate/low noise
L_1=0.3 nH	Input bias feeding and 50 Ω gain/return loss/noise impedance matching together with C_1
C_1=270 fF	Input DC blocking (DC independent from interface circuits) and 50 Ω gain/return loss/noise impedance matching together with L_1, stability at low frequencies (high-pass effect)
L_2=0.25 nH	Output bias feeding and 50 Ω gain/return loss impedance matching together with C_2
C_2=80 fF	Output DC blocking and 50 Ω gain/return loss impedance matching together with L_2, stability at low frequencies
$C_{RF\text{-}shunt}$ =15 pF	Definition of RF ground, shunts undesired RF signals from DC supply
R_{bias}=5 kΩ	RF isolating feeding of V_{g2}, very low DC gate current
$R_{stability1}$=50 Ω	Prevents resonances associated with the unknown supply connection at the input
$R_{stability2}$=17 Ω	Prevents resonances associated with the unknown supply connection at the output

In similar fashion, we can proceed with the output matching. In first order approximation, the output impedance can be described by

$$\underline{Z}_{out} = \frac{1}{j\omega C_2} + \cfrac{1}{\cfrac{1}{j\omega L_2} + \cfrac{1}{\underline{Z}_{out,CC}}} \tag{8.3}$$

where $\underline{Z}_{out,CC}$ denotes the output impedance of the cascode circuit. Referring to Eq. (7.60), we may approximate $\underline{Z}_{out,CC}$ by $g_m \cdot r_o^2$. Even with these simple equivalent models, we get relatively complex terms when separating into the real and imaginary terms. A much faster and more convenient approach is the matching in the Smith chart. In this context, we learn to appreciate the Smith chart. The corresponding matching of the cascode circuit is explained in Fig. 8.4a. By means of the matching curves and a detailed Smith chart as shown in Fig. 3.4, we can easily read out the required values of the matching elements. The matching centre frequency is 35 GHz.

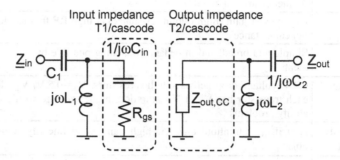

Fig. 8.3. First order equivalent networks for estimation of port impedances required for impedance matching

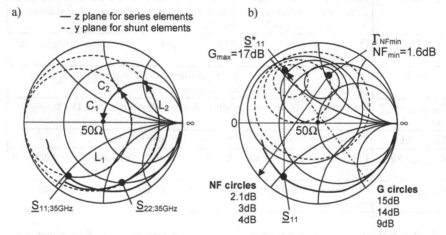

Fig. 8.4a,b. Impedance matching: **a** maximum gain/return loss; **b** source impedances for maximum gain/return loss and minimum noise of input, Γ_{NFmin} : source reflection coefficient where NF_{min} is achieved

According to Fig. 8.4b, and as expected, the optimum source impedances for maximum gain and noise of the transistor are not equal. Optimisation for maximum gain amounting to 17 dB would result in an NF of 2.2 dB, whereas matching for NF_{min} amounting to 1.6 dB would yield a lower gain of around 14 dB. In these values, the impact of the parasitics inherent in the matching networks is not incorporated yet. Including the pads, we have to take into account losses in the order of 1 dB at the input and the output leading to a NF degradation of 1 dB and a gain reduction of 2 dB. For this design, a compromise between maximum gain and NF was aimed for. Accordingly, fine-tuning was performed for the element values of the input matching network. Transmission lines are used for the realisation of the inductors, which require small inductive values. Due to the high operation frequency with respect to f_t and the corresponding moderate S_{21}, stability is relatively easy to achieve. However, for stability reasons at lower frequencies and to achieve a Rollet's factor K>1 irrespective of frequency, resistors are incorporated in the bias connections. They avoid resonances between the RF shunt capacitance and potential inductance of the DC probe tips. The gate of the common gate circuit is terminated by a RF shunt capacitor and biased by a high-ohmic resistor.

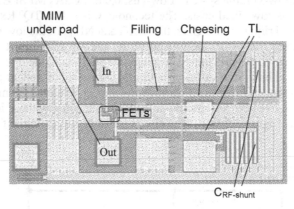

Fig. 8.5. Photograph of MMIC with overall chip size of 0.6×0.3 mm^2, compare [Ell15] © IEEE 2004

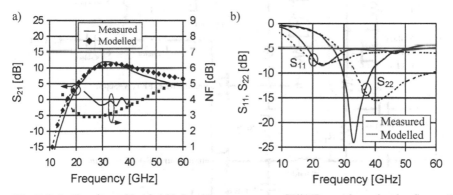

Fig. 8.6a,b. Results at V_{dc}=2.4 V, I_{dc}=17 mA, compare [Ell15]: **a** gain and noise figure; **b** return losses, compare [Ell15] © IEEE 2004

A photograph of the compact MMIC is shown in Fig. 8.5. To minimise parasitic connections, the circuit has been characterised on-wafer employing measurement tips. In Fig. 8.6a, the gain and noise figure versus frequency are shown. The corresponding return losses are illustrated in Fig. 8.6b. At a frequency of 35 GHz, a gain of 11.9 dB, a low noise figure of 3.6 dB, an input return loss of 6 dB and an output return loss of 18 dB are measured. Between the 3 dB frequency bandwidth ranging from 26 to 42 GHz, output compression points above 4 dBm are achieved.

8.2.2 Source Degenerated 4.2-5.7 GHz GaAs MESFET LNA

An LNA for wireless LAN applications at 4.2–5.7 GHz is presented in this section [Ell47]. Since the circuit is intended for mobile applications, the minimisation of the power consumption has been a major design goal. Furthermore, the circuit is targeted for a low-volume system. Low-cost 0.6 µm GaAs MESFET technology is used, which has low fixed costs. The technology is called TQTRx and is offered by the foundry Triquint. At 5 GHz, the MAG and NF_{min} of the low power consuming enhancement transistors are around 11 dB and 1.6 dB, respectively. To achieve high gain, as for the previously presented circuit, a cascode circuit is employed. The cascode yields a MAG of nearly 18 dB. Due to the excellent performance of the GaAs substrate in terms of passive devices, a relatively complex matching network can be used for the reactive impedance matching without introducing significant losses. Figure 8.7 shows the circuit schematic of the circuit.

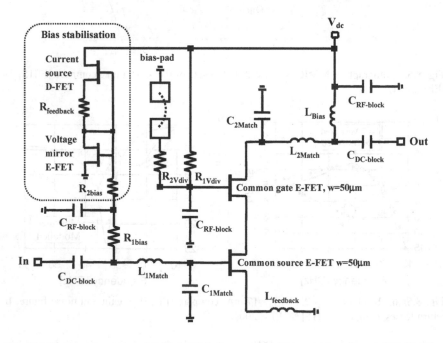

Fig. 8.7. Circuit schematic of LNA MMIC, for supply voltages higher than 1.2 V, the bias pad can be connected to ground to activate a voltage divider optimising the bias, compare [Ell47] © IEEE 2000

Since a gain of 12 dB is sufficient according to the specifications from our customer for this circuit, we have some margin. Thus, we can apply the source degeneration technique, which simplifies impedance matching at the expense of lower gain. This approach is well suited for low to moderate frequency LNAs, where gain is not the limiting factor. Source degeneration means that we include a series feedback inductance $L_{feedback}$ into the source of the input transistor. In correspondence to Sect. 7.1.2, we can approximate the real part of the input impedance by

$$\text{Re}\{\underline{Z}_{in}\} \approx \frac{g_m \cdot L_{feedback}}{C_{gs}}, \tag{8.4}$$

With given values of g_m and C_{gs}, we can obtain our desired interface impedance of 50 Ω by setting $L_{feedback}$ accordingly. In a second step, we have to compensate the imaginary part of $\underline{Z}_{in}$ with an input matching inductance L_{1Match}. The corresponding resonance frequency can be approximated by

$$\omega_m^2 \approx \frac{1}{\left(L_{feedback} + L_{1Match}\right) \cdot C_{gs}}. \tag{8.5}$$

In addition to simple impedance matching, the source degeneration has one additional advantage for LNAs. The impedances for minimum noise figure and maximum gain move closer together, thereby increasing the overall performance [Leh85]. It has been found that $\underline{S}_{11}^*$ varies with both the output load and the series feedback, whereas the source reflection coefficient for minimum noise figure Γ_{NFmin} is mainly influenced by the series feedback. Thus, it is possible to find a combination of the feedback impedance and the output load, which decreases the distance between $\underline{S}_{11}^*$ and Γ_{NFmin}.

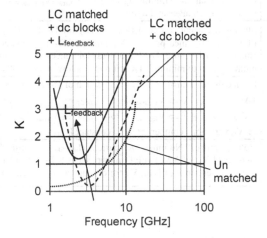

Fig. 8.8. Stability of LNA represented by Rollet's factor K for different conditions

Table 8.2. Discussion of circuit elements

Parameter	Discussion
0.6-μm GaAs MESFET	Well suited for low-cost low-volume applications with moderate frequency around 5 GHz. Excellent passive devices due to high substrate resistivity
Cascode	High power gain (high output resistance) at RF frequencies (low input capacitance)
E-FET	Very good gain per current consumption ratio
w_g=50 μm	Tradeoff: low DC current ($I \sim w_g$) and acceptable impedance matching ($Z \sim w_g^{-1}$).
Supply voltage	High performance mode (bias pad shorted): V_{dc}=2.7 V, V_{gs}=0.4 V ($I_d \approx 0.5 I_{max}$: class A), V_{g2}=1.8 V, V_{dss}=1.4 V (common source in saturation), V_{dsg}=1.3 V (common gate in saturation) Low power consumption mode (bias pad open): V_{dc}=1 V, V_{gs}=0.4 V ($I_d \approx 0.5 I_{max}$: class A), V_{g2}=1 V, V_{dss}=0.6 V (common source in saturation), V_{dsg}=0.4 V (common gate almost in saturation)
Supply current I_{dc}=1.2..1.3 mA	$I_d \approx 0.5 I_{max}$, saturation. High gain, good linearity, moderate/low noise
L_{1Match}=4.5 nH C_{1Match}=50 fF	Input matching. High input impedance transformed to lower impedance to reach low NF and good impedance matching at 50 Ω
L_{2Match}=5 nH C_{2Match}=74 fF	Output matching. High output impedance transformed to lower impedance to reach high gain and good impedance matching at 50 Ω
$L_{feedback}$= 0.5 nH	Guarantees unconditional stability (K>1), makes input more resistive to simplify input impedance matching
$C_{DC-block}$=5 pF	DC blocking, stability at low frequencies (highpass effect)
$C_{RF-block}$=7 pF	RF shunting, definition of RF ground
L_{Bias}=8 nH	RF isolation. Fed at point having lowest impedance, thus minimising required bias element value, parasitics and size
R_{1bias}=5 kΩ R_{2bias}=10 kΩ	RF isolation. FET gate current very small, high-ohmic resistors
R_{1Vdiv}=3 kΩ R_{2Vdiv}=6 kΩ	RF isolation, DC voltage divider
Bias circuit	Bias stabilisation. Minimising of $\Delta V_{th}=V_{th,real}-V_{th,nominal}$ due to process tolerance, temperature changes and aging
CS D-FET	Stabilised current source. Effect of ΔV_{th} for D-FETs (V_{th}<0) smaller than for E-FETs (V_{th}>0), since possible V_{gs} swing of D-FETs higher and effect of $\Delta V_{th}/V_{gs}$ smaller. Furthermore, D-FET process tolerances smaller since $a_{D-FET}>a_{E-FET}$. Minimum w_g used to minimise DC current to a low value of 60 μA
$R_{feedback}$=900 Ω	DC feedback, adjustment and stabilisation of D-FET current $+\Delta V_{th} \Rightarrow I_{d,DFET} \downarrow \Rightarrow V_{feedback} \downarrow \Rightarrow V_{gs,DFET} = - V_{feedback} \uparrow \Leftrightarrow I_{d,DFET} \uparrow$ $-\Delta V_{th} \Rightarrow I_{d,DFET} \uparrow \Rightarrow V_{feedback} \uparrow \Rightarrow V_{gs,DFET} = - V_{feedback} \downarrow \Leftrightarrow I_{d,DFET} \downarrow$ $\Rightarrow V_{th}$ variations compensated $\Rightarrow$ constant $I_{d,DFET}$
VM E-FET	Voltage mirror. Nominal DC current ($I_{max,EFET}/2$) defined by stabilised D-FET. Generated V_{gs} of VM E-FET mirrored to CS E-FET (identical E-FETs except of w_g) $\Rightarrow$ dependence of I_d on V_{th} for CS and cascode circuit strongly reduced

Together with increased bandwidth and better return loss matching, the series feedback improves stability as illustrated in Fig. 8.8. We can see that the un-matched cascode is only conditionally stable for frequencies below 6 GHz. Matching of the circuits worsens the stability around the matching since the gain is raised. Because of the lowpass effect of the matching networks and the highpass properties of the DC block capacitor, the stability is improved towards higher and lower frequencies, respectively. Unfortunately, according to Sect. 7.1.2, series feedback decreases the effective g_m and the gain. Therefore, the level of source degeneration should be carefully dosed. The output is reactively matched for high gain. Fine-tuning of the elements is performed by CAD software with large signal MESFET models [Ell53] and equivalent circuits of the passive devices. The elements are discussed in Table 8.2.

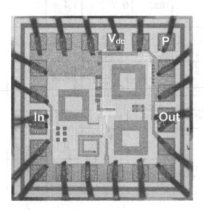

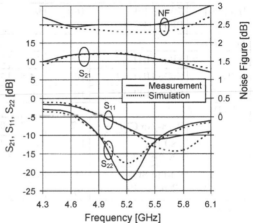

Fig. 8.9. Photo of LNA MMIC, total chip size is 1×1 mm^2. All non-marked pads are ground pads, compare [Ell47] © IEEE 2000

Fig. 8.10. Gain, noise figure and return losses of LNA in operation mode with ultra low power consumption, $V_{dc}=1$ V, $I_{dc}=1.2$ mA, compare [Ell47] © IEEE 2000

A photograph of the wire-bonded IC is depicted in Fig. 8.9. For measurements, the chip was mounted on a test substrate. Figure 8.10 shows the gain, noise figure and return losses at a supply voltage of only 1 V drawing a low current of 1.2 mA. At 5.2 GHz, the gain is 12.3 dB and the noise figure is 2.4 dB. The return loss of the noise-matched input is 7 dB and the output return loss is 14 dB. The 1 dB output compression point is approximately −8.3 dBm. A third order output intercept point of -0.8 dBm was measured. At the standard WLAN supply voltage of 2.7 V, a lower noise figure of 2 dB and a slightly higher gain of 13.1 dB were measured at a current of 1.3 mA. The 1 dB compression point is approximately 5 dB higher than for the low-power mode.

8.2.3 Comparison of Narrowband LNAs

A comparison with other reported LNAs at microwave and millimetre wave frequencies is presented in Table 8.3.

Table 8.3. Comparison of state-of-the-art LNAs

Ref	Technology/ f_t [GHz]	f_{center} [GHz]	S_{21} [dB]	NF [dB]	$P_{1dB,out}$ [dBm]	OIP3 [dBm]	V_{dc} [V]	I_{dc} [mA]
III/V								
[Ell47] – Sect. 8.2.2	0.6-µm GaAs MESFET/18	5.2	12.3	2.4	−8.3	−0.8	1	1.2
			13.1	2	−3.4	5.9	2.7	1.3
[Mim00]	150-nm InP PHEMT/n.a	26.5	14.5	1.7	n.a.	0	n.a.	n.a
[Kok99]	70-nm InP PHEMT/500	160	9	6	n.a.	0	1.4	33
SiGe HBT								
[Ell34]	SiGe HBT/47	5.2	12.7	2.4	n.a	0	1.2	1
[Ell17]	SiGe HBT/47	16	14.5	3.8	n.a.	n.a.	1.5	1.5
[Böc02]	SiGe HBT/155	19	26	2.2	n.a.	n.a.	3	8.7
CMOS								
[Cas03]	0.18-µm CMOS	5.75	14.2	0.9	n.a.	15	1V	16
[Yan01]	180-nm CMOS/50	16	9	4	n.a.	n.a.	3	15.5
[Ell15] – Sect. 8.2.1	90-nm SOI CMOS/149	35	11.9	3.6	4	n.a.	2.4	17
		40	9.5	4				

8.3 Wideband Design

Two wideband LNAs are treated in this section serving as fruitful design examples.

8.3.1 SiGe Common Base/Collector LNA with 7.8-GHz Bandwidth

In this section, a 7.8-GHz broadband LNA is presented, which has been fabricated in commercial IBM 6HP SiGe BiCMOS technology [Ell37]. The HBTs yield an f_t of up to 47 GHz and an NF_{min} of approximately 1.2 dB around 5 GHz [Ong99]. At the latter frequency, the process features inductors with quality factors up to 19. Figures 8.11 and 8.12 show the schematic and the chip photograph of the actively matched circuit. A common base input and a common collector output stage are applied for active matching to 50 Ω. The impedances can be approximated by

$$Z_{in} = \frac{1}{g_{m,CB}} \| R_1 \tag{8.6}$$

and

$$Z_{out} = \frac{1}{g_{m,CC}} \| R_3 , \tag{8.7}$$

where $g_{m,CB}$ and $g_{m,CC}$ are the transconductance of the CB (Common Base) and the CC (Common Collector) stage, respectively. The transconductance values are matched by suitable choices of the transistor sizes and the bias. The CB transistor has an emitter size of $10 \times 0.32 \ \mu m^2$, operates in class AB and draws a supply current of 1.8 mA at the applied DC voltage of -1.8 V. The CC device has an emitter size of $5 \times 0.32 \ \mu m^2$, operates in class A and consumes a supply current of 1.7 mA. No band limiting LC resonance or feedback is required for impedance matching. A further advantage is that also no area consuming inductors are necessarily needed for the impedance matching allowing a very compact layout. The two stages are DC coupled to save DC block capacitors, which would limit the bandwidth towards lower frequencies. The resistors R_1, R_2 and R_3 feed the bias. R_1 and R_3 generate a DC feedback for the CB and CC stage, respectively, thereby lowering the influence of supply voltage, temperature and process variations. Their values are a tradeoff between minimum DC voltage drop and maximum signal gain. At the bias port, a large capacitance is used as an RF shunt. The circuit operates with a single voltage supply.

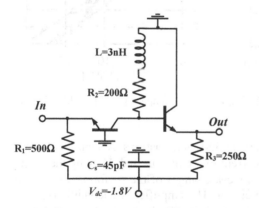
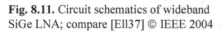
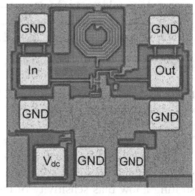

Fig. 8.11. Circuit schematics of wideband SiGe LNA; compare [Ell37] © IEEE 2004

Fig. 8.12. MMIC photograph, overall chip size is only 0.67 mm × 0.67 mm; compare [Ell37] © IEEE 2004

The CC stage acts as an output buffer and exhibits a very high input impedance. Thus, the low frequency voltage gain of the CB stage is approximately given by

$$a_v = g_{m,CB} \cdot R_2 , \tag{8.8}$$

where R_2 is the load resistance. The voltage gain of the CB stage determines the total voltage and power gain of the circuit. Inductive peaking is employed to compensate the gain drop towards high frequencies, which is mainly caused by the capacitive parasitics of the transistors and the signal pads. The frequency response can be optimised by suitable choice of the values of R_2 and the peaking inductor L. Figure 8.13a,b shows the influence of R_2 and L, respectively. High values of R_2 increase the low frequency gain but decrease the high frequency gain. Low values of R_2 increase the effect of L as well as the corresponding resonance peak. Large values of L increase the maximum gain and move the resonance peak towards low frequencies. The main design goal of this work was to achieve a flat frequency response. Accordingly, values of R_2=200 Ω and L=3 nH were chosen because they provide a good tradeoff between high gain and flat frequency response. However, for narrowband applications higher values of L and lower values of R_2 would be superior. With R_2=150 Ω and L=4.5 nH a flat gain of approximately 14.4±0.2 dB was simulated from 5–6 GHz indicating the excellent suitability of the circuit for wireless systems operating in accordance to the 802.11a or high performance radio local area network (HIPERLAN) standards.

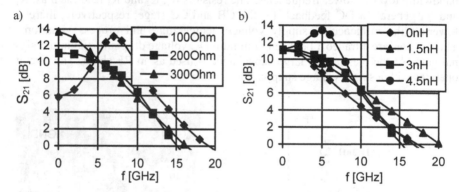

Fig. 8.13a,b. Simulated gain at V_{dc}=-1.8 V, I_{dc}=3.6 mA, compare [Ell37] © IEEE 2004: **a** vs R_2 with L=3 nH; **b** vs L with R_2=200 Ω

The circuit was measured on-wafer and at 50-Ω terminations. A supply voltage of −1.8 V and a corresponding supply current of 3.6 mA are fed. Varying the supply voltage from −1.7 to −2.2 V result in weak changes of the performance, demonstrating a low bias sensitivity of the circuit. The amplifier is unconditionally stable. In Fig. 8.14a, the gain and the noise figure are shown versus frequency. A gain of 10.6 dB and a −3 dB frequency of 7.8 GHz were measured. The measured gain cutoff frequency is 17 GHz. Up to 10 GHz, the noise figure is below 4.4 dB. The return losses versus frequency are shown in Fig. 8.14b. Up to 20 GHz the input and output return losses are above 7.8 dB and 9 dB, respectively. The circuit has an input compression point of −12 dBm.

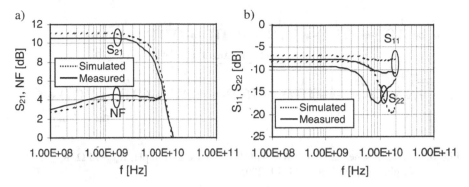

Fig. 8.14a,b. Measured results at V_{dc}=-1.8 V, I_{dc}=3.6 mA, compare [Ell37] © IEEE 2004; **a** gain and noise; **b** return losses

8.3.2 60-GHz CMOS Broadband Travelling Wave Amplifier

As outlined in Sect. 7.4.6, large bandwidth can be obtained with travelling wave amplifiers. The band-limiting capacitive parasitics are incorporated into artificial transmission lines. A 60-GHz low noise CMOS TWA (Travelling Wave Amplifier) is presented now [Ell7], which has been implemented in the same 90-nm SOI technology as the LNA presented in Sect. 8.2.1.

8.3.2.1 Design

Figure 8.15 depicts the schematics of the TWA employing cascode stages as plotted in Fig. 8.16. The input signal travels down the input line and feeds each amplifier stage. Signal reflections are absorbed in the termination resistors R_{abg} and R_{abd}. Given that the phases of the input and output lines are equal, the amplified signals are constructively added at the output line. This is, e.g. the case if the values for the inductances L and the capacitances C of the input and output lines are equal. The input capacitance is generally higher than the output capacitance. An additional shunt capacitance can be added at the output of the amplifier stages to obtain equal capacitances and phase conditions. To achieve a good tradeoff in terms of gain, noise, larges signal performance and power consumption, a typical class-A bias point with V_{gs}=0.5 V and V_{ds}=1 V is applied for the transistors. For the first design iteration, the feedback S_{12} from the input to the output of the amplifiers is neglected. Thus, the capacitance of the distributed lines is determined by the input capacitance C_{gs} of the amplifier. The characteristic impedance and the 3 dB cut-off frequency of a distributed input line section can be approximated by $Z_w = \sqrt{\dfrac{L}{C_{gs}}}$

and $f_c = \dfrac{1}{\pi\sqrt{LC_{gs}}}$ with L as the line inductance.

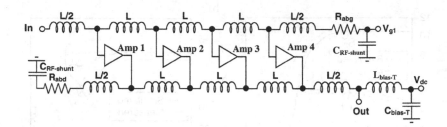

Fig. 8.15. Circuit schematics of TWA with four stages, L=170 pH, R_{abg}=75 Ω, R_{abd}=50 Ω, $C_{RF\text{-}shunt}$=25 pF, compare [Ell7] © IEEE 2005

Gain up to 60 GHz has been one major design goal of this circuit. Consequently, the f_c of the transmission line sections has to be above this frequency. With an f_c of 70 GHz and a Z_0 of 50 Ω, we obtain first estimations for C_{gs} and L with values of 90 fF and 225 pH. The value of C_{gs} corresponds a w_g of 64 μm. The impact of the parameters will be investigated systematically by simulations.

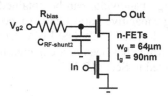

Fig. 8.16. Simplified circuit schematics of amplifier stage, $C_{RF\text{-}shunt2}$=5 pF, R_{bias}=6 kΩ, compare [Ell7] © IEEE 2005

a. Power Gain

The power gain of TWAs is limited by the losses of the gate and drain lines, which include the parasitics of the active devices. It has been shown that the losses of FET based TWAs are mainly determined by the gate line losses [Ava82]. This is especially the case for TWAs using cascode amplifier stages providing high output resistances. If the losses of the drain line and the inductors are neglected, the small signal power gain can be approximated by

$$G = G_0(1 - nA_g)^2 \tag{8.9}$$

with the low frequency gain

$$G_0 = \left(\frac{n \cdot g_m \cdot Z_0}{2}\right)^2 \tag{8.10}$$

and the gate line loss factor

$$A_g = \frac{1}{4}R_g\omega^2 C_{gs}^2 Z_0 . \tag{8.11}$$

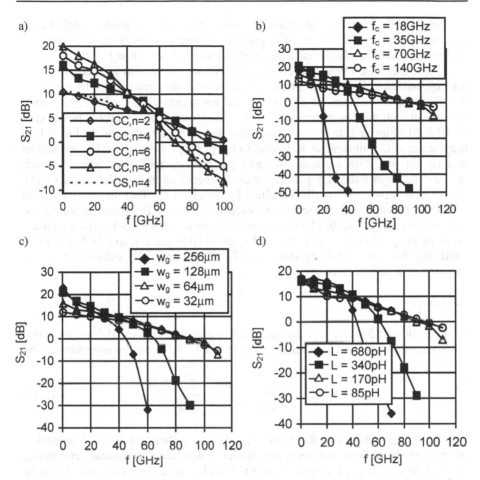

Fig. 8.17a–d. Simulated gain of different TWAs vs frequency, bias of each FET with w_g=64 μm: V_{gs}=0.5 V, V_{ds}=1 V, I_d=17 mA, compare [Ell7] © IEEE 2005: **a** cascode (CC) and common source amplifier (CS) stages with different number of stages (n), L=170 pH, f_c per section is 70 GHz; **b** different f_c of line sections, cascode TWA with four stages; **c** different FET gate width w_g, cascode TWA with four stages; **d** different L of line sections, cascode TWA with four stages

The derivations of Eqs. (8.9)–(8.11) can be found in [Ava82]. These equations show that for a given frequency, the maximum gain is achieved for an optimum number of stages of

$$n_{oG} = \frac{1}{2A_g}. \tag{8.12}$$

With given device parameters and operation frequency of 40 GHz, where according to the design goal optimum performance should be reached, we obtain A_g=0.0895, n_{oG}=5.6, G_0=21 dB and G=15 dB. Due to the additional losses generated by the drain line and the inductors, the total line losses are slightly higher than assumed. In reality, the values of n_{oG} and G are upper limits. The simulated gain vs frequency and number of cascode amplifier stages is shown in Fig. 8.17a.

The results of a TWA with common source stages are also included for comparison verifying the superior properties of the cascode circuit. For a frequency of 40 GHz, the simulations predict an n_{oG} of approximately 5, which is in good agreement with the first order calculations. As expected, the simulated power gain of 10.5 dB is lower than the one using the first order calculations. In Fig. 8.17b, the impact of f_c is illustrated. Referring to Eq. (7.66), f_c depends on C_{gs} and consequently on w_g. With increased w_g and g_m, the low frequency gain is raised, whereas the high frequency gain is lowered. According to Eqs. (7.63) and (7.64), the relation between w_g and L is defined by the given Z_w. This means that w_g and L have to be varied simultaneously to keep Z_w constant. The investigations show that the highest gain up to 60 GHz is achieved for an f_c of 70 GHz. The characteristics of the gain vs w_g and L at fixed f_c of 70 GHz are depicted in Fig. 8.17c,d verifying that a w_g of 64 um and L of 170 pH are optimally suited to meet the targeted gain performance.

b. Noise

With the device dependent drain and gate noise coefficients γ and δ, respectively, as discussed in Sect. 5.4.7, the noise figure of FET TWAs can be approximated by [Ait85]:

$$F = 1 + \frac{4\gamma}{n \cdot g_m \cdot Z_0} + \frac{n \cdot \omega^2 \cdot C_{gs}^2 \cdot Z_0 \cdot \delta}{3 g_m}. \tag{8.13}$$

The second term describes the drain noise, which is dominant at low frequencies, whereas the third term represents the frequency dependent gate noise determining the high frequency performance. From the last identity, a minimum noise figure of

$$F_{min} = 1 + \frac{2\omega C_{gs}}{g_m} \sqrt{\frac{4\gamma\delta}{3}}. \tag{8.14}$$

can be derived for a number of stages of

$$n_{oF} = \frac{2}{\omega C_{gs} Z_0} \sqrt{\frac{3\gamma}{\delta}}. \tag{8.15}$$

At an operation frequency of 40 GHz, we obtain n_{oF}=3.7 and F_{min}=3.3 dB. For comparison, noise figure simulations are performed. As depicted in Fig. 8.18a, at the target frequency of 40 GHz, the lowest noise figure of 3 dB is achieved for a n_{oF} of approximately 4. This is in excellent agreement with the theoretical results. The simulations show that the best low frequency noise performance is achieved at high n_{oF}, whereas for high frequencies, the lowest noise figures are reached for low values of n_{oF}. In accordance with Eq. (8.13), this is attributed to the fact that the drain noise is inversely proportional to n, whereas the gate noise is proportional to n. At 40 GHz, the calculated values for n_{oF} and n_{oG} are close together. Thus, a value of n=4 is used for the final realisation of the TWA. According to Eq. (8.14), F_{min} of the distributed amplifier is supposed to increase continuously with frequency, with an NF of zero towards DC. The simulations and measurements reveal deviations from this ideal characteristic. This can be explained as follows: Towards DC, the circuit is not a distributed structure any more. The circuit behaves like a single transistor with all amplifier stages connected in parallel. Furthermore, the input and output are directly terminated by the absorption resistors. Thus, at low frequencies, the contribution of the gate line termination resistor R_{abg} has to be considered. As clearly shown in Fig. 8.18b, low values of R_{abg} significantly increase the noise towards low frequencies. Thus, the low frequency noise performance can be improved by increasing the input termination resistor. However, this decreases the input return loss. A value of 75 Ω is chosen since this provides a reasonable tradeoff between enhanced noise performance and acceptable input return loss. Further increase of R_{abg} decreases the noise figure only up to a certain limit, since at high R_{abg}, gate leakages dominates the low frequency noise characteristics.

a)   b)

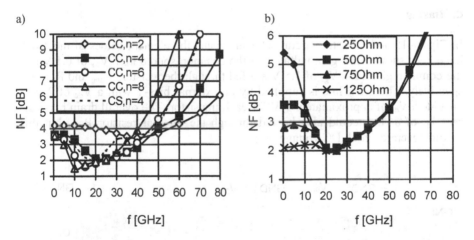

Fig. 8.18a,b. Simulated noise figure, bias of each FET: V_{gs}=0.5 V, V_{ds}=1 V, I_d=17 mA, compare [Ell7] © IEEE 2005: **a** different TWAs with cascode and common source amplifier stages, n: number of stages; **b** four stage cascode TWA with different gate line terminations R_{abg}

c. Output Power

The transistors are operated in class-A in ideal conditions providing a maximum efficiency

$$\eta = \frac{RF \text{ ouput power}}{dc \text{ power}} \tag{8.16}$$

of 50%. Neglecting the losses of the drain line and the inductors, the maximum output power of the TWA can be approximated by

$$P_{max} = \frac{1}{2} n \cdot V_{dc,eff} \cdot I_{dc} . \tag{8.17}$$

The effective supply voltage of the cascode stages $V_{dc,eff}=2\times(V_{ds}-V_{ds,sat})$ considers the voltages losses due to the saturation voltage $V_{ds,sat}$. With n=4, V_{ds}=1 V, $V_{ds,sat}$=0.3 V and $I_{dc \ per \ stage}$=16.5 mA, an upper limit of P_{max}=16.6 dBm can be calculated. The corresponding maximum η equals 35%. Since the frequency dependent losses of the inductors and the drain line are not taken into account, the expected values will be lower.

d. Biasing

In Figs. 8.15 and 8.16, the circuit schematics and the element values of the TWA have been shown. The gate voltages of the common source FET V_{g1} of 0.5 V and the common gate FET V_{g2} of 1.5 V are fed through the resistors R_{abg} and R_{bias}, respectively. The drain line is supplied by an external bias-T with a V_{dc} of 2 V to provide a V_{ds} of approximately 1 V for each transistor. The overall drain bias current I_{dc} of 66 mA could also be fed through the internal resistor R_{abd}, but this would generate a high DC voltage drop.

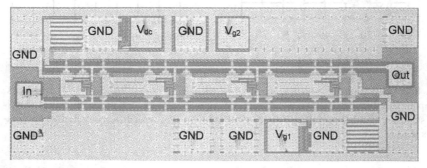

Fig. 8.19. Photograph of compact TWA MMIC with chip size of 0.89 mm × 0.33 mm, compare [Ell7] © IEEE 2005

8.3.2.2 Results

A photograph of the compact TWA MMIC is shown in Fig. 8.19. Within the full frequency bandwidth, the circuit is unconditional stable with K-factor well above unity.

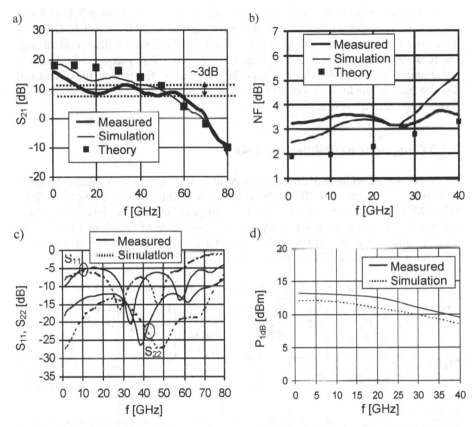

Fig. 8.20a–d. Measured results at V_{gs}=0.5 V, V_{dc}=2 V, I_{dc}=66 mA, compare [Ell7] © IEEE 2005: **a** gain; **b** noise figure; **c** return losses; **d** 1 dB output compression point

Figure 8.20a shows the measured gain. A gain of 9.7 dB±1.6 dB was measured from 10 GHz to 59 GHz. Towards DC the gain increases up to 16 dB. The gain cutoff frequency is 71 GHz. The measured wafer is based on experimental IC hardware that showed significant process variations. Better agreement between simulations and measurements and less pronounced gain ripples are expected for circuits from a wafer with more nominal parameters. The noise figure is depicted in Fig. 8.20b. Up to 40 GHz, the noise figure is below 3.8 dB. In Fig. 8.20c, the return losses are shown. From DC to 60 GHz, the measured input and output return losses are higher than 5 dB and 12 dB, respectively. In Fig. 3.20d, the 1 dB output power compression points are illustrated. Due to the increased losses, the

1 dB compression point decreases with increased frequency. The maximum 1 dB compression point of 13.3 dBm yields a power added efficiency

$$PAE = \eta\left(1 - \frac{1}{G}\right) \tag{8.18}$$

of 16% at low frequencies. The measured 1 dB compression point is 3.3 dB lower than the ideal output power calculated in paragraph c. At 20 GHz and 40 GHz, the measured 1 dB output compression points are 12.5 dBm and 9.5 dBm, corresponding to a PAE of 13.5% and 6.7%, respectively. The TWA was optimised as low noise amplifier. However, due the good large signal performances, the circuit can also be used as medium power amplifier. The output power may be sufficient for short-range WLAN systems.

8.3.3 Comparison of Wideband LNAs

A comparison of wideband amplifiers is compiled in Table 8.4. A bandwidth up to 150 GHz and a gain of 4 dB has been reported in InP HEMT technology. With SOI CMOS, a bandwidth of 91 GHz and 5 dB gain has been achieved. The TWA presented in the preceding section yields a noise figure of less than 3.8 dB up to 40 GHz and a gain beyond 8 dB up to 60 GHz. This performance is similar to TWAs using advanced HEMT technology.

Table 8.4. Comparison with state-of-the-art wideband amplifiers

Technology /f_{max} [GHz]	Band-width [GHz]	S_{21} [dB]	P_{1dB} [dBm] /up to f [GHz]	NF [dB] /up to f [GHz]	P_{dc} [V×mA]	Area [mm²]	Ref.
III–V based							
0.6-μm GaAs MESFET/18	0.1–12	7	n.a.	n.a.	4×19	1.36	[Orz01]
0.1-μm GaAs HEMT/n.a.	0.25–40	14	>4 /20	5–3.7 /40	3.5×143	6.3	[Leo01]
InP HBT/370	6–77	7.5	n.a.	n.a.	250	0.84	[Aga98]
0.1-μm InP HEMT/300	5–112	7	n.a.	n.a.	n.a.	2.2	[Agb98]
	5–147	4	n.a.	n.a.	n.a.		
Silicon Bipolar							
SiGe/65	7.8	10.6	−12	4.4 /7.8	1.8×3.6	0.14	[Ell37] Sec. 8.3.1
BJT/70	3–15	8.7	n.a.	<8.3 /8	n.a.	7.5	[Kaw01]

Table 8.4. Continued

Technology /f_{max} [GHz]	Band-width [GHz]	S_{21} [dB]	P_{1dB} [dBm] /up to f [GHz]	NF [dB] /up to f [GHz]	P_{dc} [V×mA]	Area [mm²]	Ref.
CMOS							
0.5-μm CMOS/n.a.	0.5–4	5.3	>12 /5	<8 /2	3×27	0.79	[Ball00]
0.18-μm CMOS/n.a.	1–10	8	n.a.	n.a.	n.a.	2.3	[Fra02]
90-nm CMOS/140	0–23	18	n.a.	n.a.	1.2×80	0.5	[Toi03]
0.12-μm SOI CMOS/200	4–91	5	n.a.	n.a.	2.6×35	0.82	[Plo04]
	5–86	9	n.a.	n.a.	2.6×50	1	
90-nm SOI CMOS/160	0.1–59	8	>12.5 /20	<3.8 /40	2×66	0.3	[Ell15] Section 8.3.2

8.4 Tutorials

1. Explain the design flow for an LNA.
2. What are the design goals and tradeoffs for LNAs? Outline the impact of an LNA on the system noise figure in receivers.
3. Why are low power consumption and low noise figure more important for an LNA than for a power amplifier?
4. What is the advantage and disadvantage of a cascode topology for an LNA in terms of the low frequency and RF gain, and the DC power consumption?
5. Explain all elements required for the narrowband-matched cascode LNA according to Sect. 8.2.1. Assume reasonable transistor port impedances and match the LNA at the input and the output using a Smith chart. Why can the reuse of elements for different tasks be advantageous? Which elements can be used for multifunctional tasks?
6. What is the impact of the inductive source degeneration used in the 5 GHz LNA described in Sect. 8.2.2 in terms of gain, noise, impedance matching and stability? Why is probably no inductive source degeneration employed for the 40-GHz LNA treated in the preceding section?
7. Explain the strategy of the active matched LNA presented in Sect. 8.3.1. How can we trade off gain and bandwidth?

8.   We refer to the TWA LNA treated in Sect. 8.3.2. What is the impact of the
      input termination resistor? Do we really need this resistor? Which parameters
      determine the gain and the bandwidth? Why is there an optimum number of
      stages? If we would like to sacrifice bandwidth for gain, how do we have to
      adapt the design?
9.   Which gain and NF performances are approximately achieved for leading-
      edge narrowband and wideband LNAs?

References

[Aga98]   B. Agarwal, Q. Lee, D. Mensa, R. Pullela, J. Guthrie and M. J. W. Rodwell,
           "80-GHz distributed amplifier with transferred substrate heterojunction
           bipolar transistors", IEEE Transactions on Microwave Theory and
           Techniques, Vol. 46, pp. 2302–2307, Dec. 1998.
[Agb98]   B. Agarwal, A. E. Schmitz, J. J. Brown, M. Matloubian, M. G. Case,
           M. Le, M. Lui and M. J. W. Rodwell, "112-GHz, 157 GHz, and 180-
           GHz InP HEMT traveling-wave amplifiers", IEEE Transactions on
           Microwave Theory and Techniques, Vol. 46, No. 12, pp. 2553–2559,
           Dec. 1998.
[Ait85]   C. S. Aitchison, "The intrinsic noise figure of the MESFET distributed
           amplifier", IEEE Transactions on Microwave Theory and Techniques,
           Vol. MTT-33, No. 6, pp. 460–466, June 1985.
[Ava82]   Y. Ayasli, R. L. Mozzi, J. L. Vorhaus, L. D. Reynolds and R. A. Pucel,
           "A monolithic GaAs 1–13-GHz traveling-wave amplifier", IEEE
           Transactions on Microwave Theory and Techniques, Vol. MTT-30, No.
           7, pp. 976–981, July 1982.
[Ball00]   B. Ballweber, R. Gupta and D. Allstot, "A fully integrated 0.5-5.5 GHz
           CMOS distributed amplifier", IEEE Journal on Solid-State Circuits,
           Vol. 35, pp. 231–239, Feb. 2000.
[Ber04]   www-device.eecs.berkeley.edu/~bsimsoi/
[Böc02]   J. Böck, H. Schäfer, D. Zöschg, K. Aufinger, M. Wurzer, S. Boguth,
           M. Rest, R. Stengl, T. F. Meister, "Sub 5 ps SiGe bipolar technology",
           Electron Device Meeting, pp. 763-766, Dec. 2002.
[Cas03]   D. J. Cassan and J. R. Long, "A 1-V transformer-feedback low-noise
           amplifier for 5-GHz wireless LAN in 0.18μm CMOS", IEEE Journal of
           Solid-State Circuits, Vol. 38 No. 3 pp. 427–435, March 2003.
[Ell7]   F. Ellinger, "SOI CMOS traveling wave amplifier with NF below
           3.8 dB from 0.1-40 GHz", IEEE Journal of Solid-State Circuits, Vol.
           40, No. 2, pp. 553–558, Feb. 2005.
[Ell15]   F. Ellinger, "26–42 GHz low noise amplifier MMIC fabricated on
           digital SOI CMOS technology", IEEE Journal of Solid-State Circuits,
           Vol. 39, No. 3, pp. 522–528, March 2004.

[Ell17]   F. Ellinger, H. Jäckel, "Low cost BiCMOS variable Gain LNA at Ku-band with ultra low power consumption for adaptive antenna combining", IEEE Transactions on Microwave Theory and Techniques, Vol. 52, No. 2, pp. 702-709, Feb. 2004.

[Ell34]   F. Ellinger, C. Carta, L. Rodoni, G. von Büren, D. Barras, M. Schmatz and H. Jäckel, "BiCMOS variable gain LNA at C-band with ultra low power consumption for WLAN", IEEE International Telecommunication Conference, CD ROM, Aug. 2004.

[Ell37]   F. Ellinger, D. Barras, M. Schmatz and H. Jäckel, "Low power DC-7.8 GHz BiCMOS LNA for UWB and optical communication", IEEE Microwave Theory and Techniques Symposium, Vol. 1, pp. 13–16, June 2004.

[Ell47]   F. Ellinger, U. Lott and W. Bächtold, "Ultra low power GaAs MMIC low noise amplifier for smart antenna combining at 5.2 GHz", IEEE Radio Frequency Integrated Circuit Symposium, Boston, pp. 157–159, June 2000.

[Ell53]   F. Ellinger, J. Kucera and W. Bächtold, "Improvements on a nonlinear GaAs MESFET model", IEEE MTT-S International Microwave Symposium, Baltimore, pp. 1623–1626, June 1998.

[Fra02]   B. M. Frank, P. Freundorfer and Y. M. M. Antar, "Performance of 1-10-GHz traveling wave amplifiers in 0.18 µm CMOS", IEEE Microwave and Components Letters, Vol. 12, No. 9, pp. 327–329, Sept. 2002.

[Kaw01]   M. Kawashima, H. Hazashi, T. Nakagawa and K. Araki, "A low-noise distributed amplifier using cascode-connected BJTs terminal circuit", Asia Pacific Microwave Conference, pp. 21–24, Dec. 2001.

[Kok99]   Y. L. Kok, "160–190 GHz monolithic low-noise amplifiers", Microwave and Guided Wave Letters, Vol. 9, No. 8, pp. 311–313, Aug. 1999.

[Leh85]   R. E. Lehmann et al., "X-band monolithic series feedback LNA2", Transaction on Microwave Theory and Techniques, pp. 1560–1566, 1985.

[Leo01]   R. E. Leoni III, S. J. Lichwala, J. G. Hunt, C. S. Whelan, P.F. Marsh, W. E. Hoke and T. E. Kazior, "A DC-45 GHz metamorphic HEMT traveling wave amplifier", IEEE GaAs Symposium, pp. 133–136, 2001.

[Mim00]   Y. Mimino, M. Hirata, K. Nakamura, K. Sakamoto, Y. Aoki and S. Kuroda, "High gain-density K-band P-Hemt LNA MMIC for LMDS and satellite communication", IEEE MTT-S International Microwave Symposium, Vol. 1, pp. 17–20, June 2000.

[Ong99]   S. A. St. Onge, et al., "A 0.24 µm SiGe BiCMOS mixed-signal RF production technology featuring a 47 GHz f_t HBT and 0.18 µm L_{eff} CMOS" Bipolar/BiCMOS Circuits and Technology Meeting, pp. 170–120, Sept. 1999.

[Orz01]   A. Orzati and W. Bächtold, "Monolithically integrated traveling-wave amplifier for low-cost broadband optical receiver", Workshop on Compound Semiconductor Devices and Integrated Circuits Europe, pp. 9–10, May 2001.

[Plo04] J.-O. Plouchart, J. Kim, N. Zamdmer, L-H. Lu, M. Sherony, Y. Tan, R. Groves, R. Trzcinski, M. Talbi, A. Ray and L. Wagner, "A 4–91 GHz distributed amplifier in a standard 0.12 μm SOI CMOS microprocessor technology", IEEE Journal of Solid-State-Circuits, Vol. 39, No. 9, pp. 1455–1461, Sept. 2004.

[Toi03] T. Toifl, M. Kossel, C. Menolfi, T. Morf and M. Schmatz, M.;A "23 GHz differential amplifier with monolithically integrated T-coils in 90 nm CMOS technology" IEEE MTT-S International Microwave Symposium, Vol. 1, pp. 239–242, June 2003.

[Yan01] H. Yano, Y. Nakahara, T. Hirayama, N. Matsuno, Y. Suzuki and A. Furukawa, "Performance of Ku-band on-chip matched Si monolithic amplifiers using 0.18-μm-gatelength MOSFETs", IEEE Transactions on Microwave Theory and Techniques, Vol. 49, No. 6, pp. 1086-1093, June. 2001.

9 Power Amplifiers

The best thing this world is offering is the yearning for another world.
Martin Kessel

In the preceding chapter, we treated small signal amplifiers handling weak input and output signals. Now we discuss power amplifiers applied in transmitters. For typical wireless applications, power amplifiers must provide output powers in the region of several 100 mW for mobile systems, and up to several watts for basestations. Relevant is the output power at the fundamental frequency, which is fed into the load. We denote this power by P_{L1}. The harmonic power does not contribute to the desired output power.

Power amplifier-specific figures of merit such as the p_n (Normalised Power Factor), PUF (Power Utilisation Factor), η (Efficiency) and PAE (Power Added Efficiency) are used for the comparison of approaches. There is the aim to maximise those figures of merit. The p_n is given by the ratio between P_{L1} and the product of the maximum possible drain source voltage V_{max} and drain current I_{max}:

$$p_n = \frac{P_{L1max}}{V_{max} \cdot I_{max}}, \tag{9.1}$$

where P_{L1max} is the maximum P_{L1}. The values of V_{max} and I_{max} are determined by breakdown effects, which can destroy the device. p_n describes how much power we can get out of a transistor when considering its given IV boundaries. According to Fig. 7.14b, the load line has a strong impact on the amplifier properties. The PUF describes the P_{L1max} ratio between an amplifier under test and a class-A reference amplifier:

$$PUF = \frac{P_{L1max}}{P_{L1max\ class-A}}. \tag{9.2}$$

The amplifier classes will be discussed in the next sections. To obtain maximum output power, amplifiers have to be driven close to their large signal limitations leading to saturation effects. Consequently, linearity plays an important role. The requirements for the linearity depend on the applied modulation types. Both phase and amplitude modulations can be used for data transmission. Amplitude modulation is strongly degraded by saturation effects demanding better large signal performance and higher linearity than phase modulation.

Power amplifiers consume significant DC power P_{dc}. Hence, they have a strong impact on the stand-by time of battery-driven devices. To maximise the stand-by time, the efficiency defined by

$$\eta = \frac{P_{L1}}{P_{dc}}, \tag{9.3}$$

must be as high as possible. At high frequencies, the gain of circuits is limited. Thus, the input power P_{in} can't be neglected and should be considered in the efficiency leading to the power added efficiency

$$PAE = \frac{P_{L1} - P_{in}}{P_{dc}} = \eta \left(1 - \frac{1}{G} \right). \tag{9.4}$$

For $G \rightarrow \infty$, PAE=η. Excellent books in the area of power amplifiers have been published [Cri99, Kra80], which may serve as references beyond this manuscript.

If not otherwise mentioned, the following considerations relate to n-channel FETs or npn BJTs.

9.1 Choice of Basic Amplifier Topology

As a first step, we have to choose the optimum basic amplifier topology. Because of the low available supply voltages in the order of some volts for mobile systems, power amplifiers have to handle very high currents, which may generate high DC voltage losses across the parasitic resistances in the bias path. Assume a current of 100 mA and a resistance of 10 Ω leading to a voltage drop of 1 V. Such a high voltage loss reduces the effective supply voltage of the transistors and consequently the efficiency. Given that the required bandwidth does not exceed one octave, high Q bias inductors can be applied since they provide low DC and RF losses. An efficient way to reduce the DC losses is to minimise the number of high current bias paths.

We review the biasing concepts of the three basic amplifier configurations depicted in Fig. 9.1. The common source and drain circuits have only one high current DC path associated with the feeding of the drain and the source, respectively. However, the use of a bias inductor in the source of the common drain topology may cause severe stability problems since the inductor may act as a constructive input to output feedback. Unfortunately, two high current paths, namely in the drain and the source have to be considered for the common gate circuit. Moreover, the common gate circuit does not provide current gain. In this context, we have to point out that current gain is important to achieve power gain at low load impedances required for high power. Exhibiting both current and voltage gain, the common source circuit offers high power gain. Very large transistors are required for power amplifiers. In aggressively scaled CMOS technologies, the R_{ds} of large transistors may even be lower than 50 Ω. Thus, in contradiction to small signal amplifiers featuring compact transistors, impedance matching strategies on the basis of the common gate and common drain amplifier may not be beneficial.

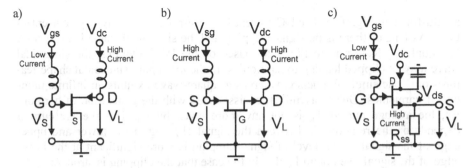

Fig. 9.1a–c. Biasing of basic power amplifier topologies drawing high currents: **a** common source; **b** common gate; **c** common drain

Because of their high output resistance, cascode circuits are capable of reducing the output power losses compared to the common source topology. The output losses can be significant for amplifiers implemented in aggressively scaled technologies. Nevertheless, cascode circuits are rarely used for power amplifiers. The involved common gate stage can lead to instabilities associated with the large RF shunt capacitor at the gate resonating the inductance of the non-ideal ground connection. For small signal amplifiers featuring compact transistors, this resonance is usually well above the operation frequency and thus not as severe as for power amplifiers. Moreover, given that one transistor is temporarily switched off, the entire supply voltage appears across one single transistor, which may cause damage if the supply voltage is higher than the drain source breakdown voltage of a single device. In addition, the designer has to consider that for the cascode power amplifier the losses associated with the saturation voltage appear twice.

We can conclude that the common source topology is the best choice for most power amplifiers. Thus, in the following sections, we will focus on the common source amplifier.

9.2 Classical Current Source Based Amplifiers

We start with the classical current source based amplifiers. Suppose that the amplifier represent an ideal voltage controlled current source with infinite output impedance. Especially for large devices in aggressively scaled CMOS we have to consider the relatively low output impedance determined by R_{ds}, which generates significant losses. Moreover, a small fraction of the signal swing appears in the triode region, where the transistor exhibits resistive characteristics. Similar conclusions can be drawn for BJTs. Nevertheless, for the sake of simplicity, we continue to assume ideal current source properties. With I_{dq} and I_{dp} as the quiescent bias current and the RF current amplitude, respectively, a linear amplifier has a sinusoidal drain current of

$$I_d = I_{dq} + I_{dp} \cos(\omega_0 t), \qquad (9.5)$$

as illustrated in Fig. 9.2a. The DC current I_{dc} corresponds to the average current vs time. As long as there is no signal clipping and the sinusoidal shape is preserved, I_{dc} equals I_{dq}. Referring to Fig. 9.2b, based on the IV limitations, the sinusoidal wave may be clipped leading to harmonics. In theory, a pure sinusoidal shape features only one single frequency, whereas a square-wave exhibits an infinite number of harmonics. Due to asymmetries associated with the generation of harmonics, the average current I_{dc} is not any more given by I_{dq}. The average current depends on the relationship between the signal clipping at the lower and upper edge of the signal. If the level of signal clipping is more significant at the lower edge of the signal, we obtain $I_{dc} > I_{dq}$. In the case that the clipping is stronger at the upper edge of the signal, we get $I_{dc} < I_{dq}$.

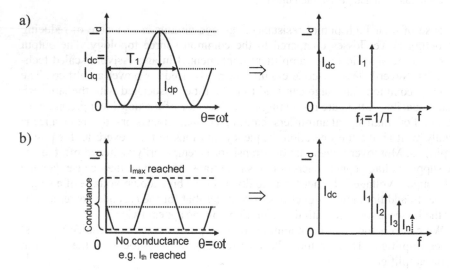

Fig. 9.2a,b. Correlation between drain current in time and frequency domain: **a** linear sinusoidal wave; **b** nonlinear wave generated by signal clipping caused by the transistor IV boundaries. I_{th} denotes the threshold current

By means of the applied bias, the portion the device spends in its active region can be influenced. Accordingly, we define a conductance angle α. If the full, the half and less than the half of the signal period is amplified, the conductance angles are 360°, 180° and less than 180°, respectively. The corresponding amplifier classes are referred to as class-A, class-B and class-C, respectively. In Fig. 9.3, the corresponding bias points are depicted. The load line determined by the load resistor describes the relation between RF load current and voltage. Based on the drain current, the DC and harmonic currents can be calculated by Fourier analysis yielding

$$I_{dc} = \frac{1}{2\pi} \int_{-\alpha/2}^{\alpha/2} I_d(\theta)d\theta, \qquad (9.6)$$

and

$$I_{Ln} = \frac{1}{\pi} \int_{-\alpha/2}^{\alpha/2} I_d(\theta)\cos(n\theta)d\theta, \qquad (9.7)$$

where I_{Ln} represents the magnitude of the nth harmonic load current, and $\theta = \omega t$. Note that the 1st harmonic equals the fundamental signal component I_{L1}. If we talk about harmonics in a general sense, we mean those beyond second order. Equations (9.6) and (9.7) are instrumental for the calculation of key measures such as the fundamental-, harmonic- and DC-power, efficiency and linearity.

a)   b)

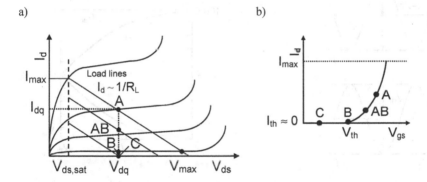

Fig. 9.3a,b. Class-A, class-AB, class-B and class-C amplifier: **a** drain current vs drain source and gate source voltage; **b** drain current vs gate source voltage in saturation region

9.2.1 Class-A

In class-A, the signal is amplified during the full signal period as sketched in Fig. 9.4a. Given that the sinusoidal waveform is preserved without signal clipping, only the fundamental signal component exists yielding a power in the load of

$$P_{L1} = \frac{V_{dp}^2}{R_L} \cdot \frac{1}{2\pi} \int_{-\pi}^{\pi} \cos^2(\theta)d\theta = \frac{V_{dp}^2}{2R_L}, \qquad (9.8)$$

with V_{dp} as the magnitude of the RF drain voltage. Assuming that $V_{ds,sat}$ is relatively small[1] the maximum V_{dp} is determined by V_{dc}. With Eq. (9.8), we get

$P_{L1max} = \dfrac{V_{dc}^2}{2 \cdot R_L}$. We can verify this result by consideration of the RMS (Root

Mean Square) signal swing amplitudes $\Delta V = \dfrac{V_{max}}{2\sqrt{2}}$ and $\Delta I = \dfrac{I_{max}}{2\sqrt{2}}$. With

$P_{Lmax} = \Delta V \cdot \Delta I = \dfrac{1}{8} \cdot V_{max} \cdot I_{max}$, $I_{max} = \dfrac{V_{max}}{R_L}$ and $V_{max} = 2 \cdot V_{dc}$, we obtain

$P_{L1max} = \dfrac{V_{dc}^2}{2 \cdot R_L}$ as well.

a)   b)

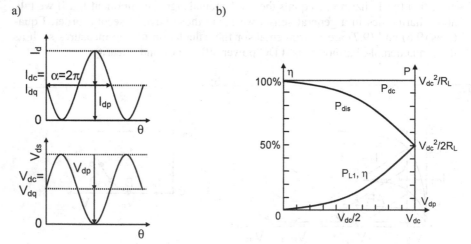

Fig. 9.4a,b. Class-A amplifier: **a** waveforms; **b** output power and efficiency vs output voltage swing

According to Eq. (9.1) we obtain $p_n = \dfrac{1}{8}$. Recall that per definition PUF=1 for a

class-A amplifier. Theoretically, up to a voltage swing of $V_{dp}=V_{dc}$, there is no signal clipping. Consequently, no harmonics or intermodulation products are generated leading to maximum linearity. The dissipated large signal DC power equals the product of the average DC current and voltage. Due the symmetrical swing, we simply get

$$P_{dc} = \frac{V_{dc}^2}{R_L} = \frac{V_{dq}^2}{R_L}. \qquad (9.9)$$

---

[1] If not mentioned otherwise, we make this assumption throughout this chapter.

Equation (9.6) yields the same results. The maximum DC to RF efficiency is given by

$$\eta_{max} = \frac{P_{L1max}}{P_{dc}} \cdot 100\% = 50\% \; . \tag{9.10}$$

The efficiency of 50% can only be exploited at maximum uncompressed signal swing. At lower voltage swing, the efficiency is reduced as indicated in Eq. (9.8) since the consumed DC power remains constant regardless of the RF signal amplitude. The characteristics are depicted in Fig. 9.4b.

The dissipated power $P_{dis} = P_{dc} - P_{L1}$ is converted into thermal energy and can severely impact circuits due to the heating. It is noted that P_{dis} exhibits a minimum at maximum drain voltage swing. We can deduce that the heating peaks at minimum input power- a result, which may be surprising at the first glance.

9.2.2 Class-B

By reduction of the bias current resulting in lower DC current, the efficiency can be increased. In class-B, the quiescence bias current is set close to the threshold region with threshold current I_{th}. Consequently, the drain current in class-B is sinusoidal only for one half wave as illustrated in Fig. 9.5a. Similarly to the class-A amplifier, we can calculate the fundamental load power by considering the RF voltage swings. There are two major differences: Since the signal power appears only for half of a period, we have to consider a power reduction of 1/2. On the other hand, the RF current swing is doubled leading to $\Delta I = \frac{I_{max}}{\sqrt{2}}$, while the voltage swing remains $\Delta V = \frac{V_{max}}{2\sqrt{2}}$. Hence, we get the same output power as for class-A of $p_n = \frac{1}{8}$ with PUF=1, $P_{L1max} = \frac{1}{8} \cdot V_{max} \cdot I_{max} = \frac{V_{dc}^2}{2 \cdot R_L}$ and $P_{L1} = \frac{V_{dp}^2}{2 \cdot R_L}$.

Since the class-B amplifier is biased at the threshold voltage, the DC power, which represents the average drawn DC power, is lower than for the class-A amplifier. Referring to Eq. (9.6), we obtain an average DC current of

$$I_{dc} = \frac{1}{2\pi} \int_{-\pi/2}^{\pi/2} \frac{2 \cdot V_{dp}}{R_L} \cdot \cos(\theta) \, d\theta = \frac{2 \cdot V_{dp}}{\pi \cdot R_L} \; . \tag{9.11}$$

yielding a DC power of

$$P_{dc} = \frac{2}{\pi} \cdot \frac{V_{dc} \cdot V_{dp}}{R_L} . \tag{9.12}$$

At maximum swing of $V_{dp} = V_{dc}$ resulting in $P_{L1} = \frac{2}{\pi} \frac{V_{dc}^2}{2 \cdot R_L}$ we can calculate the maximum efficiency amounting to

$$\eta_{max} = 78.5\% . \tag{9.13}$$

The output power, DC power, efficiency and the dissipated power versus the drain voltage swing are shown in Fig. 9.5b. Compared to class-A, the consumed DC power in class-B operation falls with decreased RF signal amplitude. No DC current is drawn in class-B as long as there is no RF input signal. Consequently, the degradation of the efficiency at low signal amplitudes is not as significant as for the class-A amplifier. Moreover, the energy dissipation and the heating are reduced. The latter property can be very important for integrated circuits, where cooling is difficult.

a)   b)

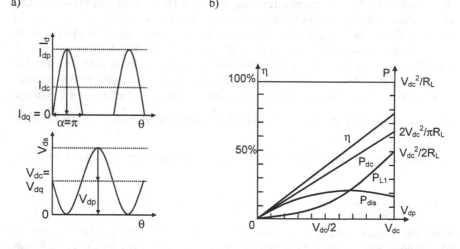

Fig. 9.5a,b. Class-B amplifier: **a** waveforms; **b** output power and efficiency vs output voltage swing

Benefits frequently come together with drawbacks. Since one half of the signal is cut, strong undesired harmonics are generated. The suppression of these harmonic mandates filters, which introduce additional losses. Of course, these losses must be lower than the achieved benefits. Otherwise class-B is not worthwhile. Since only half of the signal voltage is amplified, class-B amplifiers have 6 dB less power gain limiting the capabilities at highest frequencies. Furthermore, due to the low bias current, the transistor provides only low f_t and f_{max} resulting in a further reduction of gain.

So far, we have treated single-ended class-B amplifiers. To allow the amplification of both signal polarities, two complementary amplifiers can be employed. Corresponding operation is possible using complementary FETs or by means of transformer-based push-pull principles. Regarding the latter approach, the size and the losses of the required transformers have to be considered. The major drawback of the approach with the complementary transistors is the reduced operation frequency due to the p-channel FET.

9.2.3 Class-A/B

As its name suggests, the class-A/B amplifier has a conductance angle between 180° and 360°. Hence, its properties are between the ones of the class-A and class-B amplifiers yielding a good tradeoff between efficiency and linearity. Hence, the class-A/B amplifier is popular for various applications.

9.2.4 Class-C

It is straightforward that additional efficiency improvement can be achieved by further reduction of the quiescence bias current. In class-C amplifiers, the transistor is on for less than half a duty cycle since the extrapolated I_{dq} is below the threshold current. The waveforms are plotted in Fig. 9.6. Below the threshold current, the transistors may not provide any negative current since the current typically remains zero. This has no effect on our calculations because the signal fractions appearing below the threshold are not considered. Thus, for the calculations, it does not matter if I_{dq} is negative or zero. However, the assumption of a virtual negative I_{dq} simplifies our calculations.

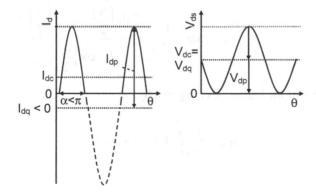

Fig. 9.6. Class-C amplifier waveforms

For the specific cases of the class-A and class-B amplifiers, the calculation of the fundamental load and DC power was relatively easy. Now we derive universal relations holding for all conductance angles α. With $\theta = \omega_0 t$ and within $-\dfrac{\alpha}{2} < \theta < \dfrac{\alpha}{2}$, the drain current is described by

$$I_d = I_{dq} + I_{dp} \cos\theta. \tag{9.14}$$

Note that I_{dq} is positive for $\alpha > \pi$ and supposed to be negative for $\alpha < \pi$. The boundary where I_d equals zero yields $I_{dq} = -I_{dp} \cos\theta$. With respect to α, we deduce

$$I_{dq} = -I_{dp} \cos\frac{\alpha}{2}. \tag{9.15}$$

The latter relation can be verified. In class-B with $\dfrac{\alpha}{2} = \dfrac{\pi}{2}$ we get $I_{dq}=0$, which is correct. For $\dfrac{\alpha}{2} < \dfrac{\pi}{2}$, the negative I_{dq} increases as expected for class-C. Equation (9.15) holds also for class-A, where $\dfrac{\alpha}{2} = \pi$ yields $I_{dq} = I_{dp} = I_{dc}$. With $I_{dp} = I_{max} - I_{dq}$ and Eqs. (9.14) and (9.15), we can write for the drain current vs α:

$$I_d = \frac{I_{max}}{1-\cos\left(\dfrac{\alpha}{2}\right)}\left[\cos\theta - \cos\left(\frac{\alpha}{2}\right)\right]. \tag{9.16}$$

The DC component and harmonics of the latter equation can be computed by Fourier analysis based on Eqs. (9.6), (9.7) and (9.16). For the DC current we get

$$I_{dc} = \frac{1}{2\pi}\int\limits_{-\alpha/2}^{\alpha/2} I_d(\theta)d\theta = \frac{I_{max}}{2\pi}\cdot\frac{2\sin\left(\dfrac{\alpha}{2}\right) - \alpha\cos\left(\dfrac{\alpha}{2}\right)}{1-\cos\left(\dfrac{\alpha}{2}\right)}. \tag{9.17}$$

The amplitudes of the harmonics are given by

$$I_{Ln} = \frac{1}{\pi}\int\limits_{-\alpha/2}^{\alpha/2} I_d(\theta)\cos(n\theta)\,d\theta \tag{9.18}$$

yielding

$$I_{L1} = \frac{I_{max}}{2\pi} \cdot \frac{\alpha - \sin\alpha}{1 - \cos\left(\dfrac{\alpha}{2}\right)} \qquad (9.19)$$

for the fundamental load current. The components are illustrated in Fig. 9.7 vs the conductance angle. We can draw first conclusions:

- The RF and DC powers decrease with lowered α and approach zero at $\alpha=0$.
- The amplitude of the fundamental does not change much between 2π (class-A)$>\alpha>\pi$ (class-B). A maximum appears in class-AB.
- With respect to the amplitude of the fundamental, the amplitudes of the harmonics tend to get higher towards zero α leading to worsened linearity.
- The amplitudes of the harmonics tend to lower with order, which is a phenomenon also observed for mixers.

The maximum fundamental load power can be computed by

$$P_{L1} = \frac{V_{max}}{2\sqrt{2}} \frac{I_{max}}{\sqrt{2}} \frac{1}{2\pi} \cdot \frac{\alpha - \sin\alpha}{1 - \cos\left(\dfrac{\alpha}{2}\right)} = \frac{1}{8\pi} \cdot \frac{\alpha - \sin\alpha}{1 - \cos\left(\dfrac{\alpha}{2}\right)} \cdot V_{max} \cdot I_{max} . \qquad (9.20)$$

resulting in a PUF of 1 and 0 for α of π and 0, respectively. Based on Eqs. (9.17) and (9.20), we can calculate the maximum efficiency:

$$\eta_{max} = \frac{\alpha - \sin\alpha}{2\left(2\sin\dfrac{\alpha}{2} - \alpha\cos\dfrac{\alpha}{2}\right)} . \qquad (9.21)$$

Figure 9.8 illustrates P_{L1} and η. As calculated in the previous two sections, 50% and 78.5% maximum efficiency are achieved in class-A and class-B, respectively. The efficiency improves towards smaller conductance angles leading to 100% efficiency at zero α. Unfortunately, at zero α, the transistor is switched off for the whole cycle and the output power and the PUF vanish to zero. Therefore, a reasonable compromise between efficiency and output power has to be made.

Generally, due to the low α, class-C amplifiers have a relatively low output power. Since more than half of the signal is clipped, class-C amplifiers behave very nonlinear. Similarly to the class-B amplifier, the harmonics have to be filtered to allow proper system operation. The losses of these filters have to be taken into account for the overall efficiency. Especially on silicon-based substrates, the losses may be more significant than the benefit of the high efficiency achievable in theory. In this context, filter losses in the order of 1–2 dB may be taken into account. Moreover, the RF gain is low for two reasons: first, because of the low bias current, g_m and f_t are very low demanding for a very fast technology with respect to the operation frequency. Second, more than 6 dB gain is wasted since less than the half of the input voltage is amplified. Consequently, the class-C approach is rarely used for silicon-based ICs operating at frequencies above 5 GHz.

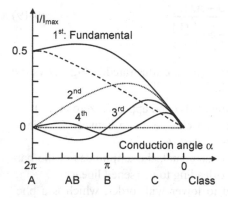

Fig. 9.7. Current amplitudes up to 4th harmonic obtained by Fourier transformation of drain current vs conductance angle defining the operation class, compare [Cri99]

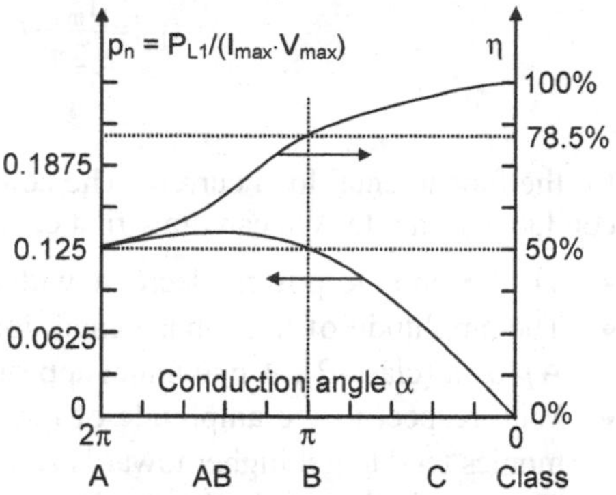

Fig. 9.8. Normalised fundamental load power and efficiency vs conductance angle defining the operation class

9.2.5 Loss Mechanisms

Up to now, idealised assumptions have been made with respect to the amplifier properties. In practice we have to consider the following loss mechanisms:

- The non-zero drain source saturation voltage $V_{ds,sat}$, also known as knee voltage, reduces the possible voltage swing $\Delta V_{eff} = V_{max} - V_{ds,sat}$. Since V_{max} is scaled together with the gate length, the impact of $V_{ds,sat}$ increases for advanced CMOS technologies.
- To get high power at limited drain source supply voltage, we have to feed high current. To yield 500 mW output power, efficiency in class-A of 50% and $V_{dc} = 3$ V, we have to deliver an I_{dc} of around 334 mA. This results in a load resistance in the order of 10 Ω, which has to be matched to the reference impedance, which in most cases amounts to 50 Ω. In the required matching inductance we can easily pick up a resistance in the order of 1–5 Ω reducing the output power and efficiency of more than 10%.
- Further losses have to be considered for the filters of the harmonics mandated in systems where the allowed emitted power in adjacent channels is strongly restricted. In this context, we have to bear in mind that the harmonics can generate intermodulation products being very close to the carrier, e.g. the third order intercept products, which are difficult to filter.
- Ground connections have resistive and inductive impedances. The latter impedance becomes severe at high frequencies leading to voltage drops and power degradations. The reduced gain lowers the PAE.
- Non-ideal current source behaviour and non-infinite output impedances can introduce further losses. For a small fraction of time, the signal appears in the triode region, where we may observe a further reduction of the output impedance.

9.2.6 Power Matching and Load Pull Investigations

In the sections concerned with small signal amplifiers, we have observed that maximum power transfer is obtained at conjugate matching. One may assume that conjugate matching is also the favourite matching goal for power amplifiers. However, for large signal power matching, we have to take into account the following issues:

- As illustrated in Fig. 9.3, the load impedance defines the relation between output voltage and current swings, and consequently the output power. The load impedance is limited by the large signal boundaries in the IV curves of the used transistors.
- A small load resistance is advantageous for two reasons. First, the losses associated with the non-infinite output impedance of the transistor are lowered. Second, at given voltage, the output power is high. However, a small load resistance increases the losses of the parasitic series resistance of the matching network as outlined in the previous section. Thus, a tradeoff has to be found.
- Moreover, we have to consider the termination for both the fundamental frequency and the harmonics. A slight mismatch of the fundamental frequency may allow a significant mismatch of the non-desired harmonics. Thus, more RF power can remain in the fundamental frequency resulting in higher output power. A low-pass like matching network may be used for this task.

We can conclude that large signal power matching and conjugate matching are not necessarily equal. The optimum load resistance can be evaluated by load pull measurements, where the output impedance is swept and the associated power is measured yielding contours with equal output power as shown in Fig. 9.9.

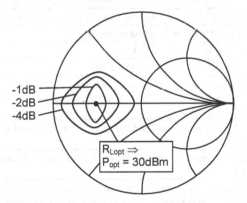

Fig. 9.9. Typical load pull measurements of a power transistor showing the load impedance R_{Lopt} for maximum output power P_{opt}, and contours with equal output power in the Smith chart

9.2.7 Class A/B Power Amplifier Design Example

A high efficiency two-stage common source amplifier using low-cost MOSFET technology has been presented in [Fall01]. The circuit schematic is shown in Fig. 9.10. The first stage is optimised for high gain, whereas the second is optimized for high efficiency. Each required element is discussed in Table 9.1. At 1.75 GHz, an output power of 25 dBm and a PAE of 48% have been measured at a low supply voltage of 1.5 V as plotted in Fig. 9.11.

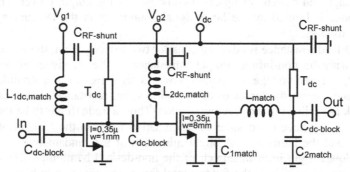

Fig. 9.10. Simplified schematic of class-AB MOSFET power amplifier

Table 9.1. Circuit discussion

Two stage common source	Two cascaded stages are used to reach sufficient gain. The first stage is optimised for high gain, whereas the second is optimised for high efficiency
0.35-µm CMOS	Low-cost, allows systems with high circuit complexity, system on a chip
w_1=1 mm	Sufficient width for adequate large signal performance and high gain
w_2=8 mm	Large width for high output power
V_{g1}	Class AB to reach high gain
V_{g2}	Class AB/B to reach good efficiency
V_{dc}	>1 V to operate in saturation. The higher V_{dc}, the higher the efficiency, but the breakdown has to be considered, which may be around 5 V for this technology
$L_{1dc, match}$	50 Ω input matching for high gain and RF isolated gate biasing of first stage
$L_{2dc, match}$	Interstage matching for high gain and RF isolated gate biasing of second stage
L_{match}, C_{1match}, C_{2match}	50 Ω output matching for maximum efficiency. The Π lowpass attenuates harmonics and intermodulation, which would decrease the efficiency and lead to interferences with other communication channels
T_{dc}	RF isolated drain biasing. A thick transmission line with small DC resistance is used to minimise the DC voltage drop. The drain supply current of power amplifiers is very high

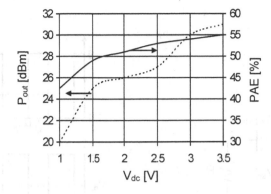

Fig. 9.11. Output power and PAE vs supply voltage at 1.75 GHz

9.3 Switched Amplifiers

The amplifiers presented so far employ a current source to transfer power into a load. One disadvantage is the static power consumption, which lowers the efficiency. A further approach applies a transistor, which is instantaneously switching between the on- and off-state as illustrated in the IV curves in Fig. 9.12. In an ideal case, these two states can be represented by impedances of zero and infinity, respectively. Consequently, either the voltage across the switch V_{sw} or the current through the switch I_{sw} are zero resulting in zero switching power in both states. Theoretically, 100% of the DC energy can be converted into RF energy. Of course, in practice the accuracy of this assumption is limited, e.g. by the nonzero $V_{ds,sat}$, non-zero switching time, resistive losses and the parasitic power transferred to undesired harmonics.

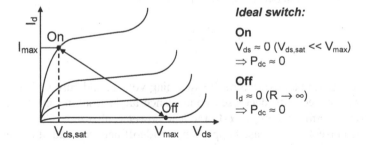

Fig. 9.12. Switch states of transistor including conditions for ideal switch

A simple switched amplifier is illustrated in Fig. 9.13 featuring square-waveforms for both V_{sw} and I_{sw}. If no filtering is applied, a substantial part of the RF energy is transferred into undesired harmonics. By means of Fourier transformation we can calculate a maximum efficiency of 81%.

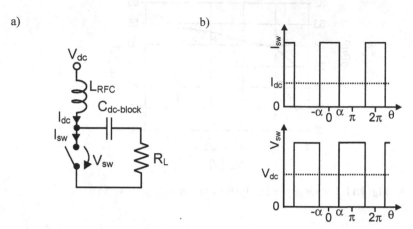

Fig. 9.13a,b. Simple switched amplifier: **a** schematic; **b** switch states in IV curves

The efficiency can be increased by filtering of the harmonics. Two simple approaches can be applied, namely a parallel LC shunt (see Fig. 9.14a) or a series LC tuned circuit (see Fig. 9.15a) connected between the drain and the load. At the fundamental frequency, these circuits yield an admittance to ground and a series impedance of zero, respectively. Consequently, when neglecting any resistive losses in the elements, the fundamental frequency is not attenuated. For the harmonics, these resonance circuits present a high admittance to ground and a large series impedance, respectively. Thus, the undesired harmonics are filtered and either the switch voltage (see Fig. 9.14b) or current (see Fig. 9.15b) becomes sinusoidal resulting in enhanced efficiency. In this context, we refer to the current and voltage switch topology, respectively. The optimum choice of the topology depends on the parasitics inherent in the bandstop and bandpass filters. The design equation of these tuned networks are simply given by

$$\omega_1 = \frac{1}{\sqrt{L_1 \cdot C_1}} \, . \tag{9.22}$$

However, we can find an overlap of the switching voltage and current. Maximum efficiency is achieved with decreased conductance angle resulting in 100% efficiency for $\alpha=0$. Unfortunately, if $\alpha=0$, the load power is also zero. Consequently, similar to the current source based approach, a tradeoff between output power and efficiency has to be made.

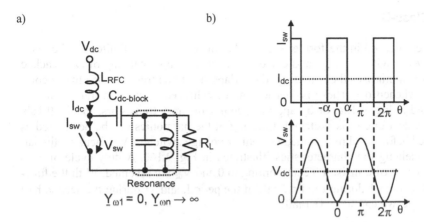

Fig. 9.14a,b. Tuned current switch with bandstop filter: **a** schematic; **b** waveforms

To generate the square-wave output characteristics required for switching, the input signal of the transistor must be large enough to drive the device into the saturation. Thus, the input to output characteristics are strongly nonlinear and the gain is low since a part of the input signal is clipped and not amplified. Typically, the associated gain reduction is well above 3–6 dB. Moreover, in our considerations we assume ideal switches with zero on and infinite off resistance. Especially the latter property does not hold towards high operation frequencies. This can be attributed to the parasitic gate and channel capacitances. In practice, we observe that fast technologies with approximately ten times higher f_t than the operation frequency are required for switched amplifiers. For class-A amplifiers, a rule of thumb may yield an f_t of at least three times the operation frequency.

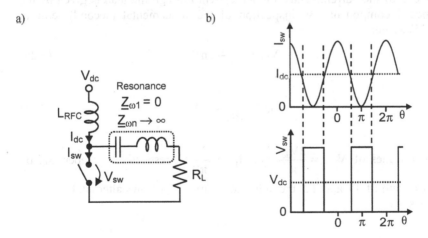

Fig. 9.15a,b. Tuned voltage switch with bandpass filter: **a** schematic; **b** waveforms

9.3.1 Class-D

We have observed in the former section that the efficiency is limited by the over-lap between switch voltage and current. The alternating switching of two stacked transistors minimises these undesired overlaps since one transistor is always completely switched off. These properties can be achieved by using a transformer as illustrated in Fig. 9.16a, or by applying complementary transistors. In Fig. 9.16b, the equivalent circuit is sketched. Filtering of the harmonics can be performed as explained before leading to tuned current or voltage switches. We focus on the latter one yielding the characteristics illustrated in Fig. 9.16c. A duty cycle of 50% and V_{ds} levels of transistor T2 amounting to 0 and V_{dc} are assumed. With the function $s(\theta)$ being +1 during the first half of the period, and −1 during the second half of the cycle, we can write [Kra80]

$$V_{T2} = V_{dc}\left[\frac{1}{2} + \frac{1}{2}s(\theta)\right].$$

(9.23)

Fourier analysis yields

$$s(\theta) = \frac{4}{\pi}\left(\sin\theta + \frac{1}{3}\sin 3\theta + ...\right)$$

(9.24)

and consequently

$$V_{T2} = V_{dc}\left(\frac{1}{2} + \frac{2}{\pi}\sin\theta + \frac{2}{3\pi}\sin 3\theta + ...\right).$$

(9.25)

By supposing that the LC resonance has a reactance of zero at the fundamental frequency and a high impedance at the harmonics, we may neglect the harmonic components of the current. Thus, the RF current through the load is given by the fundamental component. By inspection of the fundamental (second) term in Eq. (9.25) we get

$$V_{L1} = \frac{2V_{dc}}{\pi}\sin\theta.$$

(9.26)

and

$$I_{L1} = \frac{2V_{dc}}{\pi R_L}\sin\theta.$$

(9.27)

with peak values of $V_{L1p} = \frac{2V_{dc}}{\pi}$ and $I_{L1p} = \frac{2V_{dc}}{\pi R_L}$, respectively. Depending on the on/off state of T1 and T2, the sinusoidal current I_{L1} flows alternately through the transistors.

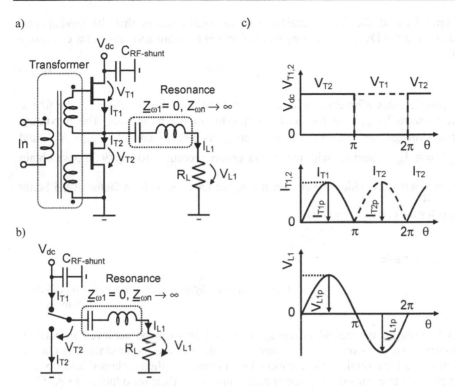

Fig. 9.16a–c. Voltage switching class-D amplifier: **a** schematic; **b** simplified schematic; **c** waveforms

The maximum output power at the fundamental frequency is given by

$$P_{L1} = \frac{1}{2} \cdot V_{L1p} \cdot I_{L1p} = \frac{2}{\pi^2} \frac{V_{dc}^2}{R_L} . \qquad (9.28)$$

What about the DC power? The DC current is given by the average of I_{L1}. Since the associated Fourier transformation of a sinus function adds a factor of $\frac{1}{\pi}$ we get

$$I_{dc} = \frac{I_{max}}{\pi} = \frac{2V_{dc}}{\pi^2 R_L} . \qquad (9.29)$$

finally yielding a DC power consumption of

$$P_{dc} = V_{dc} \cdot I_{dc} = \frac{2}{\pi^2} \frac{V_{dc}^2}{R_L} . \qquad (9.30)$$

Comparison of the latter relation with Eq. (9.28) shows that the fundamental power and the DC power are equal resulting in a theoretical maximum efficiency of

$$\eta_{max} = 100\% . \tag{9.31}$$

In practice, this efficiency is limited by the losses in the switches and the filters, the nonzero $V_{ds,sat}$, and the nonideal filtering of the harmonics. The maximum drain voltage and current of the transistors are given by $V_{max} = V_{dc}$ and $I_{max} = \pi \cdot I_{dc}$. Together with the output power according to Eq. (9.28), we obtain $p_n = \dfrac{1}{\pi} \approx 0.318$, which means that the power handling is by a factor of 2.5 better than for the class-A topology.

9.3.2 Class-E

If a problem is defined, the solution is on half-way.
Julian Huxley

As in class-D, theoretical efficiencies up to 100% can be achieved in class-E. However, no transformer or complementary devices are required making class-E well suited for monolithic integration. Due to their simplicity, class-E tuned power amplifiers have gained widespread acceptance since their introduction [Sok75]. A comprehensive theory has been presented in [Raa77].

Referring to Fig. 9.17a,b, the least complex configuration includes a single transistor switch, a series tuned output circuit and an RF choke. As common for switched based amplifiers, the input has to be driven hard enough to allow an input switching signal with square-wave shape. A duty cycle of 50% is assumed, which has been demonstrated to be the optimum [Raa77].

Similar to other switching type amplifiers, the idea of a class-E amplifier is as follows. When the transistor switch is closed, the drain voltage is zero and a large drain current can exist. If the switch is open, no current flows, but the drain voltage is high. Thus, the simultaneous appearance of nonzero DC voltage and current is avoided eliminating transistor power losses in both the open and the closed state. Therefore, the total real power is transferred to the load. Figure 9.17c depicts the corresponding waveforms.

The circuit conditions are determined by the reactive output network, which is designed such that the drain voltage falls back to zero just before the device is switched on and current is drawn. The output network consists of an inductor L_x, and a charge capacitance C_s which is due to what is inherent in the transistor and added by the load network. This capacitance keeps the drain voltage at zero volts during the switch is closed to avoid switching losses.

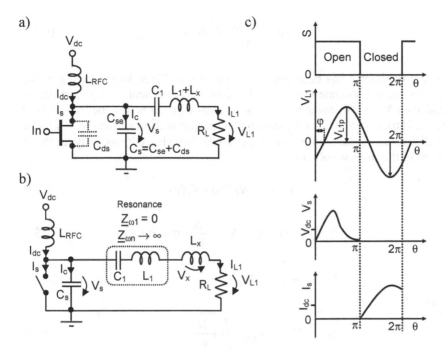

Fig. 9.17a–c. Class-E amplifier: **a** circuit schematics; **b** simple circuit schematic, transistor acts as switch; **c** waveforms

Filtering of the harmonics is necessary to achieve sinusoidal output waveforms and maximum efficiency. The class-E output network by itself acts already as a lowpass filter, thereby providing some level of filtering. Additional filtering, e.g. by means of the series tuned LC filter may be performed to improve the attenuation of the harmonics. It is obvious that this filter has to have a high Q, otherwise the filtering is not worthwhile because of the additional resistive losses.

Compared to the amplifier types presented before, the analysis of the class-E amplifier is tedious since all parameters are interrelated. E.g. the drain voltage is a function of the current charging C_s, which depends on the load voltage and in turn is determined by the drain voltage. The reader may not worry about the understanding of the following derivations. They are indeed complex. The practical guys may switch directly to Table 9.2, where the major design equations are summarised. However, the persistent theoreticians may appreciate the derivations. Let's start with the calculation of the sinusoidal load voltage V_{L1} and current I_{L1} having peak amplitudes of V_{L1p} and I_{L1p}, respectively. Following the analysis presented in [Raa77], we can write

$$V_{L1}(\theta) = V_{L1p} \cdot \sin(\omega t + \varphi) = V_{L1p} \cdot \sin(\theta + \varphi) \qquad (9.32)$$

and

$$I_{L1}(\theta) = \frac{V_{L1p}}{R_L} \cdot \sin(\theta + \varphi). \tag{9.33}$$

The phase constant φ has been defined in Fig. 9.17c. Given ideal filtering of the harmonics, we can assume that the voltage of the fundamental frequency across the capacitor V_{s1} equals the sum of the sinusoidal voltages across the load and the inductor. Thus, also V_s must be a sinusoidal function with an amplitude V_{s1p} and a phase difference ϕ generated by the inductor with impedance $X = \omega L_x$. By means of theorems of the sums we obtain

$$V_{s1}(\theta) = V_{L1}(\theta) + V_x(\theta) \tag{9.34}$$

$$= V_{L1p} \cdot \sin(\theta + \varphi) + V_{L1p} \cdot \frac{X}{R_L} \cdot \cos(\theta + \varphi) \tag{9.35}$$

$$= V_{s1p} \cdot \sin(\theta + \varphi_1) \tag{9.36}$$

with amplitude

$$V_{s1p} = V_{L1p} \cdot \sqrt{1 + \frac{X^2}{R_L^2}} \tag{9.37}$$

and phase

$$\varphi_1 = \varphi + \phi = \varphi + \arctan\left(\frac{X}{R_L}\right). \tag{9.38}$$

In the time frame $0 < \theta < \pi$, where the switch is off, we can deduce an expression for the total voltage across C_S:

$$V_s(\theta) = \frac{1}{\omega_0 \cdot C_s} \int_0^\theta I_c(u) du \tag{9.39}$$

with u as an equivalent of θ used for integration. For the current through the capacitor I_C we can write

$$I_C(u) = I_{dc} - I_{L1}(u) = I_{dc} - \frac{V_{L1p}}{R_L} \sin(u + \varphi). \tag{9.40}$$

By putting Eq. (9.40) into Eq. (9.39) we obtain

$$V_s(\theta) = \frac{1}{\omega_0 \cdot C_s} \cdot \left[\frac{V_{L1p}}{R_L} \cdot \sin\left(\varphi - \frac{\pi}{2}\right) + \theta \cdot I_{dc} + \frac{V_{L1p}}{R_L} \cdot \cos(\theta + \varphi) \right]. \tag{9.41}$$

Up to now we have too many unknowns calling for boundary conditions. By inspection of the desired waveforms we observe that

$$V_s(\pi) = 0 \tag{9.42}$$

and

$$\frac{dV_s(\pi)}{d\theta} = 0. \tag{9.43}$$

Based on Eqs. (9.41)–(9.43), we can write for V_{L1p} and φ:

$$V_{L1p} = \sqrt{1 + \frac{\pi^2}{4}} \cdot I_{dc} \cdot R_L = g \cdot I_{dc} \cdot R_L \tag{9.44}$$

and

$$\varphi = \tan^{-1}\left(\frac{-2}{\pi}\right) = -32.482° \tag{9.45}$$

with

$$g = \sqrt{1 + \frac{\pi^2}{4}} = 1.8621, \tag{9.46}$$

being an important design constant. Since V_{dc} is the time average of V_s within one time period, we can write

$$V_{dc} = \frac{1}{2\pi} \int_0^{2\pi} V_s(\theta) d\theta = \frac{I_{dc}}{2 \cdot \pi \cdot \omega_0 \cdot C_s} \cdot \left[\frac{\pi^2}{2} - \pi \cdot g \cdot \cos(\varphi) - 2 \cdot g \cdot \sin(\varphi)\right]. \tag{9.47}$$

Applying $\frac{\pi^2}{2} - \pi \cdot g \cdot \cos(\varphi) = 0$ and $-g \cdot \sin(\varphi) = 1$ allows us to simplify Eq. (9.47) to

$$V_{dc} = \frac{I_{dc}}{\pi \cdot \omega_0 \cdot C_s}. \tag{9.48}$$

With Eq. (9.44), we get for the fundamental load power

$$P_{L1} = \frac{V_{L1p}^2}{2 \cdot R_L} = \frac{I_{dc}^2 \cdot g^2 \cdot R_L}{2}. \tag{9.49}$$

The DC power P_{dc} is simply

$$P_{dc} = I_{dc}^2 \cdot R_{dc}. \tag{9.50}$$

With Eqs. (9.49) and (9.50), we obtain the efficiency

$$\eta = \frac{P_{L1}}{P_{dc}} = \frac{g^2 \cdot R_L}{2 \cdot R_{dc}} .$$ (9.51)

It is our goal to exploit the maximum efficiency of 100%. Hence, we demand

$$\eta_{max} = 1 .$$ (9.52)

Equations (9.51) and (9.52) yield

$$R_L = \frac{2}{g^2} \cdot R_{dc} = 0.578 \cdot R_{dc} .$$ (9.53)

We can deduce that

$$I_{dc} = \frac{V_{dc}}{R_L} \cdot \frac{2}{g^2} \approx \frac{V_{dc}}{1.734 \cdot R_L} ,$$ (9.54)

$$P_{dc} = P_{L1} = \frac{I_{dc}^2}{R_L} = \frac{2}{g^2} \cdot \frac{V_{dc}^2}{R_L} \approx 0.578 \cdot \frac{V_{dc}^2}{R_L}$$ (9.55)

and in turn

$$V_{L1p} = \frac{2}{g} \cdot V_{dc} \approx 1.075 \cdot V_{dc} .$$ (9.56)

On the basis of Eqs. (9.48) and (9.53), the operation frequency at which the efficiency yields 100% can be evaluated:

$$\omega_0 = \frac{2}{\pi \cdot g^2 \cdot R_L \cdot C_s} .$$ (9.57)

We observe that the maximum possible ω_0 is limited by the minimum C_s, which is impacted by the parasitic output capacitance of the transistor in turn being a function of the gate width and subsequently the maximum current I_{max}. The latter parameter can be computed by employing Eqs. (9.40) and (9.44):

$$I_{max} = I_{dc} \cdot (1+g) \approx 2.86 I_{dc} .$$ (9.58)

According to Eq. (9.41), the maximum voltage across the transistor switch V_{max} can be found by $\frac{dV_s}{d\theta} = 0$ and appears at $\theta_{Vsmax} = -2\varphi$. Together with $\frac{1}{\omega_0 C} = \frac{1}{\pi} \cdot \frac{1}{R_{dc}}$ we get for V_{max}:

$$V_{max} = V_{dc} \cdot \pi \cdot (-2\varphi) \approx 3.56 V_{dc} . \tag{9.59}$$

Consequently, $p_n=0.0981$ and PUF=0.785. We can conclude that the power capability of the class-E amplifier is around 20% lower than the one of the class-A amplifier. Fourier transformation of Eq. (9.41) yields an expression for V_{slp}, which is related by means of Eq. (9.37). This operation yields the angle φ_1, leading to the missing reactance X. We can write

$$V_{slp} = \frac{1}{\pi} \int_0^{2\pi} V_s(\theta) \cdot \sin(\theta + \varphi_1) \cdot d\theta = V_{L1p} \sqrt{1 + \frac{X^2}{R_L^2}} . \tag{9.60}$$

Together with Eq. (9.38) we get

$$\phi = \arctan\left[\frac{\pi}{8}\left(\frac{\pi^2}{2} - 2\right)\right] = 49.052° \tag{9.61}$$

and finally

$$X = \tan(\phi) \cdot R_L = 1.1525 \cdot R_L . \tag{9.62}$$

Indeed, these calculations were long and complex since the parameters are interrelated. The most important design equations are summarised in Table 9.2. It is clear that the presented results are only achievable in the ideal case. In practice, losses due to nonideal switching and parasitics of the passive elements have to be taken into account. In this regard, the reader is referred to [Raa78]. CAD large signal simulations and optimisations are required to achieve optimum performance.

Table 9.2. Summary of key design equations for class-E amplifier

V_{dc}	I_{dc}	C_s	L_x	P_{L1}	g
$V_{max}/3.56$	$I_{max}/(1+g)$	$\dfrac{2}{\omega_0 \cdot \pi \cdot g^2 \cdot R_L}$	$\dfrac{1.153 \cdot R_L}{\omega_0}$	$V_{dc} \cdot I_{dc}$	1.862

$$g = 1.8621.$$

9.3.3 Class-F

Compared to sinusoidal signal characteristics, we have observed that square-waves can lead to enhanced efficiency. This is attributed to the fact that the overlap between the voltage and current in transistors can be minimised resulting in lower parasitic power consumption since more power can be transferred into the load. Given a hard input drive, square-wave characteristics can be obtained by employing the transistor as a switch rather than a current source.

a)   b)

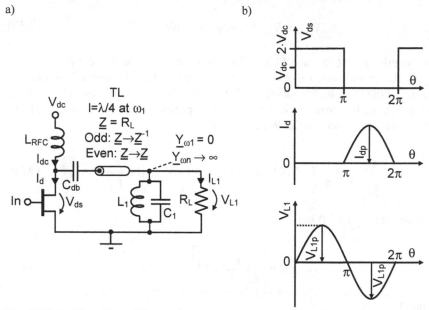

Fig. 9.18a,b. Class-F amplifier with quarter wave transmission line: **a** circuit schematics, C_{db}: DC decoupling capacitor; **b** waveforms

Square-like properties are achieved for the class-F amplifier by directly manipulating the harmonics. Hence, much less input power is required thereby decreasing the gain loss compared to the switch type amplifiers presented in the preceding sections. An efficient class-F approach consisting of a transistor, a transmission line with lengths $1 = \frac{\lambda}{4}$ and impedance being equal to the one of the load, and a LC filter is illustrated in Fig. 9.18a. The parallel tuned LC filter may produce an open at the fundamental frequency and a short for all harmonic frequencies. That means that only current at the fundamental frequency can flow into the load resulting in sinusoidal output characteristics. To evaluate the impedance seen at the drain, we have to review the transformation properties of the $\frac{\lambda}{4}$ line. Referring back to Sect. 3.6.4, the impedance seen at the input of the line is given by

$$Z_{in} = \frac{Z_w^2}{Z_L},$$
(9.63)

with line impedance Z_w and load impedance Z_L. The equation reveals that Z_{in} and Z_L are inversely proportional. We can draw the following conclusions concerning the impedance seen at the drain:

- Since $\underline{Z}_L = \underline{Z}_w = R_L$, the impedance seen at the fundamental frequency remains R_L.
- Due to the reciprocal characteristics of the line, at all odd frequencies, the shorts determined by the filter are transformed into opens.
- At all even frequencies, the shorts remain shorts since the line length becomes a multiple of $\dfrac{\lambda}{2}$ preserving impedances. Or in other words two $\dfrac{\lambda}{4}$ lines are connected in series, where two times the reciprocal value results in the original impedance.

Consequently, due to the impedance conditions, the drain voltage features only odd harmonics including the fundamental component, which is odd as well. Note that rectangular functions with positive amplitude and 50% duty cycle feature only odd harmonics as described in Eq. (9.25). Thus, similar to the switched amplifiers a square-wave drain voltage results. What about the drain current? After inspection of the impedances, we can conclude that currents appear only at the fundamental component and the even harmonics. Consequently, power is solely generated at the fundamental frequency leading to an efficiency of 100% and the waveforms depicted in Fig. 9.18b.

Biasing with $V_{dc}=1/2\ V_{max}$ allows a peak-to-peak drain voltage of $2\ V_{dc}$. Comparison with Eqs. (9.26) and (9.27) reveals peak fundamental components of the square-wave of $V_{L1p} = \dfrac{4}{\pi} \cdot V_{dc}$ and $I_{L1p} = \dfrac{4}{\pi} \cdot \dfrac{V_{dc}}{R_L}$ leading to

$$P_{L1} = \frac{V_{L1p}^2}{2 \cdot R_L} = \frac{\left[\left(\dfrac{4}{\pi}\right) \cdot V_{dc}\right]^2}{2 \cdot R_L}. \tag{9.64}$$

With $V_{max}=2\ V_{dc}$ and $I_{max}=2\ I_{L1p}$, the normalised power handling capacity yields

$$p_n = \frac{\left[\left(\dfrac{4}{\pi}\right) \cdot V_{dc}\right]^2}{2 \cdot R_L} \cdot \frac{1}{2 \cdot V_{dc} \cdot \left(\dfrac{8}{\pi} \cdot \dfrac{V_{dc}}{R_L}\right)} = \frac{1}{2 \cdot \pi} \approx 0.159 \tag{9.65}$$

resulting in PUF=1.272. We can conclude that the maximum output power with respect to the device power limitations is nearly 30% higher than the one of a class-A amplifier.

9.3.4 Class-F with Simple 3rd Harmonic Resonator

The transmission line based class-F amplifier shows promising results, which have been verified by hybrid implementations providing transmission lines and filters with excellent Q. However, for integrated solutions, the size and the losses of the $\frac{\lambda}{4}$ transmission line can be problematic. In silicon-based approaches, the losses of this line can easily be in the range of 2 dB.

a)   b)

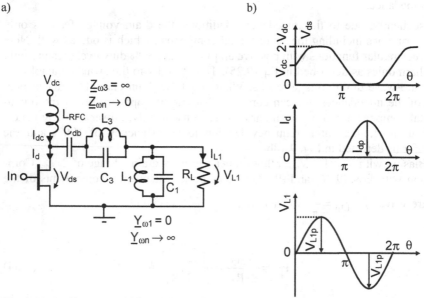

Fig. 9.19a,b. Class-F amplifier with simple lumped element resonator: **a** circuit schematics; **b** waveforms

Thus, for monolithic integration, transmission line based class-F amplifiers are not well suited. This motivates for solutions, which do not require transmission lines. The $\frac{\lambda}{4}$ line can be substituted by lumped element resonators. Compared to the distributed line, these lumped element resonators act at only one frequency. Thus, multiple resonators would be required, e.g. to provide opens for the odd frequencies. Is a manipulation of all harmonics really required? Implied by the fact that the level of harmonics tend to decrease with order, a limited number of lumped element resonators may be a good compromise between wave-shaping and output network complexity. It has been shown that the control of the first odd harmonic, namely the third harmonic is efficient leading to the circuit shown in Fig. 9.19a. The corresponding waveforms sketched in Fig. 9.19b demonstrate that the efficiency is above that of a class-B amplifier since the overlap between transistor voltage and current is decreased.

For this circuit having only one harmonic resonator, an open at the 3rd harmonic is not necessarily the optimum impedance. The corresponding wave characteristics represented by

$$V(\theta) = V_1 \cos(\theta) - V_3 \cos(3\theta) \qquad (9.66)$$

with the amplitude V_1 and V_3 of the fundamental and the third harmonic, respectively, are illustrated in Fig. 9.20 for different levels of V_3. It can be demonstrated that the most square-like characteristics are achieved for $V_3 = 1/9$ leading to an optimum trade-off between flat response and low harmonic content. For larger amplitudes of V_3, the overshoot and the efficiency decreases. The manipulation of V_3 can be accomplished by tuning the series LC tank.

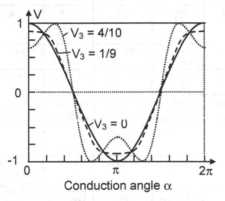

Fig. 9.20. Squaring of sinusoidal wave by means of subtracting third harmonic resulting in $V(\theta)$ as presented in Eq. (9.66), compare [Cri99]

The maximum efficiency of this simple class-F amplifier should be higher than the one of a class-B amplifier providing $\eta_{max}=78.5\%$. On the other hand, the maximum efficiency must be lower than 100% achieved by the ideal multi-resonator approach. It has been deduced theoretically that an maximum efficiency of 88.4% can be achieved with $V_3=1/9$ [Raa97].

9.4 Summary

In Table 9.3, the key characteristics of the discussed amplifiers are listed. We can distinguish between the classical current source based amplifiers and the switched approaches. The optimum choice depends on the application and the available technology. By sacrificing the linearity, the RF gain and the output power, the efficiency of current source amplifiers can be improved by decreasing the bias current and conductance angle from class-A over class-B to class-C. The class-A and class-B amplifiers are capable of operating with amplitude, phase and frequency modulation schemes. In class-A, the highest operation frequency of around a third of the f_t is achieved. Thus, millimetre wave amplifiers are frequently operated in class-A. A good overall performance is achieved with class-AB amplifiers, which are successfully employed in various RF systems.

Table 9.3. Comparison of amplifier approaches

Type	Classical Current Source Amplifier				Switched Amplifiers		
Sub-type	Class-A[a]	Class-AB[a]	Class-B[a]	Class-C[a]	Class-D[b]	Class-E[a]	Class-F[c]
Bias point	$I_{dc}=I_{max}/2$ $V_{dc}=V_{max}/2$	Between A and B	$I_{dc}=I_{max}/\pi$ $V_{dc}=V_{max}/2$	$I_{dc}<I_{max}/\pi$ $V_{dc}=V_{max}/2$	$I_{dc}=I_{max}/\pi$ $V_{dc}=V_{max}$	$I_{dc}=I_{max}/2.86$ $V_{dc}=V_{max}/3.56$	$I_{dc}=I_{max}/\pi$ $V_{dc}=V_{max}/2$
Output waveform	Both polarities of sinusoidal (α=360°)	More than one polarity of sinusoidal (α>180°)	Positive part of sinusoidal (α=180°)	Peak wave (α<180°)	Both polarities of sinusoidal	Both polarities of sinusoidal	Both polarities of sinusoidal
p_n	1/8	Up to 0.1375	1/8	<1/8 0.112 at α=147°	$1/\pi$	0.0981	$1/2\pi$
η_{max} at P_{max}	≤50%	50%–78.5%	78.5%	78.5%–100%	100%	100%	100%
Heat power/P_{dc}	≥50%	21.5%–50%	≥21.5%	≤21.5%	0%	0%	0%
Maximum RF gain	MAG	Between A and B	6 dB below MAG	Between B and 0	Well below MAG, input saturation required for switching		
Typ. max. operation freq.	~1/3 f_t	~1/4 f_t	~1/5 f_t	~1/6 f_t	<1/6 f_t		
Possible bandwidth	High	High	Moderate	Low	Depends on transformer bandwidth	Small, class-E resonance	Small due to filter bandwidth
Content of harmonics and distortion	Low	Moderate	High	Very high	High	High	Low
Demands for filtering	Low	Moderate	High	Very high	Very high	Very high	High
Modulation	AM, PM, FM	AM, PM, FM	AM, PM, FM	PM, FM	PM, FM	PM, FM	PM, FM
Comments	Highest gain at highest frequencies	Good tradeoff	Push-pull and complementary replace topology allow amplification of both polaries	Weak RF properties	Push-pull yields higher output power but requiers two transformers	Low design flexibility	Strong impact on filter performance and Q

[a]Single transistor
[b]Complementary switching according to Fig. 9.16
[c]Basically a current source based amplifier with switch-like characteristics, $\lambda/4$ transmission line type resonator according to Fig. 9.18 is considered

The switched-type topologies may provide enhanced efficiency compared to the current source approaches. However, the high efficiency is dearly purchased by lower linearity and RF gain due to the hard input drive required for switch operation. Because of the input saturation, amplitude modulation by means of the gate voltage is not possible. Amplitude modulation on basis of supply voltage variations may be an option. However, this makes the drive of the amplifier difficult since high control power is necessary. Additionally, from system perspective, supply voltage modulation may introduce critical feedback effects for other components. The push-pull class-D amplifier provides superior output power but mandates a transformer or complementary devices.

A good trade-off between both basic concepts is provided by the class-F approach with a 3rd order resonator. Square-like drain characteristics are achieved by harmonic termination rather than by saturation. Thus, high efficiency is achieved at moderate input drive making the approach well suited for RF applications.

Key results of state-of-the-art power amplifier MMICs are listed in Table 9.4.

Table 9.4. Key performances of state-of-the-art power amplifier MMICs

Transistor/f_{max}	Operation frequency	Class	V_{dc}	P_{1dB}	Gain	PAE	Ref.
0.3-μm PHEMT/n.a.	10–11 GHz	E	5 V	24 dBm	10 dB	63%	[Tay01]
0.35-μm MOSFET/10 GHz	690–710 MHz	E	2.2 V	29.5 dBm	16 dB	62%	[Mert02]
0.35-μm MOSFET/10 GHz	1.7–1.8 GHz	B	1 V/ 3.4 V	25 dBm/ 30 dBm	20/ 23 dB	43/ 55%	[Fall01]
Bipolar/50 GHz	1.8–2 GHz	C	3 V	1.4 dBm	28 dB	55%	[Sim00]
0.5-μm MESFET/ 14 GHz	5.15–5.3 GHz	AB	3.3 V	20 dBm	19 dB	15%	[Bös96]
0.1-μm GaAs PHEMT/220 GHz	90–97 GHz	A	2.2 V	25 dBm	25 dB	7%	[Wan01]

9.5 Efficiency and Linearity Enhancement Techniques

As observed in previous sections and illustrated in Fig. 9.21, there is a tradeoff between efficiency and linearity. Linearisation techniques have been proposed, which relax this trade-off. Corresponding approaches will be outlined in the following sections. Elaborate discussions can be found in the specific literature [Gre05, Cri99, Ken00].

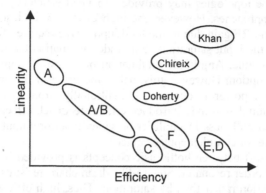

Fig. 9.21. Trade-off between efficiency and linearity

By the way, we can define two different kinds of linearities, which are partly inter-related:

1. Linearity as the level relation between a desired harmonic (e.g. the fundamental) and undesired harmonics or intermodulation products. Useful measures are the IM3 and the IP3. This type of linearity has a strong impact on adjacent communication channels.
2. Input to output linearity demanding for constant gain versus input power. A relevant measure is the 1 dB compression point. For amplitude-modulated systems, this kind of linearity plays a dominant role. In this context, switch-based amplifiers are quite non-linear since they are operated in saturation. In this regard, switch-based amplifiers may be termed as DC to RF converters rather than amplifiers.

9.5.1 Backoff Optimisation for Efficiency

According to Fig. 9.22a, maximum efficiencies are only achieved at maximum output power P_{max}, which is determined by the maximum possible swing within the IV curves. Consider, e.g. a class-A amplifier, where the DC power is drawn independent on the output power. Referring to Eqs. (9.8) and (9.9), it becomes clear that the efficiency is reduced if the output power is below P_{max}. The relation between P_{max} and the output power P_{L1} is represented by the backoff power ratio:

$$\text{backoff} = \frac{P_{max}}{P_{L1}} \geq 1 . \qquad (9.67)$$

Because of varying properties within the communication channel, coverage ranges and system requirements, the backoff ratio can easily be above 10 dB.

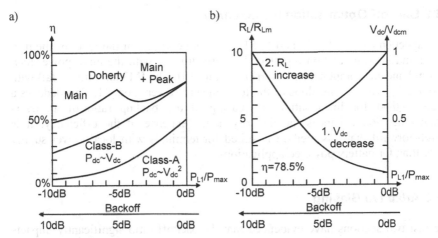

Fig. 9.22a,b. Backoff characteristics: **a** class-A, class-B and Doherty amplifier; **b** Two possible approaches maintaining maximum efficiency of class-B amplifier, V_{dcm}: V_{dc} at P_{max}, R_{Lm}: R_L at P_{max}

In Fig. 9.22a, the maximum efficiencies of class-A and class-B amplifiers are depicted versus backoff ratio. At 10 dB backoff, the efficiencies fall from 50% to 5% and from 78.5 % to 25 %, respectively. As expected, the worsening of the class-A amplifier is more significant than for the class-B amplifier. In class A, the DC power stays constant at raised backoff, whereas in class-B, the DC power is reduced at increased backoff. The efficiency of the class-A amplifier decreases by the same amount as the magnitude of the backoff ratio. At 10 dB backoff, we observe that 95% of the DC power is converted into heat, which may overstrain the amplifier and the system. The backoff ratio can be lowered by reducing P_{max} according to the system requirements. By doing so, the maximum efficiency can be preserved. Three approaches can be applied to maintain efficiency at varying backoff:

1. Decrease of supply voltage
Obviously, as deduced in preceding sections, P_{max} decreases with lowered V_{dc}. Hence, as demonstrated for a class-B amplifier in Fig. 9.22b, adequate reduction of V_{dc} allows to keep the backoff at 0 dB thereby preserving the maximum efficiency of 78.5%.

2. Increase of load resistance
Moreover, we have observed that P_{max} can be decreased by raising of R_L. In this context, refer also to Fig. 9.22b. A dynamically adjustable load is difficult to realise as physical passive element. The Doherty amplifier employs this idea by providing a virtual load determined by the relation between the effective voltage and current seen by the amplifier.

3. Switching of power cells
Also this idea is included in the Doherty amplifier.

9.5.2 Backoff Optimisation for Linearity

By inspection of Eqs. (4.47) and (5.51), we can observe that the level of the harmonic and intermodulation products decrease stronger with the input power than the fundamental components. Referring to Fig. 4.16, the IM3 amplitudes fall with a slope of 3, whereas the slope of the fundamental power is only 1. This deduces a crude method for linearisation. Increasing of the backup ratio of an over-dimensioned PA enhances the linearity at the expense of the efficiency. It is straightforward that this method is suited for terminals with local power supply rather than for battery powered applications.

9.5.3 Adaptive Biasing

The last two sections have evidenced that the backoff ratio significantly impacts the tradeoff between efficiency and linearity. Adaptive drain bias can be applied to optimise this tradeoff with respect to specific requirements, e.g. by moving from bias points B_1 to B_2 as depicted in Fig. 9.23a. In multi-channel systems with large data traffic, maximum linearity may be required demanding for class-A operation and subsequently higher bias current. In the latter case, the bias point may be shifted from B_1 to A.

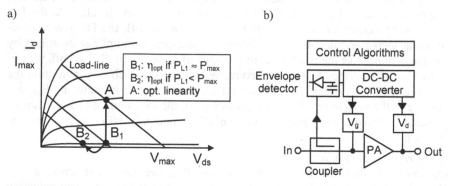

Fig. 9.23a,b. Adaptive bias taking into account the input power and requirements in terms of linearity and efficiency: **a** IV curve, η_{opt}: optimum efficiency; **b** system with self controlling envelope tracking

We can conclude that an adaptive approach with flexible supply voltage and conductance angle can be very effective [Han99]. The required output power may be known in the baseband. By means of DC converters, the supply voltages can be adjusted. Evidently, the power consumption of these DC converters must be well below that of the power amplifier. Otherwise the method would not be effective. Self-controlling amplifiers may be realised on basis of envelope detectors and power couplers preceding the amplifier, and control logic. A corresponding circuitry is sketched in Fig. 9.23b. For further information, the reader is referred to [Sta02].

9.5.4 Doherty Amplifier

The Doherty amplifier [Doh36] employs the latter two ideas outlined in Sect. 9.5.1. Depending on the backoff ratio, the virtual R_L and the number of power cells are adapted to maximise the efficiency and the output power. Referring to Fig. 9.24a,b, the input power is split between the main amplifier operating at all conditions, and the peak amplifier only contributing at high power starting at a backoff ratio of around 6 dB. The threshold of the amplifiers can be defined by the bias and or the threshold of the employed transistors. IC technologies feature devices with different threshold voltages. Since the peak amplifier should conduct at higher input power than the main amplifier, the conductance angle of the peak amplifier is lower than the one of the main amplifier. Moreover, the gate width of the peak device is typically larger to provide high peak output power.

In Fig. 9.22a, the typical performance of a Doherty circuit with a class-AB main amplifier and a class-B/C peak cell has been illustrated. The superior performance under conditions with high backoff ratio and varying output power is demonstrated. Since the Doherty amplifier is based on current source amplifiers, the linearity is relatively high. Thus, in addition to phase and frequency modulation, amplitude modulation may be possible. The two amplifiers generate a load pulling effect, where the loads $\underline{Z}_m$ and $\underline{Z}_p$ seen by the main and peak amplifier depend on the currents $\underline{I}_m$ and $\underline{I}_p$ fed into R_L. Referring to Fig. 9.24c, we find by the Kirchhoff circuit theory that

$$\underline{Z}_m \cdot \underline{I}_m = R_L \cdot \left(\underline{I}_m + \underline{I}_p \right) \tag{9.68}$$

resulting in

$$\underline{Z}_m = R_L \cdot \left[1 + \frac{\underline{I}_p}{\underline{I}_m} \right] \tag{9.69}$$

and

$$\underline{Z}_p = R_L \cdot \left[1 + \frac{\underline{I}_m}{\underline{I}_p} \right]. \tag{9.70}$$

Let's discuss a typical example, where the phases of $\underline{I}_m$ and $\underline{I}_p$ are equal. To optimise the efficiency and the backoff conditions at low input power, a high resistive value of $\underline{Z}_m$ is desired. Unfortunately, at weak output power, the magnitude of $\frac{I_p}{I_m}$ is low resulting in a small resistive value of $\underline{Z}_m$. Opposite characteristics are desired. An impedance inverter can be employed, which can be realised by a $\lambda/4$ transmission line incorporated between the main amplifier and the load node. To enable constructive in-phase combining, a second $\lambda/4$ line is added at the input of the peak amplifier. For details concerning the impedance conversion properties of $\lambda/4$ transmission lines, the reader is referred back to Sect. 3.6.4. At high output power, the magnitude of $\frac{I_p}{I_m}$ is high leading to a small $\underline{Z}_p$, which is fruitful to maximise the contribution of the peak amplifier. The reader may derive the characteristics of the circuit in the case that $\underline{I}_m$ and $\underline{I}_p$ have opposite phases.

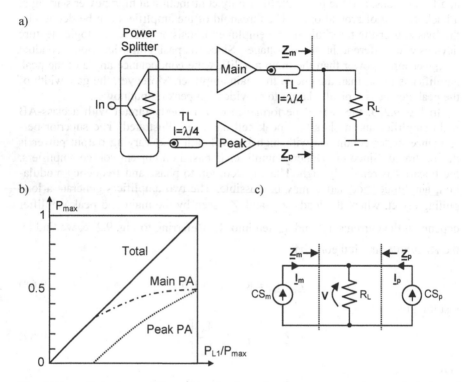

Fig. 9.24a–c. Doherty amplifier: **a** circuit schematics; **b** power properties; **c** calculation of the load seen by the amplifiers, CS: complex current source

9.5.5 LINC and Chireix Outphasing Amplifier

Mixed AM and PM modulation schemes provide high data rates. Unfortunately, amplitude modulation is not possible for switched or saturated amplifiers such as class-C, class-D or class-E. This is a pity because these amplifiers provide the highest efficiency. Various methods have been presented, which solve this problem by varying the phase in the combiner of amplifier systems [Ken00]. Depending on the combiner phase, the system amplitude can be modulated. These circuits are known as LINC (Linear Amplification with Nonlinear Components) systems.

One example of an LINC approach is the Chireix outphasing amplifier [Chi35] illustrated in Fig. 9.25. The differential input signal is split into two equal paths which are modulated by $+\phi$ and $-\phi$, amplified by two identical amplifiers and finally combined. With peak output power at $\phi=90°$ and zero output power at $\phi=0°$, 100% amplitude modulation depth can be achieved by varying the phase. It is clear that the efficiency degrades towards $\phi=0°$ since the DC dissipation stays more or less constant. This problem can be mitigated by adding phase dependent reactance in the outputs of the amplifiers [Gre05]. The manipulation of the impedance seen by the amplifiers reduces the DC power and the degradation in efficiency. Despite its architectural appeal LINC amplifiers are not widespread today. The reason is the challenging realisation of a linear combiner with sufficient bandwidth, low loss, and high isolation required to suppress load pulling effects. Given that the two input phases would be equal, these properties could be achieved with a simple Wilkinson combiner. However, the fact that the input phases vary significantly with frequency, the theoretical benefit of the concept is limited.

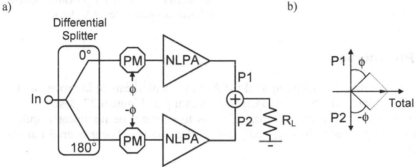

Fig. 9.25a,b. Chireix amplifier: **a** circuit schematics; NLPA: Nonlinear Power Amplifier **b** amplitude combining

9.5.6 Kahn Envelope Elimination and Restoration Amplifier

If an idea does not seem to be absurd at the first time, the idea is not good.
A. Einstein

L. R. Kahn has demonstrated a further method allowing both AM and PM modulation on basis of nonlinear amplifiers [Kah52]. According to Fig. 9.26, the mixed AM/PM signal is fed into a limiter preserving the PM information. A coupler and detector located in front of the circuit track the AM envelope modulating the drain supply voltage of the PA. Recall that the amplitude of the PA can be varied with supply voltage. Consequently, efficient mixed AM/PM modulation is possible using nonlinear switch type amplifiers. Since they always operate at P_{max} due to the input saturation, a high efficiency can be achieved. However, the approach is only efficient as long as the power consumption of the control chain is well below that of the amplifier.

a)

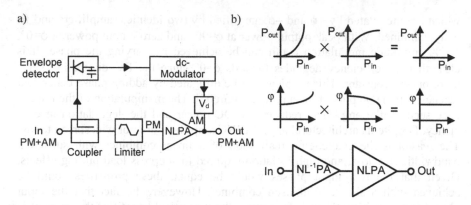

Fig. 9.26. Kahn envelope elimination and restoration amplifier

Fig. 9.27. Illustration of predistortion, cascading of NLPA (Nonlinear Power Amplifier) together with circuit providing inverse nonlinear characteristics (NL^{-1}PA)

9.5.7 Predistortion

The input to output linearity required for AM modulation can be improved by using predistortion, which is an open-loop technique. Figure 9.27 describes the properties of the approach. The basic idea is to cascade the nonlinear amplifier with a circuit providing inverse nonlinearities leading to a linear overall transfer function.

9.5.8 Envelope Tracking

The input and output power can be determined and compared as depicted in Fig. 9.28 allowing for bias corrections and consequently linear gain up to higher input power, and an enhanced 1 dB compression point. This approach is similar to the smart biasing presented in Sect. 9.5.3.

a)   b)

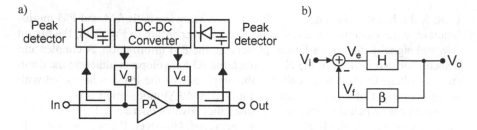

Fig. 9.28a,b. Feedback techniques: a envelope tracking; b negative feedback

9.5.9 Negative Feedback

Negative feedback means that the original signal and the feedback signal have opposite relative phases. The corresponding schematic is illustrated in Fig. 9.28b. In a common source amplifier, the opposite phase is generated by its inherent phase shift of 180°. Hence, a simple resistor between the gate and the drain may be used for the feedback. With magnitude transfer function H, feedback β, and $V_0 = H \cdot V_e = H \cdot (V_i - V_f) = H \cdot (V_i - V_0 \cdot \beta)$, the open loop voltage gain can be described by

$$a_v = \frac{V_0}{V_i} = \frac{H}{1 + \beta \cdot H}.$$
(9.71)

We get

$$\frac{da_v}{dH} = \frac{1}{(1 + \beta \cdot H)^2},$$
(9.72)

and subsequently

$$\frac{da_v}{a_v} = \frac{1}{1 + \beta \cdot H} \cdot \frac{dH}{H}.$$
(9.73)

From the latter equation we can deduce that the impact of variations of H on a_v are reduced by a factor of $1 + \beta \cdot H > 1$. Thus, the properties in terms of compression, nonlinearities, process variations and aging are improved. This comes at the expense of decreased gain. Hence, negative feedback is not well suited for high operation frequencies.

9.5.10 Feedforward

Figure 9.29. illustrates the typical schematics of a feedforward system. The input signal is split in two equal parts. One path is amplified yielding a distorted signal, which is subtracted by the non-amplified linear signal available in the second path. The resulting error signal is fed through a second amplifier. A much more linear signal is achieved after subtracting the distorted signal from the error signal. To allow maximum cancellation of the distortion, the phases are tuned in phase shifters and the amplitude of the error signal is adjusted in the second amplifier. Evidently, the second amplifier introduces distortion as well. However, since the amplitude of the error signal is much lower than the one at the fundamental frequency, the linearity of the error path is sufficient. One major drawback of the approach is the additional power consumption in the second amplifier reducing the efficiency. Moreover, the circuit size is relatively large.

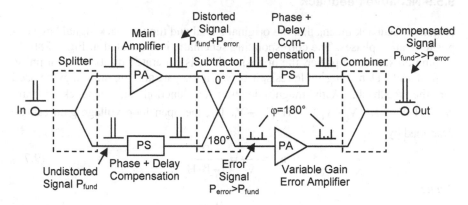

Fig. 9.29. Feedforward linearisation

9.6 Power Combining

As illustrated in Fig. 9.30, the supply voltage of CMOS technologies have been scaled down from around 5 V for 1-µm technology in 1990 to around 1 V for to-day's 90-nm processes. Major reasons for this scaling are the speed improvements and the power saving for logic circuits. The relation between the supply voltage and the gate lengths may be approximated by

$$V_{dc} \approx 1V \sqrt{\frac{l_g [nm]}{90nm}} \ . \tag{9.74}$$

Unfortunately, the scaling of the gate length significantly decreases the potential for high output power, which is more or less proportional to V_{dc}^2. Hence, the supply voltage of the PA can't be reduced at the same amount as the one of the other components. Due to the limited breakdown properties of scaled CMOS technologies, III/V rather than silicon-based technologies are used for the PA. Thus, many mobile systems comprise at least two chips. Needless to say that a single chip solution would be preferred from economic point of view.

A further power and efficiency limitation is associated with the channel resistance, which scales down together with the gate length. Typical RF CMOS PA devices exhibit parasitic drain source resistances with values well below 500 Ω. The corresponding current drop with respect to the load connected in parallel can be significant. Moreover, due to the low output impedance, low loss impedance matching becomes more and more challenging, since the parasitic series resistance becomes significant.

The next sections are devoted to power combining techniques, which can be employed to achieve high output power at low device supply voltages. By the way, we have already treated some power combining concepts, e.g. the travelling wave, balanced, push-pull and Doherty amplifier.

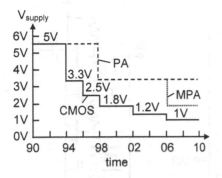

Fig. 9.30. Supply voltage maturation, MPA: moderate power amplifier

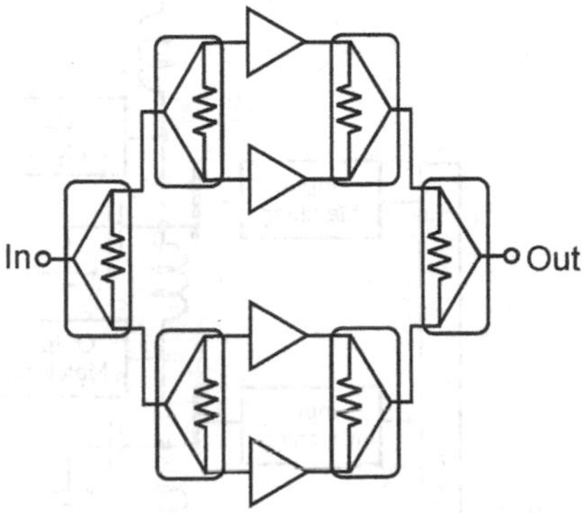

Fig. 9.31. Power combining using Wilkinson splitters/combiners

9.6.1 Amplifier Arrays

Wilkinson splitters/combiners can be applied to built up power clusters as shown in Fig. 9.31. In hybrid form, this method is easy to realise and exhibits efficient and scalable performance. However, monolithic integration introduces several drawbacks. Wilkinson splitters/combiners consume lots of costly circuits area and have significant losses due to the required $\lambda/4$ transmission lines. For details about Wilkinson splitters/combiners, the reader is referred back to Sect. 6.10. Size reduction is possible by using lumped element equivalents of the lines. However, also these couplers are still bulky and can easily have losses in the order of 1–2 dB per device. Both size and losses reduce with frequency. Thus, at frequencies well above 10 GHz and up to a certain complexity, lumped element based power splitters/combiners can do a good job. In the presented circuit schematics, the input and output impedances are preserved. Thus, common amplifiers with 50 Ω terminations can be employed. Depending on the desired impedance conditions, the input splitters are not necessarily required and may be replaced by a common input node. Due to the relatively large circuit area and handled power, substrate coupling may generate undesired impacts on other components. Substrate shielding techniques are required to mitigate this effect.

9.6.2 FET Stacking

We have already got to know one FET stacking topology, namely the cascode circuit benefiting from increased output resistance and higher RF gain leading to enhanced efficiency and PAE. However, problems associated with the RF grounding of the gate of the common gate stage may appear.

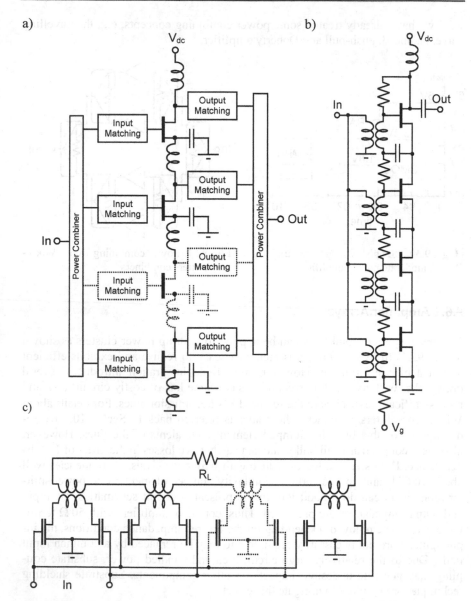

Fig. 9.32a–c. Transistor stacking: **a** parallel RF isolated stages with series DC output;
b DC coupling using transformer, parallel input, series output; **c** differential input coupled
by transformer to single ended output

An architecture stacking n common source circuits is depicted in Fig. 9.32a
[Ezz85]. The signal is divided at the input, amplified and combined at the output.
Choke inductors connect the sources and drains of the individual FETs. Thus, the
FETs are DC connected in series, whereas RF-wise the inputs/outputs are in parallel.

To provide RF isolation between the FETs, the source is RF grounded by a capacitance. The output power is n times that of a single stage. Major drawback is the loss generated by the inductors, which must have a large value to provide sufficient RF impedance. Moreover, the inductors require lots of circuit size and must have a high Q to avoid DC losses. For the latter reasons, this topology has been implemented into III/V technology.

A further topology employing stacked common source circuits, where the inputs are in parallel and the outputs are in series is illustrated in Fig. 9.32b [Ror99]. The output voltages are added increasing the output resistance. This lowers the losses associated with impedance transformations of the transistors with very large gate widths. DC isolation between the input nodes is performed by transformers allowing RF grounding of the source nodes by means of capacitors. A ladder of high-ohmic resistors acts as voltage divider determining the gate DC voltages. Compared to the preceding approach, no combiner is required at the output.

The circuit schematic in Fig. 9.32c illustrates a smart topology where the power of differential amplifiers is added by means of transformers feeding a single ended load [Aok02]. As in the former approach, the output resistance is increased avoiding extreme impedance transformations. Due to the constructive combining of the differential signals the double voltage swing is achieved per amplifier.

9.6.3 Travelling Wave Amplifier

Referring to Sects. 7.4.6 and 8.3.2, the travelling wave amplifier adds the currents of the individual stages. Very large bandwidth can be achieved. Concerning the maximum output power and efficiency, we have to identify one drawback, which is the output line termination resistor. Significant RF power is dissipated in this resistor. The output power can be increased using a high number of stages and/or using large transistors. However, this lowers the maximum operation frequency and bandwidth. Typically, the RF power is limited by the voltage swing in the last stage, where the signal amplitude is maximal.

9.7 Tutorials

1. Which basic circuit topology would you prefer for power amplifiers? Why?
2. What are the design tradeoffs for power amplifiers? Compare the class-A, class-B, and class-C characteristics in terms of bias point, conductance angle, output waveform, maximum gain, operation frequency, maximum output power, efficiency and linearity. Generally, how can we increase the efficiency?
3. In an ideal case, could we achieve 100% efficiency in class-C? What is the major drawback?
4. Derive the efficiency for a class-A amplifier.

5. Which are the key sources of losses inherent in transistors impacting the power and efficiency of amplifiers?

6. You design a single transistor amplifier, which is unstable. How do you stabilise it? Which stabilisation principles should be avoided?

7. Explain each element of the power amplifier example according to Sect. 9.2.7. Why is the output FET larger than the preceding FET? Why do we bias the first FET towards class-A and the output FET towards class-B? Calculate the approximate RF and DC drain current in the input and output FET.

8. What is the idea behind switched amplifiers in terms of increased efficiency? Illustrate the desired waveforms across the transistors. What are the drawbacks?

9. Explain the basic idea of a class-E amplifier by a simplified circuit schematic. Why are the analytic calculations complex? What about the design flexibility compared to other approaches? What is the theoretical maximum efficiency? What limits the operation frequency? If you would like to design an amplifier at highest frequencies, how do you choose C_s? Why do we need this capacitance? What is the task of L_x? What are the practical limitations of class-E amplifiers?

10. Explain how class-F amplifiers work. Why do we need less input power and saturation compared to other switched amplifier approaches? How would you realise a fully integrated class-F amplifier? What are the options?

11. How can we realise a class-D amplifier providing both half-waves? Explain two principles. Discuss the pros and cons.

12. What is the difference between the drain/emitter efficiency and the PAE? Derive the equation for the PAE.

13. Why do we have to trade off linearity and efficiency?

14. What is backoff ratio and how does it influence the efficiency and linearity? Is backoff an issue for mobile systems? Compare the impact of backoff on class-A, class-B and class-C amplifiers.

15. How could we detect and monitor the backoff in a circuit?

16. What is smart biasing? How could we optimise the backoff for efficiency and how for linearity?

17. Are the gain matching and the output power matching the same under large signal conditions? Which other kind of impedance matching do you know? Which ones are relevant for power amplifiers?

18. What does load-pull mean? How can we use load-pull data for amplifier optimisations?

19. Your amplifier must be very simple and has to handle an amplitude modulation scheme. Which kind of amplifiers do you use? What about the case when the system applies phase modulation only?

20. What happens if an amplifier is nonlinear and generates undesired harmonics and intermodulation products? Consider neighbouring channels. You have to design a power amplifier at 1.8–2.4 GHz and at 9–10 GHz. In comparison, what are the qualitatively requirements in terms of permitted adjacent power and consequently linearity?

21. Consider both the current source and switched amplifier topologies. Which f_t performances do we approximately need for amplifiers specified according to the 802.11a WLAN standards?

22. How can we enhance the linearity of power amplifiers?

23. Could we apply nonlinear amplifiers for linear amplification? Explain corresponding principles. What are the drawbacks of these principles with respect to monolithic integration?

24. What are the basic ideas behind the Doherty amplifier?

25. Why does negative feedback improve the linearity? What about the impact on process variations?

26. Explain the feedforward principle. What are the pros and cons? Is it suited for low-cost systems?

27. What are the two major drawbacks of the gate length scaling with respect to the achievable output power? How did the supply voltage mature during the last decade?

28. Your boss wants you to design a 1-W single transistor amplifier in 90-nm CMOS technology. Estimate the required output resistance and deduce the corresponding problems in terms of output matching and efficiency. What is your opinion concerning this design task?

29. How could you increase the load resistance at given output power?

30. Comment on the use of integrated transformers. What are the benefits and disadvantages?

31. How could you benefit from differential power amplifiers?

32. Explain methods for power combining. How could you realise an integrated divider and splitter? What do you have to consider in terms of input phases? How does this effect the layout?

33. What are typical performances of leading-edge power amplifiers?

References

[Aok02] I. Aoki, S. D. Kee, D. B. Rutledge, A. Hajimiri, "Fully integrated CMOS power amplifier design using the distributed active-transformer architecture", IEEE Journal on Solid-State Circuits, Vol. 37, No. 3, pp. 371–383, March 2002.

[Bös96] T. Bös et al., "A monolithic integrated, on chip matched GaAs power amplifier for HIPERLAN with a single 3.3V supply", Microwave and Millimeter-Wave Monolithic Circuits Symposium, San Francisco, pp. 81–84, June 1996.

[Chi35] H. Chireix, "High power outphasing modulation" Proc. IRE, Vol. 23, No. 11, Nov. 1935, pp. 1370–1392.

[Cri99] S. C. Cripps, RF Power Amplifiers for Wireless Communication, Artech House, Boston, 1999.

[Doh36] W. H. Doherty, "A new high efficiency power amplifier for modulated waves", Proc. IRE, Vol. 24, No. 9, Sept. 1936, pp. 1163–1182.

[Ezz85]   A. Ezzeddine, H. L. Hung, H. C. Huang, "High voltage FET amplifiers and phased-array applications", IEEE Microwave Theory and Techniques Symposium, pp. 336–339, 1985.

[Fall01]   C. Fallesen, "A 1 W CMOS power amplifier for GSM-1800 with 55% PAE", Microwave Theory and Techniques Symposium, pp. 911–914, June 2001.

[Gre05]   A. Grebennikov, RF and Microwave Power Amplifier Design, McGraw-Hill, 2005.

[Han99]   G. Hanington et al., "High efficiency power amplifier using dynamic power supply voltage for CDMA applications", IEEE Transactions on Microwave Theory and Techniques, Vol. 47, No. 8, Aug. 1999.

[Kah52]   L. R. Kahn, "Single sideband transmission by envelope elimination and restoration", Proc. IRE, Vol. 40, pp. 803–806, July 1952.

[Ken00]   P. B. Kenington, High-linearity RF Amplifier Design, Artech House, 2000.

[Kra80]   H. L. Krauss, C. W. Bostian, F. H. Raab, Solid-state Radio Engineering, Wiley, New York, 1980.

[Mert02] K. L. R. Mertens, "A 700 MHz fully differential CMOS Class-E power amplifier" Journal of Solid State Circuits, Vol. 37, No. 2, pp. 137–141, Feb. 2002.

[Raa77]   F. H. Raab, "Idealized operation of the class-E tuned power amplifier", IEEE Transactions on Circuits and Systems, CAS-24, pp. 725–735, Dec. 1977.

[Raa78]   F. H. Raab, N. O. Sokal, "Transistor power losses in the class E tuned power amplifier", IEEE Journal on Solid-State Circuits, Vol. SC-13, No. 6, June 1978.

[Raa97]   F. H. Raab, "Class-F power amplifiers with maximally flat waveforms", IEEE Transactions on Microwave Theory and Techniques, Vol. 45, No. 11, pp. 2007–2011, Nov. 1997.

[Ror99]   J. G. McRory, G. G. Rabjohn, R. H. Johnston, "Transformer coupled stacked FET power amplifiers", IEEE Journal on Solid-State Circuits, Vol. 34, No. 2, pp. 157–161, Feb. 1999.

[Sim00]   W. Simburger et al., "A monolithic 2.5V, 1W silicon bipolar power amplifier with 55% PAE at 1.9GHz", Microwave Theory and Techniques Symposium, pp. 853–856, June 2000.

[Sok75]   N. O. Sokal, A. D. Sokal, "Class-E – a new class of high efficiency tuned single-ended switching power amplifiers", IEEE Journal on Solid-State Circuits, Vol. SC-10, No. 3, June 1975.

[Sta02]   J. Staudinger, "An overview of efficiency enhancements with application to linear handset power amplifiers", IEEE RFIC Symposium, pp. 45–48, June 2002.

[Tay01]   R. Tayrani, "A monolithic X-band class-E power amplifier", GaAs IC Symposium, pp. 205–208, 2001.

[Wan01]   H. Wang, L. Samoska, T. Gaier et al., "Power-amplifier modules covering 70–113 GHz using MMICs", Transactions on Microwave Theory and Techniques, Vol. 49, No. 1, pp. 9–16, Jan. 2001.

10 Mixers

I know that I know nothing.
Socrates

As discussed in Sect. 2 and illustrated in Fig. 10.1, mixers convert the high RF frequency to a low IF frequency in receivers and vice versa in transmitters. The corresponding circuits are referred to as down- and up-mixers, respectively. An LO signal provided by a VCO is required for this operation. For capacity reasons and to allow coexistence with other standards, the data has to be transmitted by means of a high RF carrier frequency, whereas in the receiver, low IF frequencies are required for simple baseband processing.

The requirements for down-mixers, e.g. regarding gain and noise, are more challenging than for up-mixers. This is attributed to the fact that the signal to noise ratio in transmitters is high because of the strong signal power being locally available. In the following sections we focus on down-mixers. However, the gained insights can be mapped to up-mixers.

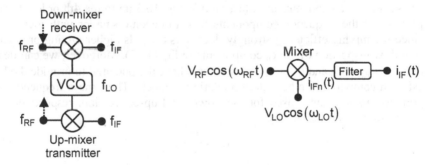

Fig. 10.1. Mixers used for frequency down- and up-conversion in receivers and transmitters

Fig. 10.2. Mixer fed by sinusoidal RF and LO signals yielding the IF signal

10.1 Nonlinearities and Mixing Products

The nonlinearities of a device as a function of a time variant LO signal are instrumental for mixing [Maa03]. Undesired frequency components have to be filtered out. Referring to the schematic depicted in Fig. 10.2, the voltage across any nonlinear device may be represented by

$$v_{IF}(t) = V_{LO}\cos(\omega_{LO}t) + V_{RF}\cos(\omega_{RF}t). \qquad (10.1)$$

In this simple model, we assume that the mixer acts as a signal adder. The current characteristics of nonlinear device can by described by a Taylor-series:

$$i_{IF}(t) = c_0 + c_1 v_{IF}(t) + c_2 v_{IF}^2(t) + c_3 v_{IF}^3(t)... + c_n v_{IF}^n(t). \qquad (10.2)$$

The constants c_n are obtained by inspection of the individual current versus voltage characteristics of a device. For FETs these properties may be based on a quadratic function, whereas for BJTs, the dependency might rather be exponential. With Eqs. (10.1) and (10.2) we get

$$i_{IF}(t) = c_0 + c_1\left[V_{RF}\cos(\omega_{RF}t) + V_{LO}\cos(\omega_{LO}t)\right] +$$

$$c_2\left[\begin{array}{l}\dfrac{V_{RF}^2}{2}\left(1+\cos(\omega_{RF}t)\right) + \dfrac{V_{LO}^2}{2}\left(1+\cos(\omega_{LO}t)\right) + \\[2mm] \dfrac{V_{RF}V_{LO}}{2}\left(\cos(\omega_{LO}t+\omega_{LO}t) + \cos(\omega_{LO}t-\omega_{LO}t)\right)\end{array}\right] + ... \qquad (10.3)$$

These mixing components are illustrated in Fig. 10.3 up to the 4th order. The amplitudes of the frequency components tend to become smaller with increased n, since the mixing efficiency strongly decreases towards higher mixing products as found by inspection of the c_n components in Eq. (10.3). Moreover, we can deduce that the $\omega_{RF}-\omega_{LO}$ and $\omega_{RF}+\omega_{LO}$ components have the potential to provide the highest gain compared to other intermodulation products. Thus, these frequency components are frequently used for the down- and up-conversion, respectively. The conversion loss is defined by

$$L_c = \frac{P_{IF}}{P_{RF}}, \qquad (10.4)$$

where P_{RF} and P_{IF} denote the RF and IF power. Conversion gain G_c can be reached for active mixers. We have to consider that the LO power is usually much higher than the RF power, because the RF signal is attenuated by propagation in the air, whereas the LO is locally fed. Thus, the $n \cdot f_{RF}$ components are relatively weak, whereas the undesired $n \cdot f_{LO}$ components are strong and have to be filtered. The suppression of undesired components is measured by means of the port to port isolations, which can be important in systems, where we want to avoid compression of circuits following the mixer. Important are high LO to IF and LO to RF isolations since the LO power is very strong. In circuits targeted for low power consuming applications, the conversion efficiency

$$\eta_c = \frac{P_{IF}}{P_{RF} + P_{LO} + P_{dc}} \qquad (10.5)$$

is an important measure, which includes also the LO power P_{LO} and the DC power P_{dc}. Similar to amplifiers, we may define the following figure of merit for mixers:

$$FOM = \frac{f_c[GHz]}{f_t[GHz]} \cdot \frac{G_c[dB]}{NF[dB]} \cdot \frac{IIP3[dBm]}{P_{dc}[dBm]}. \qquad (10.6)$$

The first term in the equation describes the centre operation frequency f_c with respect to the f_t of the used technology, the second factor considers the conversion gain to noise figure ratio and the third term relates the large signal properties to the consumed DC power. Several modifications of Eq. (10.6) can be found in the literature. Optionally, the representation of the large signal properties by means of P_{1dB} instead of the IIP3 may be reasonable as well. The use of non-logarithmic values may be an option. For wideband mixers, f_c may be substituted by the bandwidth.

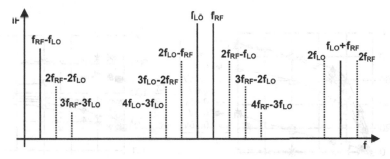

Fig. 10.3. Part of IF mixer spectrum generated by nonlinear mixing of f_{RF} with f_{LO}

Table 10.1. Element variations of 90 nm n-channel SOI MOSFET with w_g=64 μm

Element	Variation, $f(V_{gs})$	Variation, $f(V_{ds})$	Resulting nonlinear effect
R_g, R_d, R_s	Small values, negligible dependency		Very weak
R_{gs}	<20%		Weak, path dominated by series C_{gs}
C_{ds}	<10%		Weak since dominated by par. R_{ds}
C_{gd}	30%		Weak miller effect due to mixer bias, major input capacitance is C_{gs}
C_{gs}	30%	25%	Moderate
r_o	6 Ω–2 kΩ		Strong
g_m	0–80 mS		Very strong

as function of voltage variations from 0 to 1 V on condition that other voltage is fixed at optimum value between 0 and 1 V

Transistors are frequently used as the nonlinear devices in monolithically integrated mixers. VLSI IC processes feature no optimised mixing diodes as commonly applied in discrete and off-chip implementations. These mixing diodes would require special doping profiles. For cost reasons such doping profiles are rarely available.

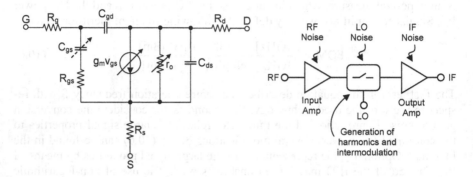

Fig. 10.4. Simplified equivalent circuit of MOSFET and dominant nonlinear elements

Fig. 10.6. Generic schematic of mixer, nonlinearities are generated by switching

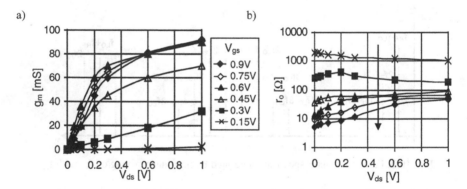

Fig. 10.5a,b. Dominant nonlinearities of 90-nm SOI n-channel FET: **a** transconductance; **b** channel resistance

The nonlinear characteristics of FETs can be a function of V_{gs} and or V_{ds}. In Fig. 10.4, the simplified equivalent circuit of a FET is shown indicating the major nonlinearities. To gain first insights concerning the potential for mixing, the relative variations of the equivalent circuit elements for 90-nm n-channel SOI MOSFET are measured and listed in Table 10.1. As expected, no nonlinearities are associated with the contact resistances R_g, R_d and R_s. Weak changes of less than 30% are observed for the elements R_{gs}, C_{ds}, C_{gd} and C_{gs}. The highest nonlinear properties are expected for g_m dominating the current source properties of the device followed by the output impedance r_o. Figure 10.5 illustrates the characteristics of the latter two parameters. According to Table 5.5, $r_o=r_{ds}\,||\,R_{ds}$. In the triode or resistive region, where $V_{ds}<V_{ds,sat}$, r_o is determined by R_{ds}, which can become quite low at high V_{gs}.

The optimum choice of the type of nonlinearity depends on the application. The g_m properties allow higher conversion gain, whereas the r_o characteristics yield higher linearity with respect to unwanted frequency components. The final choice depends on the desired tradeoff between frequency bandwidth, required LO power, DC power and the requirements in terms of gain/loss, linearity and output power.

10.2 Noise

According to Sect. 4.3.5, the system noise figure of receivers is mainly determined by the noise figure of the LNA located in front of the mixer. Given that the LNA has a high gain, the contribution of the mixer noise on the system noise figure is weak. Consequently, mixer noise is usually not critical as long as the noise is not exorbitantly high.

The analysis of mixer noise requires nonlinear noise modelling, which is complex [Maa93, Hul93]. Since the theories have to be approximated heavily to give analytical results, it is not a surprise that the agreement between theories and practice is limited. Many conclusions based on the analysis are contradicted by experimental results of other researchers. Thus, from didactic point of view, no noise modelling is treated in this book. However, the designer should keep the insights concerning linear noise in mind. Since the nonlinear noise is based on the conversion of linear noise sources, the minimisation of linear noise sources helps to lower the nonlinear mixer noise. Corresponding design strategies involve the proper choice of the transistors, the bias, the RF input matching and the avoidance of resistive parasitics. The generic schematics depicted in Fig. 10.6 may help to understand the noise sources qualitatively. In a thought experiment, we may split the mixer circuit into three parts: an RF input amplifier, the section where the nonlinearity is generated, and an IF amplifier. Sure, in practice all three parts may merge. Given that the gain of the virtual RF amplifier is high, the noise is determined by the input stage. The linearity of the active mixers may be determined by the IF stage, which has to handle the gain of the preceding stages. By the way, the noise estimation of resistive mixer is easy. In good approximation, the noise equals the resistive loss.

For details concerning mixer noise, the interested and scientifically motivated reader is referred to literature: FET gate-pumped transconductance mixers [Tie83], MOS Gilbert cell [Ter99, Hey04], SiGe Gilbert cell [Joh05], and distributed Gilbert cell mixer [Saf05].

Compared to the theoretical analysis, practical measurements can be accomplished relatively easy. As discussed in Sect. 15.4.6, both the single side band (SSB) and double side band (DSB) noise figures are used for the characterisation.

10.3 Topologies

An overview of the most important approaches used for mixers is sketched in Fig. 10.7. Either active or passive topologies can be employed [Maa03, Rob04, Maa98]. As mentioned beforehand, active topologies mainly apply nonlinear transconductances and channel resistances. The differential pair and Gilbert cell mixers are based on nonlinearities associated with hard switching. Passive mixers mainly exploit the nonlinearities of resistances and reactive elements at zero DC channel voltage yielding excellent large signal properties at zero DC power. Drawback of passive approaches is their loss, which has to be compensated by amplifiers drawing DC power.

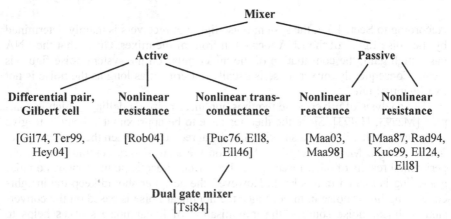

Fig. 10.7. Overview of mixer topologies and corresponding nonlinearities

10.3.1 Transconductance-Pumped Mixers

The simplified equivalent circuit of a g_m-type down-mixer FET is shown in Fig. 10.8. As outlined in Table 10.2, either the gate or the drain voltage can be pumped by the LO signal leading to a time-variant function of g_m, which generates the nonlinearities required for mixing.

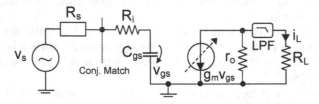

Fig. 10.8. Simplified equivalent circuit of transconductance FET down-mixer, $g_m = f(V_{gs}, V_{ds})$: nonlinear transconductance, R_i: gate resistance $f(R_g, R_{gs})$, r_o: drain source impedance, C_{gs}: gate source capacitance, R_s: source impedance (could also be complex), R_L: load, compare [Ell14] © IEEE 2004

Drain pumped down-mixers have been demonstrated in practice [Dar96, Bur76, Ell14] and analysed theoretically [Beg79]. They have a remarkable advantage compared to the gate pumped approach. The RF and the LO frequency, which are close together, are injected at different ports simplifying the filtering and improving the LO to RF and RF to LO isolation. The isolation is limited by the C_{gd} of the transistor. Biased at the transition between the linear and saturation region, the transistor provides the most pronounced level of nonlinearities. Aggressively scaled CMOS FETs have the advantage that they have very small drain source saturation voltage $V_{ds,sat}$ of below 0.2 V. Thus, no V_{ds} bias voltage is necessarily required since the applied LO power is sufficient to drive the device from zero V_{ds} to $V_{ds,sat}$.

Table 10.2. Overview of transconductance pumped mixers

	Gate pumped	Drain pumped	
Circuit schematics			
Functional principle			
V_{ds} bias	$> V_{ds,sat}$	$V_{ds,sat}$	0 V possible if $V_{ds,sat}$ small
V_{gs} bias	V_{th}	$V_{gs} > V_{th}$	
Non-linearity	$g_m = f(V_{gs})$	$g_m = f(V_{ds})$	
Reference	[Kwo93, Orz03, Puc76, Ell46]	[Dar96, Bur76, Beg79]	[Ell14]

The gate pumped mixer operates in the saturation region with V_{gs} biased close to the threshold voltage V_{th}, where maximum non-linear variations of g_m are achieved [Puc76, Kwo93, Orz03].

For both the gate- and drain-pumped mixer, the LO signal leads to a time variant g_m, which may be approximated by a cosine function with 50% duty cycle. In this context, compare the g_m waveforms according to Table 10.2. Only the positive half-wave of the cosine exists since $g_m(t)$ equals zero for $V_{gs} \leq V_{th}$ and $V_{ds} \leq 0$, respectively, assuming n-channel devices. The conversion gain mainly depends on the fundamental component of $g_m(t)$. Following the derivations of [Maa03], the fundamental component can be calculated by Fourier series [Bro91] yielding

$$g_{m1}(t) = \frac{1}{2} g_{m,max} \cos(\omega_{LO}t) \qquad (10.7)$$

where $g_{m,max}$ represents the peak value of $g_m(t)$. The RF input port of the mixer is fed with the RF voltage

$$v_{s,RF}(t) = V_{s,RF} \cos(\omega_{RF}t), \qquad (10.8)$$

where V_{sRF} denotes the RF input amplitude. For maximum power transfer we demand conjugate matching at the node between the RF input and the mixer circuit. According to Fig. 10.8, at the gate node, we can find the following identity concerning the gate currents:

$$v_{gs,RF}(t) \cdot \omega_{RF} C_{gs} = \frac{v_{s,RF}(t)}{2R_i}, \qquad (10.9)$$

where R_i is a function of R_{gs} and R_g. The overall load current including the RF, LO and mixing frequencies is given by

$$i_d(t) = -g_m(t) \cdot v_{gs,RF}(t) \cdot \frac{r_o}{r_o + R_L}. \qquad (10.10)$$

With Eqs. (10.7)–(10.10), the trigonometric theorem

$$\cos(\omega_{LO}t) \cdot \cos(\omega_{RF}t) = \frac{1}{2}\cos\left[(\omega_{RF} - \omega_{LO})t\right] + \frac{1}{2}\cos\left[(\omega_{RF} + \omega_{LO})t\right], \qquad (10.11)$$

and assuming ideal filtering of all frequency components except the IF frequency $\omega_{IF} = \omega_{RF} - \omega_{LO}$, we obtain an IF drain current of

$$i_{L,IF}(t) = -\frac{g_{m,max} V_{s,RF} \cos(\omega_{IF}t)}{8\omega_{RF} C_{gs} R_i} \cdot \frac{r_o}{r_o + R_L}. \qquad (10.12)$$

With Eqs. (10.11) and (10.12), the IF load power yields

$$P_{L,IF} = \frac{1}{2}\left|i_{L,IF}\right|^2 R_L = \frac{g_{m,max}^2 V_{s,RF}^2}{128\pi^2 \omega_{RF}^2 C_{gs}^2} \cdot \frac{R_L r_o^2}{R_i^2 (r_o + R_L)^2}. \qquad (10.13)$$

Assuming conjugate input matching, the available RF input power can be calculated by

$$P_{s,RF} = \frac{V_{s,RF}^2}{8R_i}. \qquad (10.14)$$

Finally, with Eqs. (10.13) and (10.14) we can compute a conversion gain of

$$G_c = \frac{P_{L,IF}}{P_{s,RF}} = \underbrace{\frac{g_{m,max}^2}{16\omega_{RF}^2 C_{gs}^2}}_{k_1} \cdot \underbrace{\frac{R_L r_o^2}{R_i (r_o + R_L)^2}}_{k_2} = k_1 \cdot k_2 , \qquad (10.15)$$

where the factor $\dfrac{g_{m,max}^2}{C_{gs}^2}$ is bounded by the maximum ω_t of the applied transistor.

Equation (10.15) gives first fruitful insights for optimisation. To achieve high conversion gain or low loss, an optimum bias point and LO power has to be applied providing maximum $g_{m,max}$. As expected, the conversion gain decreases with frequency. The first term k_1 is independent on the transistor gate width w_g because both $g_{m,max}$ and C_{gs} are proportional to w_g. The second term k_2 indicates the existence of a maximum with respect to w_g. Since r_o and R_i depend on w_g, an optimum w_g as a function of R_L, r_o and R_i can be found. As long as $R_L < r_o$, the gain can be enhanced by increasing R_L.

10.3.1.1 CMOS Drain Pumped Transconductance Mixer at 30–40 GHz

As an example, a passive drain pumped transconductance mixer topology with zero DC power consumption is presented, which is well suited for short channel FET technologies [Ell14]. The circuit is implemented in IBM 90 nm VSLI SOI CMOS technology.

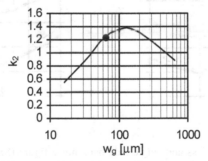

Fig. 10.9. Calculated characteristics of second term of Eq. (10.15) vs gate width at $V_{gs}=0.4$ V, $V_{ds}=0.7$ V and $R_L=50$ Ω, compare [Ell14] © IEEE 2004

Low loss is achieved by reusing the LO power to drive the device RF-wise into the active region despite the fact that the DC drain source voltage is zero. An n-channel FET with a w_g of 64 μm is used. As indicated in Fig. 10.9 and in accordance with Eq. (10.15), this size is well suited to reach a high conversion gain at 50 Ω terminations, since the coefficient k_2 is close to its maximum. The maximum k_2 is reached at twice the transistor width. However, large transistors increase the feedback capacitance C_{gd} thereby degrading the LO to RF isolation. Furthermore, the chosen w_g allows relatively simple impedance matching and filtering at the specified RF, LO and IF centre frequencies of 35 GHz, 32.5 GHz and 2.5 GHz, respectively.

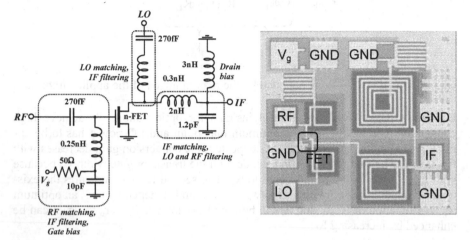

Fig. 10.10. Simplified circuit schematics of passive drain pumped transconductance mixer, compare [Ell14] © IEEE 2004

Fig. 10.11. Photograph of mixer MMIC with overall chip size of 0.5×0.47 mm^2, compare [Ell14] © IEEE 2004

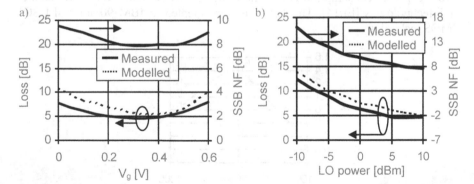

Fig. 10.12a,b. Conversion loss and single side band noise figure (SSB NF), f_{RF}=35 GHz, f_{LO}=32.5 GHz, f_{IF}=2.5 GHz, V_{ds}=0 V: **a** versus V_g, LO power=7.5 dBm; **b** versus LO power, V_g=0.3 V, compare [Ell14] © IEEE 2004

The characteristics of g_m versus V_{ds} and V_{gs} of the FET are shown in Fig. 10.5a. For V_{gs} smaller than 0.9 V, $V_{ds,sat}$ is around 0.2 V. Thus, even with zero V_{ds}, only small LO power is required to drive the transistor into the transition between linear and saturation region, where the highest level of nonlinearity is generated. The plot indicates that for low V_g, this nonlinear transition is reached with minimum LO power. However, this decreases the maximum value of g_m and increases the conversion loss as shown in Eq. (10.15). Thus, V_g is a design tradeoff between minimum LO power and conversion gain. As demonstrated later, measurements and simulations reveal an optimum V_g of approximately 0.4 V.

The simplified circuit schematics of the mixer and a photograph of the IC are shown in Figs. 10.10 and 10.11. To decrease losses and to maximise the conversion efficiency of the mixer, the number of lossy passive elements is kept as low as possible. Where feasible, the LC filter elements are reused for impedance matching, bias feeding and DC blocking. At the RF port, a highpass filter is applied for suppression of the IF frequency, RF impedance matching, DC blocking and feeding of the gate bias. The lowpass filter at the IF port allows IF impedance matching and filtering of the LO and RF frequencies. At the LO port, a bandpass filter is used for the filtering of the IF frequency, LO matching and DC blocking. A grounded bias choke inductance defines a V_{ds} DC voltage of zero.

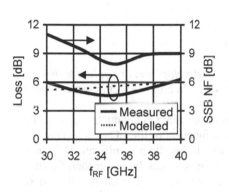

Fig. 10.13. Conversion loss and SSB NF versus RF frequency, LO power=7.5 dBm, f_{IF}=2.5 GHz, V_g=0.3 V, V_{ds}=0 V, compare [Ell14] © IEEE 2004

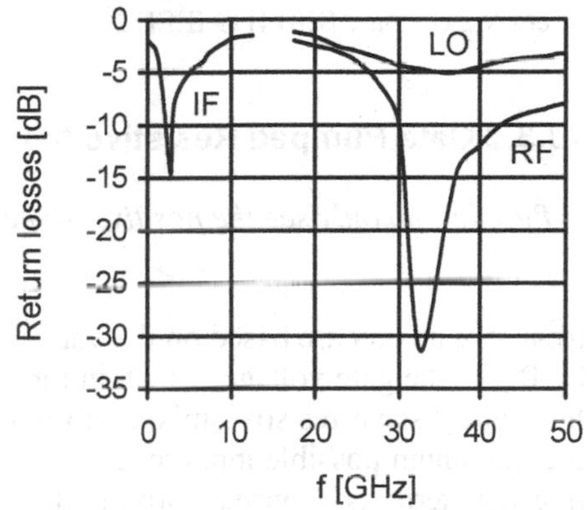

Fig. 10.14. Return losses at V_g=0.3 V and V_{ds}=0.2 V representing an average large signal point, compare [Ell14] © IEEE 2004

In Fig. 10.12, the conversion loss and the SSB NF vs V_g and LO power are depicted. The lowest loss and SSB NF are measured at V_g=0.3 V. At an RF frequency of 35 GHz, an LO frequency of 32.5 GHz, an IF frequency of 2.5 GHz and an LO power of 7.5 dBm, a conversion loss of 4.6 dB and a SSB NF of 7.9 dB are measured. The increase of the conversion loss and the SSB NF with decreased LO power is relatively weak. At an LO power of only 0 dBm, the conversion loss of 6.3 dB and the SSB NF of 9.7 dB are still low enough for low power consuming WLAN applications. The properties versus frequency are shown in Figs. 10.13 and 10.14. The measured IIP3 versus LO power is plotted in Fig. 10.15. The IIP3 rises with increased LO power since the maximum signal swing increases. At an LO power of 0 dBm and 7.5 dBm, the IIP3 is -2 dBm and 2 dBm, respectively. The 1 dB input compression points is –6 dBm at an LO power of 7.5 dBm. In Table 10.3, the minimum port isolations are listed.

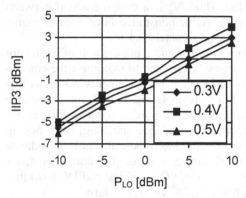

Table 10.3. Measured minimum
port isolations

LO to IF	RF to IF	LO to RF	IF to RF
45 dB	48 dB	11 dB	37 dB

LO power: 0–10 dBm

Fig. 10.15. Measured IIP3 at V_{ds}=0 V for different V_g, compare [Ell14] © IEEE 2004

10.3.2 Gate-Pumped Resistive Mixer

People who only see the positive aspect of a development are called technicians.
Werner Mitsch, German aphorist

Resistive mixers are based on the variation of the channel conductance determined by R_{ds} vs the gate voltage, which in turn is a function of the LO power. Due to the fact that passive resistive mixers do not have to handle a significant DC current, the maximum possible input level of resistive mixers can be larger than for the active counterparts. Consequently, high linearity is achievable for resistive mixers. Because of their passive and resistive nature, resistive mixers have a relatively high conversion loss. Several resistive mixers have been reported in literature [Maa87, Sch98, Zir01, Ver00]. Theoretical aspects have been addressed in the rigorous work presented in [Sal71].

Resistive mixers can be implemented in different circuit configurations. Due to their low complexity, those shown in Fig. 10.16 are well suited for monolithic integration. The overall mixer loss may be described by

$$L_{overall} \approx L_{RF-filter} \cdot L_{IF-filter} \cdot L_{switch} \cdot L_{conv}, \qquad (10.16)$$

with the insertion loss of the RF and IF filters $L_{RF\text{-filter}}$ and $L_{IF\text{-filter}}$, the insertion loss caused by the non-ideal on- and off-resistance of the FET L_{switch}, and the conversion loss due to the nonlinear mixing process L_{conv}.

The nonlinear resistance can be realised by a resistive (cold) FET, which can either be connected in series or shunt configuration. In ideal case, the nonlinear FET acts as a switch with an on-resistance Z_{on} of zero and an off-resistance Z_{off} of ∞. Note that switching is always associated with strong nonlinearities. Suppose ideal switching transitions, where an infinite number of harmonics and intermodulation products are generated due to the square-wave multiplied with any input signal.

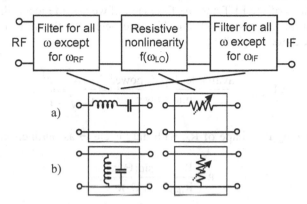

Fig. 10.16a,b. Resistive mixers, nonlinear resistance can be realised in: **a** series or; **b** shunt configuration, filters can be implemented as series bandpass (**a**) or shunted bandpass (**b**), compare [Ell8] © IEEE 2005

The simplified equivalent circuit of a resistive FET, the corresponding values of the equivalent circuit elements in on- and off-mode, and the channel resistance of a transistor with w_g=64 μm are shown in Figs. 13.3 and 13.4. Depending on the impedance conditions of the environment, the nonideal characteristics of Z_{off} or Z_{on} can dominate the losses. Thus, an optimum w_g exists as demonstrated later. Following [Sal71, Lin01], the loss associated with the non-ideal on- and off-resistances may be estimated by

$$L_{switch} = 1 + 2\left[1 + \sqrt{1 + \left(\frac{1}{\delta}\right)^2}\right]$$
(10.17)

with

$$\delta = \left|\frac{Z_{on}}{Z_{off}}\right|.$$
(10.18)

The best conversion efficiency at given LO power is achieved at a DC bias corresponding with an average value between Z_{on} and Z_{off}. By considering a sinusoidal LO signal with sufficient magnitude to drive the channel resistance from an on- to off-state, we observe the time variant characteristics of R_{ds} as illustrated in Fig. 10.17.

With the assumption that the LO power is much higher than the RF power, the nonlinear R_{ds} can be expanded into a Fourier series:

$$R_{ds}(t) = R_{ds0} + 2\sum_{n=1}^{\infty} R_{dsn} \cos(n\omega_{LO}t)$$
(10.19)

with R_{ds0} as the fundamental and R_{dsn} the nth Fourier coefficient of R_{ds}. If we neglect the parasitics of the FET and the filters and if we assume ideal filtering of all frequencies except the desired RF and IF frequencies, the theoretical conversion loss can be calculated by [Sal71]

$$L_{conv} = \frac{available\ input\ power}{output\ power} = \frac{1+\sqrt{1-\epsilon^2}}{1-\sqrt{1-\epsilon^2}} . \qquad (10.20)$$

Assuming a rectangular shape of R_{ds} with duty cycle Θ as outlined in Fig. 10.17 yields

$$\epsilon = \frac{R_{ds1}}{R_{ds0}} = \frac{\sin(\Theta \cdot \pi)}{\Theta \cdot \pi} . \qquad (10.21)$$

This identity can be found in conversion tables of Fourier series [Bro91]. Due to the passive nature of the device, and depending on the shape and the duty cycle Θ, the factor ϵ ranges from unity to zero. In theory, this results in a loss of zero and ∞, respectively. Unfortunately, an ϵ of unity can never be reached since the optimum source resistance R_{source} and output resistance R_{out} of the mixer related by [Bar67, Bäc99]

$$R_{source} = R_{out} = \sqrt{R_{ds0}^2 - R_{ds1}^2} \qquad (10.22)$$

would approach zero. Consequently, in practice, no impedance matching would be possible.

Filters are required to minimise the signal energy converted to unwanted frequencies. These filters can be implemented as series or shunted bandpass filters, in ideal case providing an open or a short, respectively, for all frequencies except the desired RF or IF frequency. Compared to series bandpass filters, the shunted bandpass filters have two significant advantages. Both the capacitor and the inductor are shunted at one port. Thus, a significant part of the substrate losses are short-circuited, thereby yielding a higher quality factor. Furthermore, due to the shunted inductance, the input and output of the resistive FET are DC grounded. Hence, a DC drain source voltage of zero is provided to keep the device within the resistive region. As a benefit, no additional area consuming and lossy bias elements are needed. Consequently, bandpass filters in shunt configuration are favourable. The filters can be designed by $\omega_{RF} = \frac{1}{\sqrt{L_{RF}C_{RF}}}$ and $\omega_{IF} = \frac{1}{\sqrt{L_{IF}C_{IF}}}$.

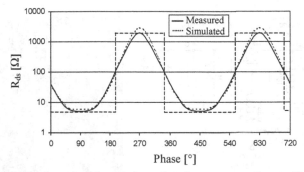

Fig. 10.17. Nonlinear characteristics of channel resistance according to Fig. 10.5b when driven by sufficiently high LO signal, V_g=0.45 V, compare [Ell8] © IEEE 2005

For many system applications, differential mixers are advantageous. Figure 10.18 depicts the topology of a differential resistive mixer. The double balanced structure consists of four passive FETs. In the simplified schematic, the impedance matching, bias and filter networks are not included. For details concerning such mixers, the reader is referred to the literature [Lee04, Cir05]. A loss and SSB NF of around 7 dB have been demonstrated [Pih01].

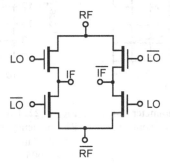

Fig. 10.18. Simplified schematic of double balanced resistive mixer

10.3.2.1 Gate-Pumped CMOS Resistive Mixer at 26.5–30 GHz

In Figs. 10.19 and 10.20, the simplified circuit schematic and chip photograph of a gate-pumped resistive CMOS mixer are shown, which has been implemented in IBM 90-nm SOI technology [Ell8]. The circuit is optimised for RF and IF centre frequencies of 28 GHz and 2.5 GHz, respectively. Figure 10.21 depicts the simulated characteristics of the filters. The RF filter has a loss $L_{RF\text{-filter}}$ of 1.1 dB at the RF frequency and a suppression of the IF frequency of 18 dB. A loss $L_{IF\text{-filter}}$ of 2.5 dB at the IF frequency and an IF to RF and LO isolation of higher than 22 dB are simulated for the IF filter.

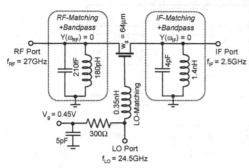

Fig. 10.19. Simplified circuit schematic of resistive mixer, compare [Ell8] © IEEE 2005

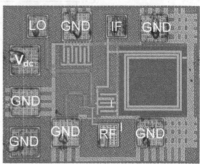

Fig. 10.20. Photograph of mixer with chip size of 0.4 × 0.3 mm², compare [Ell8] © IEEE 2005

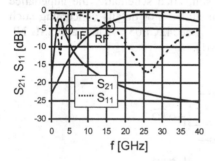

Fig. 10.21. Simulated S-parameters of filters including parasitics, compare [Ell8] © IEEE 2005

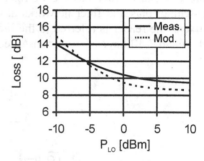

Fig. 10.22. Conversion loss vs LO power at RF frequency of 27 GHz, IF frequency of 2.5 GHz and V_g=0.45 V, compare [Ell8] © IEEE 2005

By using Eqs. (10.17), (10.18), (13.1) and (13.2), and the values of the FET according to Fig. 13.4a, we can estimate a loss of L_{switch}≈0.5 dB at 27 GHz. Due to the small l_g and the low associated parasitics, this value is relatively low, even at such high frequencies. As illustrated in Fig. 10.17, we can extract a duty cycle of θ=35%. Thus, from Eqs. (10.20) and (10.21), a conversion loss of 5.9 dB can be calculated. From Eq. (10.16) we can now estimate the theoretical overall mixer loss amounting to

$$L_{overall} \text{ [dB]} \approx 1.1 \text{ dB}+2.5 \text{ dB}+0.5 \text{ dB}+5.9 \text{ dB}=10 \text{ dB}. \tag{10.23}$$

We can conclude that the nonlinear mixing has the highest contribution to the overall loss, followed by the insertion loss of the IF filter. For a wide LO power range, minimum loss is achieved for a DC voltage of approximately 0.45 V. Hence this bias is fed.

In Fig. 10.22, the conversion loss is plotted versus LO power. The minimum measured loss is 9.7 dB at an LO power of 10 dB, which is close to the theoretical value of 10 dB obtained from the analytical calculations. The loss raises only slightly with decreased LO power. At a low LO power of 0 dB, a conversion loss of 10.3 dB is measured making the circuit also well suited for low power applications. Usually, resistive mixers require high LO power to achieve low loss. The measured 3 dB RF frequency bandwidth ranges approximately from 26.5 to 30 GHz. In Fig. 10.23, the return losses are shown at the small signal DC operation point. The measured SSB NF and IIP3 are plotted in Fig. 10.24. As expected from resistive mixers, the SSB NF is close to the value obtained for the conversion loss. A minimum noise figure of approximately 11 dB is measured at 5 dBm LO power. Up to 10 dBm, the IIP3 increases with LO power. At 0 dBm and 10 dBm LO power, high values of 12.7 dBm and 20 dBm, respectively, are measured. The port isolations are summarised in Table 10.4. Due to the suppression of the filters relatively high port isolations are achieved.

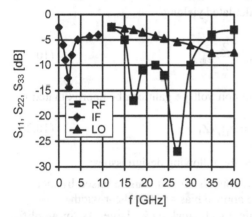

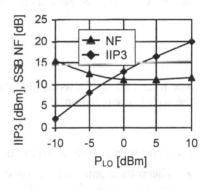

Fig. 10.23. Return losses in small signal operation point at bias of 0.45 V, compare [Ell8] © IEEE 2005

Fig. 10.24. Measured input third intercept point and single side band noise figure at bias of 0.45 V, f_{RF}=27 GHz and f_{IF}= 2.5 GHz, compare [Ell8] © IEEE 2005

Table 10.4. Port isolations

	LO to IF	LO to RF	RF to IF	IF to RF
Measured	22 dB	24 dB	33 dB	25 dB
LS model	30 dB	20 dB	26 dB	17 dB

f_{RF}=27 GHz and f_{IF}=2.5 GHz, 5 dBm LO power, 0.45 V bias

10.3.3 Differential Pair and Gilbert-Cell Mixer

The mixers presented in this section are those most frequently used in commercial integrated systems. Among the significant advantages are the differential LO and IF paths, high gain and large bandwidth. In Fig. 10.25, the simplest configuration based on a differential pair is shown. By means of an RF tranconductance stage, the RF input voltage is converted into an RF current, which is commutatively switched by the upper two transistors. We assume ideal switches with duty cycle of 50% and square-wave characteristics with magnitude of 1 and −1. Consequently, the current through the RF stage is multiplied with the function [Bef03, Bro91]

$$h(t) = \frac{4}{\pi} \cdot \sum_{n=0}^{\infty} \frac{(-1)^n}{2n+1} \cdot \sin\left[(2n+1)\omega_{LO}t\right]. \qquad (10.24)$$

By Fourier analysis, the conversion transconductance of the desired $\omega_{IF} = \omega_{RF} - \omega_{LO}$ component can be calculated yielding

$$g_{m1} = \frac{2}{\pi} \cdot g_m. \qquad (10.25)$$

The magnitude of the associated conversion voltage gain can be approximated by the well-known relation of

$$a_{v1} = g_{m1} \cdot Z_L. \qquad (10.26)$$

From the latter equations we can deduce the following design issues:

- The voltage gain and the associated conversion gain can be raised by increasing of g_m, which is a function of the applied bias and transistor width.
- We can increase the conversion gain also by making Z_L large. As for amplifiers, the maximum gain is bounded by stability constraints. High values of Z_L increase the RC constant at the output leading to a limitation of the maximum operation frequency and bandwidth. Consequently, the optimum value of Z_L has to be traded off for maximum gain on one hand, and maximum operation frequency and stability on the other hand. Since Z_L determines the signal swing in the IV curves, Z_L impacts also the linearity. By the way, how can we get a high Z_L? One solution is the to use a parallel LC network with $\omega_{IF} = \frac{1}{\sqrt{LC}}$. Neglecting resistive losses, this LC network provides an impedance of ∞. In this context, the reader is referred to Sect. 11.5.3, where a similar load is used for a cross-coupled oscillator. One disadvantage might be the large size of the inductors. An alternative solution is to use a compact parallel RC load [Bef03]. However, it is clear that the latter load provides lower RF conversion gain than the LC approach.

- Harmonics and intermodulation products are generated by switching. The mentioned LC filter is suited to suppress undesired frequencies at the output.
- The LO power has to be high enough to allow switching with a rectangular-like function. Symmetrical switching is advantageous regarding the shape of the RF signal and the associated properties in terms of linearity, noise and conversion gain.

Compared to the single balanced differential pair, the Gilbert cell illustrated in Fig. 10.26 offers one advantage. It is a doubled balanced topology [Gil74]. LO-wise the outputs of the two single-balanced mixers are connected inversely parallel, whereas IF-wise the outputs are in parallel. Consequently, at the output, the LO signal is cancelled, whereas the IF amplitude is doubled. Thus, the doubled balanced mixer provides a high level of LO to IF isolation, which can be important in systems. We may consider the potential to saturate following circuits such as the IF amplifiers or the analogue to digital converters by means of the high LO power. Disadvantages of double balanced mixers are the high power consumption. The drawback of double supply current is made up for double output power. However, we have to consider the voltage headroom required for the additional current source. All signal ports of the Gilbert cell mixer are fully differential, which is not the case for the single balanced topology. This may not necessarily be an advantage since the RF input provided by the LNA may anyway be single ended.

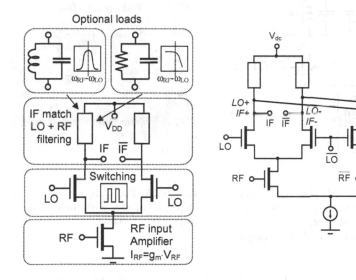

Fig. 10.25. Differential pair (single balanced)

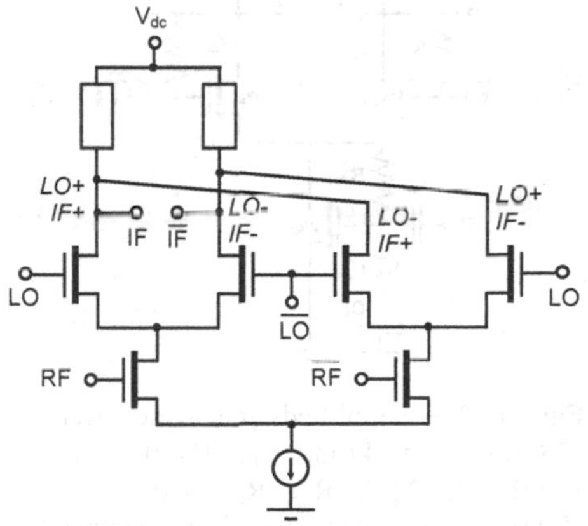

Fig. 10.26. Gilbert cell (double balanced)

10.3.3.1 CMOS Differential Pair Mixer Covering 27–33 GHz

A 27–33-GHz CMOS differential down-mixer is treated now [Ell10], which uses the same CMOS SOI technology as the mixers presented in the preceding sections. The circuit schematic and the chip photograph are shown in Figs. 10.27 and 10.28. An LC bandpass filter is applied as load and for impedance matching. The gates of the FETs are biased via high-ohmic resistors allowing the operation with single supply voltage. The corresponding gate and drain supply voltages of all FETs are approximately $V_{dd}/2$. This bias corresponds to class-A operation, where the RF buffer and the switching FETs have good gain properties. Furthermore, the FET can be symmetrically switched by a sinusoidal LO voltage. With an average g_m of 80 mS and a load resistance Z_L of 50 Ω, we can estimate an upper limit of the conversion gain of around 6 dB. With a loss of the IF filter of approximately 2.5 dB and losses due to the nonideal RF input matching in the range of 2 dB, we can predict an upper conversion gain of 1.5 dB. Due to the nonideal characteristics of the FETs at these high frequencies, in practice, the achievable conversion gain will be lower. The nonideal characteristics are mainly caused by the non-zero on-resistance and the non-infinitely off-resistance of the switch FETs, and the non-zero channel conductance of the RF buffer FET.

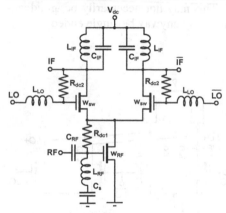

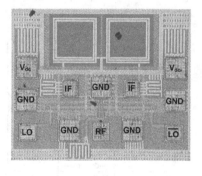

Fig. 10.27. Simplified schematic, $w_{RF}=$ 128 μm, $w_{sw}=64$ μm, $C_{RF}=100$ fF, $L_{RF}=$ 140 fH, $C_s=2$ pF, $R_{dc1}=R_{dc2}=4$ kΩ, $L_{LO}=$ 400 fH, $L_{IF}=1.2$ nH, $C_{IF}=4$ pF, compare [Ell10] © IEE 2004

Fig. 10.28. Photograph of mixer with chip size of 0.5×0.4 mm^2, compare [Ell10] © IEE 2004

To simplify measurements, the circuit has been designed for 50-Ω terminations. Much higher load impedances could be applied in systems. This would yield significantly higher gain as indicated in Eq. (10.26). The circuit is operated with RF, IF and LO frequencies of 30 GHz, 2.5 GHz and 27.5 GHz, respectively, and biased with a supply voltage of 1.2 V, a corresponding supply current of 17 mA and an LO power of 5 dBm. The return losses are depicted in Fig. 10.29. The lowest loss and SSB NF of 2.5 dB and 13 dB are measured at 29 GHz and 28 GHz, respectively. Within a broad frequency range from 26 to 34 GHz, the loss and noise figure increase is less than 3 dB. The loss and SSB NF versus LO power are illustrated in Fig. 10.30. The lowest loss is measured at an LO power of 2.5 dBm. Up to 10 dBm LO power, the SSB NF decreases with LO power. As shown in Fig. 10.31, up to 10 dBm LO power, the IIP3 increases with LO power. At 5 dB LO power and 1.2 V supply voltage, the IIP3 is 0.5 dBm. The port isolations are listed in Table 10.5.

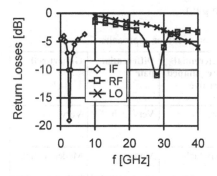

Fig. 10.29. Measured small signal return losses at V_{dc}=1.2 V and I_{dc}=17 mA, compare [Ell10] © IEE 2004

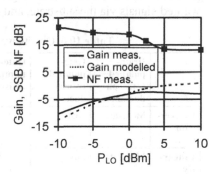

Fig. 10.30. Conversion loss and measured SSB noise figure versus RF frequency with fixed IF frequency of 2.5 GHz at 5 dBm LO power, V_{dc}=1.2 V and I_{dc}=17 mA, compare [Ell10] © IEE 2004

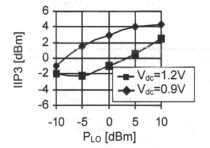

Fig. 10.31. Measured third order intercept point at input versus LO power, V_{dc}=1.2 V, I_{dc}=17 mA, f_{RF}=30 GHz, f_{IF}=2.5 GHz, compare [Ell10] © IEE 2004

Table 10.5. Port isolation

Isolation	LO to IF	LO to RF	RF to IF	IF to RF
Meas.	8 dB	12 dB	20 dB	30 dB
Sim.	17 dB	18 dB	19.5 dB	37 dB

LO power: 2.5 dBm

10.3.4 Comparison of Mixer Approaches

In Table 10.6 we find a qualitative comparison of the mixer approaches treated in the preceding sections. It is clear that passive mixers are not capable of providing gain. To compensate for the losses, the saved DC power can be used for an amplifier located in front of the mixer yielding a low system noise figure. Especially on silicon-based technologies having poor substrate isolation, the differential topologies are superior due to their high immunity against noise and pick-up of unwanted signals via the substrate, and the high gain.

Table 10.6. Qualitative comparison of mixer approaches

	Drain-pumped transconductance	Gate-pumped transconductance	Gate-pumped resistive	Differential gate-pumped resistive	Differential pair	Gilbert-cell
Conversion gain	Moderate	High	Low	Low	Very high	Very high
Linearity	Moderate	Low	High/very high	Very high	Moderate	Moderate/high
Noise	Low	Very low	Moderate	Moderate	High	High/very high
LO to IF suppression	Moderate (down-mixer) Low (up-mixer)	Low (down-mixer) Moderate (up-mixer)	Low	Low	Moderate	Very high
Power consumption	Low/zero	Moderate/low	Zero	Zero	High	Very high
Chip area	Moderate/large (inductors)	Moderate/large (inductors)	Moderate/large (inductors)	Moderate/large (inductors)	Small (RC load) Moderate/large (LC load)	Small/moderate (RC load) Large (LC load)
System integration	May be limited since single-ended	May be limited since single-ended	May be limited since single-ended	Good/very good	Good (LO, IF differential)	Very good

10.3.5 Mixer Performances

The key performances of several state-of-the-art down-mixer topologies are listed in Table 10.7.

Table 10.7. Mixer performances

	Principle	f_{RF}/f_{IF} [GHz]	G_c [dB]	SSB NF [dB]	IIP3 [mW]	P_{LO} [mW]	P_{dc} [V×mA]	Ref.
Active Mixers								
200-nm InP HEMT	Gate pumped g_m	95/1	0.9	n.a.	n.a.	1.6	n.a.	[Kwo93]
100-nm InP HEMT		64.5/16	1	n.a.	n.a.	0.7	n.a.	[Orz03]
0.5-μm SiGe HBT	Gilbert cell	24/0.1	≈10[a]	n.a.	n.a.	≈1	15 × 3.6	[Sön03]
180-nm CMOS	Cascode	10/0.2	16	n.a.	n.a.	n.a.	n.a.	[Mad01]
180-nm CMOS	Diff. pair	24/4.9	13[b]	17.5	n.a.	n.a.	1.5 × 27	[Gua04]
90-nm SOI CMOS	Diff. pair	30/2.5	−2.5	13	1	3	1.2 × 17	Section 10.3.3.1 [Ell10]
Passive Mixers								
0.5-μm GaAs MESFET	Gate pumped resistive	1.9/0.11	−7	8	6.3	3.2	0	[Kuc99]
0.5-μm GaAs MESFET		5.2/0.95	−5.5	6.5	200	10		[Ell24]
130-nm GaAs HEMT		25.7/0.2	−7.2	n.a.	n.a.	10		[Ver00]
90-nm SOI CMOS		27/2.5	−9.7	11	100	10		Section 10.3.2.1 [Ell8]
140-nm InP PHEMT		60/n.a.	−7.7	n.a.	n.a.	3.2		[Zir01]
130-nm GaAs HEMT		77/0.03	−10	n.a.	n.a.	1		[Sch98]
Schottky diode	8th harmonic	38/1.2	−23	n.a	n.a.	100		[Zha01]
90-nm SOI CMOS	Drain pumped g_m	35/2.5	−4.6	7.9	1.5	5		Section 10.3.1.1 [Ell4]

[a]Inclusive buffer [b]High-ohmic input

10.4 Tutorials

1. Why do we need mixers? Do we need frequency conversion for UWB systems?
2. What are the major design constraints for mixers? Compare the requirements for down- and up-mixers.
3. Under which conditions is the noise of mixers important?
4. What is the difference of harmonics and intermodulation products?
5. How can we mathematically describe the nonlinear characteristics (e.g. current) of a device?
6. Show analytically that the mixing product of two frequencies can give permutations of frequencies. What is the important requirement?
7. What about the amplitudes of higher harmonics? Why do they decrease with increasing order?
8. Draw the equivalent circuit of a FET. What are the most non-linear elements? Why?
9. Explain the functional principle of gate and drain pumped transconductance mixers. How can we maximise the conversion gain? What are the limitations in terms of maximum operation frequency? Which one would be better suited in terms of LO to IF/RF isolation for a down- and up-mixer?
10. What are the pros and cons concerning active transconductances and passive resistances?
11. Explain the functional principle of a differential pair and Gilbert cell mixer. How can we optimise the conversion gain? Review the tradeoffs for the loads with respect to gain, highest operation frequency, bandwidth, power consumption, stability and circuit size.
12. Which device has stronger g_m nonlinearities – the BJT or the FET? Which one should provide higher conversion gain at low and high frequencies?
13. Could we design frequency multipliers and dividers with mixers, e.g. using a differential pair?
14. Review the typical performances of state-of-the-art mixers.

References

[Bäc99]   W. Bächtold, Mikrowellen-Technik, Vieweg Verlag, Braunschweig, 1999.
[Bar67]   M. R. Barber, "Noise figure and conversion loss of the Schottky barrier mixer diode", IEEE Transactions on Microwave Theory and Techniques, MTT-15, No. 11, pp. 629–635, Nov. 1967.
[Bef03]   F. Beffa, A Low-power CMOS Bluetooth Transceiver, Diss. ETH No. 15303, 2002.
[Beg79]   G. Begemann, A. Jacob, "Conversion gain of MESFET drain mixers", Electronic Letters, pp. 567–568, Aug. 1979.

[Bro91]   I. N. Bronstein, K. A. Semedjajew, Taschenbuch der Mathematik, Teubner Verlag Stuttgart, 1991.

[Bur76]   P. Bura, and R. Dikshit, "FET mixers for communication satellite transponders", IEEE Microwave Symposium, pp. 90-92, June 1976.

[Cir05]   R. Circa, d. Pienkowski, G. Böck, R. Kakerow, M. Müller, R. Wittmann, "Resistive mixers for reconfigurable wireless front-ends", IEEE Radio Frequency Integrated Circuit Symposium, pp. 513–516, June 2005.

[Dar96]   A. H. Darsinooieh and O. Palamutcuoglu, "On the theory and design of subharmonically drain pumped microwave MESFET distributed mixers", IEEE Mediterranean Electrotechnical Conference, Vol. 1, pp. 595–598, May 1996.

[Ell8]   F. Ellinger, "26.5-30 GHz resistive mixer on 90 nm VLSI SOI CMOS technology with high linearity for WLAN", IEEE Transactions on Microwave Theory and Techniques, Vol. 53, No. 8, pp. 2559–2565, Aug. 2005.

[Ell10]   F. Ellinger, "26-34 GHz CMOS mixer", IEE Electronics Letters, Vol. 40, No. 22, pp. 1417–1418, Oct. 2004.

[Ell14]   F. Ellinger, L. C. Rodoni, G. Sialm, C. Kromer, G. von Büren, M. Schmatz, C. Menolfi, T. Toifl, T. Morf, M. Kossel, H. Jäckel, "30-40 GHz drain pumped passive down mixer MMIC fabricated on digital SOI CMOS technology", IEEE Transactions on Microwave Theory and Technique, Vol. 52, No. 5, pp. 1382–1391, May 2004.

[Ell24]   F. Ellinger, R. Vogt and W. Bächtold, "Compact, resistive monolithic integrated mixer with low distortion for HIPERLAN", IEEE Transactions on Microwave Theory and Techniques, Vol. 50, No. 1, pp. 178–182, Jan. 2002.

[Ell46]   F. Ellinger, R. Vogt, W. Bächtold, "Ultra low power, low noise GaAs up-converter MMIC for a broadband superheterodyne L-band receiver", IEEE GaAs Integrated Circuit Symposium, Seattle, pp. 103–106, Nov. 2000.

[Gil74]   B. Gilbert, "A high performance monolithic multiplier using active feedback", IEEE Journal of Solid-State Circuits, Vol. SC-9, No. 6, Dec. 1974.

[Gua04]   X. Guan and A. Hajimiri, "A 24 GHz CMOS front-end", IEEE Journal of Solid-State Circuits, Vol. 39, No. 2, pp. 155–158, Feb. 2004.

[Hey04]   P. Heydari, "High-frequency noise in RF active CMOS mixers", IEEE Design Automation Conference, pp. 57–61, Jan 2004.

[Hul93]   C. D. Hull and R. G. Meyer, "A systematic approach to the analysis of Noise in Mixers", IEEE Transactions on Microwave Theory and Techniques, Vol. 40, No. 12, pp. 909–919, Sept. 1993.

[Joh05]   T. K. Johanse, J. Vidkjaer, V. Krozer, "Analysis and design of wide-band SiGe HBT active mixers", IEEE Transcations on Microwave Theory and Techniques, Vol. 53, No. 7, pp. 2389–2397, July 2005.

[Kuc99]   J. J. Kucera and U. Lott, "A zero DC-power low-distortion mixer for wireless applications", IEEE Microwave and Guided Wave Letters, Vol. 9, No. 4, April 1999.

[Kwo93] Y. Kwon, D. Pavlidis, P. Marsh, G.-I. Ng, T. L. Brock, "Experimetal characteristics and performance analysis of monolithic InP-based HEMT mixers at W-band", IEEE Transactions on Microwave Theory and Techniques, Vol. 41, No. 1, pp. 1–8, Jan. 1993.

[Lee04] T. H. Lee, "The Design of CMOS Radio-Frequency Integrated Circuits", Cambridge, 2004.

[Lin01] E. W. Lin and W. H. Ku, "Device considerations and modeling for the design of an InP-based MODFET millimeter-wave resistive mixer with superior conversion efficiency", IEEE Transactions on Microwave Theory and Techniques, Vol. 43, No. 8, pp. 1951–1959, Aug. 2001.

[Maa87] S. A. Maas, "A GaAs MESFET Mixer with very low intermodulation", IEEE Transactions on Microwave Theory and Techniques, Vol. 35, No. 4, pp. 425–429, April 1987.

[Maa93] S. A. Maas, Microwave Mixers, Artech House, 1993.

[Maa98] S. A. Maas, The RF and Microwave Circuit Design Cookbook, Artech House, 1998.

[Maa03] S. A. Maas, Nonlinear Microwave and RF circuits, Artech House, 2003.

[Mad01] M. Madihian, H. Fujii, H. Yoshida, H. Hisamitsu, and T. Yamazaki, "A 1–10 GHz 0.18µm CMOS chipset for multi-mode wireless applications", IEEE International MTT-S Microwave Symposium, pp. 1865–1868, June 2001.

[Orz03] A. Orzati, F. Robin, H. Benedikter, W. Bächtold, "A V-band up-converting InP HEMT active mixer with low LO-power requirements", IEEE Microwave and Wireless Components Letters, Vol. 13, No. 6, pp. 202–204, June 2003.

[Pih01] J. Pihl, K. T. Christensen, E. Bruun, "Direct downconverswion with swiching CMOS mixer", IEEE International Symposium on Circuits and Systems, pp. I–117–120, May 2001.

[Puc76] R. A. Pucel, D. Massé and R. Bera, "Performance of GaAs MESFET mixers at X-band", IEEE Transactions on Microwave Theory and Techniques, Vol. MTT-24, No. 6, pp. 351–360, June 1976.

[Rad94] M. M. Radmanesh, N. A. Barakat, "State of the art S-band FET mixer design", IEEE MTT-S International Microwave Symposium Digest, pp. 1435–1438, June 1994.

[Rob04] D. Roberson, "RFIC and MMIC Design and Technology", IEE, London, 2001.

[Saf05] A. Q. Safarian, A Yazdi, P. Heydari, "Design and analysis of an ultra-wideband distributed CMOS mixer", IEEE Transactions on Very Large Scale Integration Systems, Vo. 13, No. 5, pp. 618–629, May 2005.

[Sal71] A. A. M. Saleh, Theory of resistive mixers, M.I.T press, 1971.

[Sch98] U. Schaper, A. Schaefer, A. Werthof, G. Bök, "70-90 GHz balanced resistive PHFET Mixer MMIC, Electronic Letters", Vol. 34, pp. 1377–1379, 1998.

[Sön03]   E. Sönmez, A. Trasser, P. Abele, F. Gruson, K.-B. Schad and H. Schumacher, "24 GHz high sensitivity downconverter using commercial SiGe HBT MMIC foundry technology", IEEE Topical Meeting on Silicon Monolithic Integrated Circuits in RF Systems, pp. 68–71, April 2003.

[Ter99]   M. T. Terrovits, R. G. Meyer, "Noise in current-commutating CMOS mixers", IEEE Journal of Solid-State Circuits, pp. 772–783, June 1999.

[Tie83]   G.K. Tie, C. S. Aitchinson, "Noise and associated conversion gain of a microwave MESFET gate mixer", European Microwave Conference, pp. 579–584, 1983.

[Tsi84]   C. Tsironis, R. Meierer, R. Stahlmann, "Dual-gate MESFET mixers", IEEE Transactions on Microwave Theory and Techniques, Vol. 32, No. 3, pp. 248-255, March 1984.

[Ver00]   L. Verweyen, A. Tessmann, Y. Campos-Roca, M. Hassler, A. Bessemoulin, H. Tischler, W. Liebl, T. Grave, V. Güngerich, "LMDS Up- and Down-Converter MMIC", IEEE MTT-S International Microwave Symposium, Bosten, pp. 1685–1688, June 2000.

[Zha01]   W. Zhao, C. Schöllhorn, E. Kasper amd C. Rheinfelder, "38 GHz coplanar harmonic mixer on silicon", IEEE Topical Meeting on Silicon Monolithic Integrated Circuits in RF Systems, pp. 138–141, Sept. 2001.

[Zir01]   H. Zirath, C. Fager, M. Garcia, P. Sakalas, L. Landen and A. Alping, "Analog MMICs for millimeter-wave applications based on a commercial 0.14-mm pHEMT technology", IEEE Transactions on Microwave Theory and Techniques, Vol. 49, No. 11, pp. 2086–2092, Nov. 2001.

11 Oscillators

Man muss sich wundern können.

H. Barkhausen, TU Dresden

In many transceivers, oscillators are required for the up- and down-conversion of transmitted and received signals by means of a local oscillator (LO) signal as illustrated in Fig. 11.1a. The main function of an oscillator is the generation of the LO signal, which is a periodic output signal at a specific or tuneable frequency. Similar to amplifiers, the efficiency is given by the relation between LO output power and required DC power. In this context, the reader is referred to Eq. (9.3). We can identify the following general oscillator properties:

1. Based on a start-up signal, which can be the intrinsic oscillator noise, or a switched signal, e.g. a supply voltage ramp, the signal grows every oscillation cycle. This signal amplification requires an active device providing gain. Needless to say that an oscillator can not be purely passive.

2. To prevent from self-destruction, the oscillator must have an amplitude limiting effect. Most of the integrated oscillators employ transistors with IV characteristics as sketched in Fig. 11.1b. Fortunately, given a proper design, the lower and upper limits of the amplitude are already defined, e.g. for the FET by the threshold and linear (resistive) region exhibiting zero current and high damping resistance, respectively. Starting at the applied DC bias point, the signal swings along the load line with slope determined by the load resistance. Indeed, the corresponding properties are very similar to those of power amplifiers. One major difference is the fact that the RF input power of oscillators is generated internally. This is not the case for amplifiers.

3. Many oscillators feature frequency stabilisation based on the impedance and phase characteristics of a high-Q resonator.

Oscillators operate under large signal conditions, since the RF signal swing is large approaching parts of the IV curves exhibiting varying characteristics. Harmonics and nonlinearities are generated. Precise large signal models are required for accurate predictions of oscillators. Nevertheless, simple models can be used to explain the basic operation of oscillators.

At the start-up point, oscillators show small signal behaviour, because the amplitude swing is small. The linearised time domain, feedback and negative resistance approaches will be discussed in the following sections. The first concept is well suited for general understanding; the feedback approach is rather appropriate for system considerations, whereas the negative resistance theory is frequently employed for investigations at device level. Although the three theories are based on different assumptions and simplifications, similar results are obtained. Small signal analysis does not provide any information regarding the output power. In accordance to amplifier theory, power and efficiency estimations can be performed in the IV curves based on R_L and the DC bias.

a)   b)

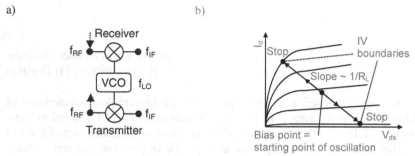

Fig. 11.1a,b. Oscillator: **a** required for frequency conversion in transceivers, VCO: voltage controlled oscillator; **b** illustration of transistor IV boundaries limiting signal swing, in this case for a FET, similar for BJTs

11.1 Simple Time Domain Model

A simple oscillator may be composed of an active device and a resonator as illustrated in Fig. 11.2a. In this example we apply a series LC resonator. Parallel resonators are feasible as well. To evaluate the start-up conditions of an oscillator, it is fruitful to describe the properties in the time domain. The voltage across any active device can be approximated by a potential series given by

$$v_A(t) = V_0 + R_A i(t) + A_A i^2(t) + \dots \qquad (11.1)$$

where V_o, R_A and A_A are specific device parameters. For the sake of simplicity, let us assume linear properties[1] with

$$v_A(t) = V_0 + r_A i(t). \qquad (11.2)$$

In a loop, the sum of the voltages equals zero. Therefore, the sum of the derivations is also zero yielding the differential equation

---

[1] Since the parameters are extracted and only valid under linear/small signal conditions, they are denoted by small letters.

$$\sum \frac{dv(t)}{dt} = L_R \cdot \frac{d^2i(t)}{dt^2} + \frac{1}{C_R}i(t) + R_R \cdot \frac{di(t)}{dt} + r_A \cdot \frac{di(t)}{dt} = 0 . \quad (11.3)$$

It is well known that such kinds of differential equations can be solved with exponential functions of the type

$$i(t) = I_0 \cdot e^{st} . \quad (11.4)$$

with I_0 as the signal amplitude at t=0. In many approaches, the noise current serves as I_0. Substituting into Eq. (11.3) yields

$$s = \sigma \pm j\omega = -\frac{R_R + r_A}{2L_R} \pm \frac{j}{\sqrt{L_R \cdot C_R}} \cdot \sqrt{1 - \frac{(R_R + r_A)^2 \cdot C_R}{4 \cdot L_R}} . \quad (11.5)$$

The latter relation reveals exponential growth as illustrated in Fig. 11.2b on the condition that

$$\sigma = -\frac{R_R + r_A}{2L_R} > 0 \quad (11.6)$$

and

$$r_A < -R_R . \quad (11.7)$$

Suppose that a resonator is purely passive with $R_R > 0$. Thus, the last equation demands that the active device must generate a negative resistance with magnitude of at least equivalent to R_R in order to compensate the losses inherent in the oscillator.

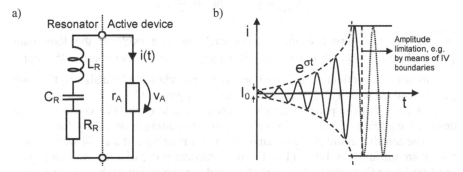

Fig. 11.2a,b. Simple oscillator model: **a** equivalent circuit with series resonator; **b** exponential growth of current until saturation occurs

By inspection of the imaginary parts of s, and by assuming $r_A = -R_R$ representing the start-up condition, we get for the oscillation frequency

$$\omega_{LO} = \frac{1}{\sqrt{L_R \cdot C_R}} , \quad (11.8)$$

a result one might have deduced by intuition. As mentioned before and as indicated in Fig. 11.2b, an amplitude limiting effect has to prevent from self-destruction.

11.2 Feedback Theory

The feedback theory defines the oscillation conditions by means of the transfer functions of the device characterised by $\underline{H}(j\omega)$ and the feedback network represented by $\underline{\beta}(j\omega)$. Both parameters are a function of the frequency and the bias.

Evidently, one or both transfer functions must comprise an active device. Otherwise, the losses of the passive elements cannot be compensated.

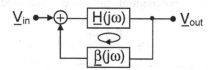

Fig. 11.3. Transfer function of feedback loop

Similar to Eq. (9.71) and in accordance with Fig. 11.3, the open loop voltage transfer function in steady-state can be derived yielding

$$\frac{\underline{V}_{out}(j\omega)}{\underline{V}_{in}(j\omega)} = \frac{\underline{H}(j\omega)}{1 - \underline{\beta}(j\omega)\cdot\underline{H}(j\omega)} . \tag{11.9}$$

In start up, the output voltage grows each periodic cycle if the loop gain $\underline{\beta}(j\omega)\cdot\underline{H}(j\omega)$ is larger than unity and if the signals combine in phase at the input.

Due to saturation effects, e.g. based on the IV boundaries of transistors, the loop gain is reduced to unity when the steady-state is approached.

It is clear that the phase determines whether the loop exhibits constructive or destructive adding of energy. Constructive energy adding is necessary for oscillation. The amplitude and phase conditions for an inverting and a non-inverting amplifier are outlined in Table 11.1. From a practical point of view, the loop gain must be larger than one to ensure start-up and compensation of the resistive parasitics within reasonable margins determined by process and temperature variations, and aging.

Known as the Barkhausen criteria, the presented conditions hold for most types of commonly used oscillators. However, it must be mentioned that these criteria are necessary but not sufficient. For some exotic types of oscillator systems, which are rarely used for IC realisations, these conditions demand further investigations [Ngu92].

Table 11.1. Oscillator conditions

1. Amplitude condition
$\left|\underline{\beta}(j\omega_{LO})\right|\left|\underline{H}(j\omega_{LO})\right| = 1$ (steady-state)
$\left|\underline{\beta}(j\omega_{LO})\right|\left|\underline{H}(j\omega_{LO})\right| > 1$ (start-up)

2. Phase condition
$\angle\underline{\beta}(j\omega_{LO}) + \angle\underline{H}(j\omega_{LO}) = n \cdot 360°$ (steady state)

a. Inverting amplifier	b. Non-inverting amplifier
$\Rightarrow$ Phase of $\underline{H}(j\omega_{LO}) = 180°$	$\Rightarrow$ Phase of $\underline{H}(j\omega_{LO}) = 0°$
E.g. common source with source feedback C_s	E.g. common gate with source feedback L_s
$\Rightarrow$ Phase of $\underline{\beta}(j\omega_{LO}) = 180° \cdot (2n - 1)$	$\Rightarrow$ Phase of $\underline{\beta}(j\omega_{LO}) = 360° \cdot n$

11.3 Negative Resistance Theory

The simple theory of negative resistance is well suited to derive the oscillation conditions of common LC oscillators at the device level [War89]. What does negative resistance mean? A device providing negative resistance is capable of compensating losses associated with positive parasitic resistances. At the node, where negative resistance appears, the outgoing signal power is larger than the incoming signal power. Transistors fed by a DC supply can achieve such characteristics. A proper feedback network ensures the instability of the active device. Matching at the output optimises the power transferred to the load. A resonator is required for the determination of the oscillation frequency. Either a parallel or series connection of an inductor L_R and a capacitor C_R can be employed to realise the resonator. In both cases, maximum energy is generated at the associated resonance frequency. At frequencies being apart from the resonance frequency, the energy vanishes in ideal case.

11.3.1 Necessary Conditions

We assume that a transistor with three ports is used as the active device. If oscillation occurs, certain oscillation conditions are met at all ports [War89]. For analysis, the circuit can either be cut between the active device and the resonator, the feedback network or the load as illustrated in Fig. 11.4. The analysis is based on the characteristics of the impedances at the interfaces, which can also be represented by the corresponding reflection coefficients.

For some circuits, the load plays an important role. One example is an oscillator, where we want to maximise the power transferred to the load. In this regard, the consideration of the plane between the active device and the load may be advantageous. In the following, we discuss this case. The impedance of the passive load $\underline{Z}_L$ is a function of ω, whereas the impedance of the active device $\underline{Z}_A$ is strongly dependent on the oscillator amplitude I and slightly dependent on the frequency. Between the load and the active device, the small signal condition for steady-state oscillation is

$$\underline{\Gamma}_A\left(I,\omega\right)\cdot\underline{\Gamma}_L\left(\omega\right)=1. \tag{11.10}$$

We can use Z-parameters yielding

$$\underline{Z}_A\left(I,\omega\right)+\underline{Z}_L\left(\omega\right)=0 \tag{11.11}$$

with

$$\underline{Z}_A\left(I,\omega\right)=r_A\left(I,\omega\right)+jX_A\left(I,\omega\right) \tag{11.12}$$

and

$$\underline{Z}_L\left(\omega\right)=R_L\left(\omega\right)+jX_L\left(\omega\right). \tag{11.13}$$

From inspection of Eqs. (11.12) and (11.13), we can derive that

$$r_A\left(I,\omega\right)=-R_L\left(\omega\right) \tag{11.14}$$

and

$$X_A\left(\omega\right)=-X_L\left(I,\omega\right). \tag{11.15}$$

The last two conditions are essential for the design of oscillators and can be summarised as follows:

1. The active device must compensate the parasitic resistance of the resonator. In start-up, where the amplitudes are very small, an over-compensation with $r_A\left(I,\omega\right)<-R_L\left(\omega\right)$ is required to provide sufficient margin with respect to process and temperature variations. In this case, the magnitude of the negative resistance of the active device must be larger than the positive resistance of the resonator. The amplitude grows until saturation occurs reducing the magnitude of the negative resistance to $r_A\left(I,\omega\right)=-R_L\left(\omega\right)$, which is the final steady-state condition. Negative resistance can be interpreted as gain, which in turn is a function of g_m.

2. The resonator must compensate the imaginary impedance of the active device at the desired oscillation frequency. The higher the total Q of the resonator, the higher the attenuation of undesired signals such as noise and harmonics located

around the oscillation frequency. From the start-up to the steady-state condition, the reactive characteristics of the active device vary. However, the impact of these variations is typically small compared to those of the resonator elements dominating the total reactive conditions. Corresponding frequency variations are well below 10% and may be compensated by frequency tuning.

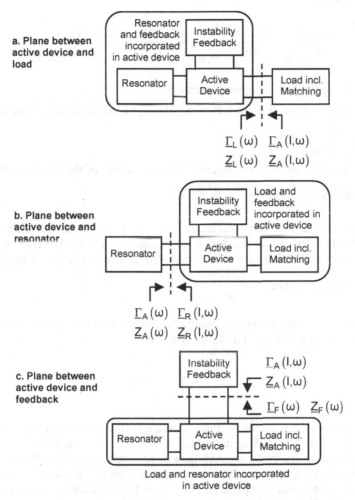

Fig. 11.4a–c. Simplified block diagrams of negative resistance model. Circuit is cut at one port of the active device, which in this case is a transistor with three ports

11.3.2 Sufficient Conditions

The conditions presented in the preceding section are mandatory but not sufficient. Since parameters such as $\underline{Z}_A(I,\omega)$ are both amplitude and frequency dependent, we may find certain conditions, where the oscillation frequency is unstable in the steady-state, since amplitude and frequency characteristics may have opposing impacts. Kurokawa has derived an equation specifying stable oscillation [Kur69]. The derivation is complex and can be found in [Gon97]. Representing a differential change of the impedances with respect to amplitude and frequency, the relation is given by

$$\frac{dr_A(I)}{dI}\bigg|_{I=I_0} \cdot \frac{dX_L(\omega)}{d\omega}\bigg|_{\omega=\omega_{LO}} - \frac{dX_A(I)}{dI}\bigg|_{I=I_0} \cdot \frac{dR_L(\omega)}{d\omega}\bigg|_{\omega=\omega_{LO}} > 0 . \quad (11.16)$$

We may make the following simplifications:

- The real part of the load R_L is typically constant with frequency. Thus $\frac{dR_L(\omega)}{d\omega}\bigg|_{\omega=\omega_{LO}} = 0$.

- In active devices, we may assume that the changes of the reactive part X_A are relatively small yielding $\frac{dX_A(I)}{dI}\bigg|_{I=I_0} = 0$.

- Assuming that the active device is biased at a DC point with high gain, the amplitude swing drives the device into regions with lower DC current and gain (e.g. threshold and linear/resistive regions). Hence, the negative magnitude of r_A is reduced leading to $\frac{dr_A(I)}{dI}\bigg|_{I=I_0} > 0$.

Based on these assumptions, we can express the condition sufficient for stable oscillation according to Eq. (11.16) simply as

$$\frac{dX_L(\omega)}{d\omega}\bigg|_{\omega=\omega_{LO}} > 0 . \quad (11.17)$$

This means that the value of X_L must increase with frequency. Such performance can be achieved by including an inductive element. The matching or output bias inductor, or an interconnection such as a bond wire may be used for this task. In many circuits, we find such inductive properties anyway. On the condition that the inductive effect is not compensated, e.g. by a DC decoupling capacitor, the designer may fulfil Eq. (11.17) without special consideration.

11.3.3 Maximum Output Power

We have derived the following necessary requirements for the operation of oscillators:

- $r_A (I, \omega) = -R_L (\omega)$ in steady-state operation, and

- $r_A (I, \omega) < -R_L (\omega)$ in start-up.

Now we estimate the optimum load for maximum output power in steady-state. As will be verified in Sect. 11.5.1, the generated negative resistance is proportional to g_m. Suppose that the oscillator is operated in a class-A bias point with maximum g_m at start-up where the RF current amplitude I is very small[2]. In this case we refer to as the small signal case. The amplitude growths continuously until the maximum RF current amplitude I_{max} is approached in steady-state, where we operate under large signal conditions. Due to the large output signal excursions, the active device enters IV regions where the g_m is smaller than at start-up as illustrated in Fig. 11.5a. This variation of g_m during oscillation must be taken into account. In the threshold region, the losses may be so high that for a certain time frame no negative resistance can be generated. Moreover, we have to take into account the lowering of the g_m towards the resistive/saturation region. We can conclude that in steady-state the magnitude of the negative resistance must be higher than in start-up. But how much higher? To simplify the calculations, we assume that g_m and subsequently the negative resistance are linearly decreasing with I as sketched in Fig. 11.5b. In accordance with [Gon97] we may write for the negative large signal (steady state) resistance

$$R_A (I) = r_A (1 - \frac{I}{I_{max}}) , \qquad (11.18)$$

where r_A is the small signal R_A at I=0 in start-up. Keep in mind that both r_A and R_A have a negative magnitude. The power delivered from R_A to R_L can be calculated by

$$P = \frac{1}{2} \cdot I^2 \cdot |R_A (I)| = \frac{1}{2} \cdot I^2 \cdot r_A \cdot \left[1 - \frac{I}{I_{max}} \right] . \qquad (11.19)$$

Maximum power is achieved if

$$\frac{dP}{dI} = \frac{1}{2} \cdot I^2 \cdot r_A \cdot \left[2I - \frac{3I^2}{I_{max}} \right] = 0 \qquad (11.20)$$

---

[2] Let us recall the impact of the operation current on the g_m of transistors. According to Eq. (5.65) bipolar transistors exhibit a linear dependence of g_m vs the collector emitter current. In FETs, the impact of the drain source current on g_m is weaker. As demonstrated in Eq. (5.16), the relation is given by a square root function. However, these characteristics hold only up to a certain gain compression. Well above class-A, the g_m may even decrease at raised supply currents.

resulting in an optimum current of $I_{LS,opt} = \dfrac{2}{3}I_{max}$. Consequently, by inspection of Eq. (11.18) we can determine the optimum R_A of

$$R_A = \frac{r_A}{3} .$$

(11.21)

That means that, in the optimum large signal case, only a third of the negative small signal resistance is effectively generated. Thus, the optimum load impedance, which has a positive magnitude since it is assumed to be passive, can be approximated by

$$R_{Lopt} = -\frac{1}{3}r_A .$$

(11.22)

The last equation is valuable since r_A is based on the small signal transistor parameters, which can be found in data sheets and can be measured relatively easy. Large signal parameters are typically not available. To obtain maximum output power, the magnitude of the negative resistance of the active device under small signal conditions must be three times larger than the magnitude of the load resistance: $r_A = -3R_L$. Fortunately, this claim is in accordance with the start-up condition $r_A(I,\omega) < -R_L(\omega)$. Despite the strong simplifications, in practice, Eq. (11.22) has proved to be a fruitful rule of thumb.

a)   b)

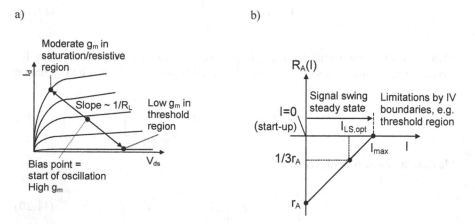

Fig. 11.5a,b. Impact of signal swing: **a** g_m characteristics along load line; **b** negative resistance assuming linear dependency

11.4 Noise

Noise is one of the most critical parameters in oscillators and has been extensively treated in the literature [Lee00, Ham01, Len04]. Due to the complexity of the problem, simplifications have to be made, which degrade the modelling accuracy. The theories and considerations presented in the next sections are well suited to understand at least the qualitative characteristics of the oscillator noise. One of the major problems is that an oscillator operates under large signal conditions. Thus, the nonlinear noise is converted into different frequencies. Analytical analyses are challenging since the noise conversions can have a variety of causes and interdependencies. Many approaches neglect these interdependencies and assume partly linear properties with parameters averaged within the large signal swing. For the sake of understanding, we will follow these simplified approaches. Designers requiring more precise analysis employ large signal simulations on the basis of CAD tools. However, also the accuracy of computer simulations is limited by the performance of the available models. Note that the accuracy of typical design kit models is still disappointing in terms of nonlinear noise modelling. Thus, also in the future, oscillator noise modelling continues to be an interesting subject of research.

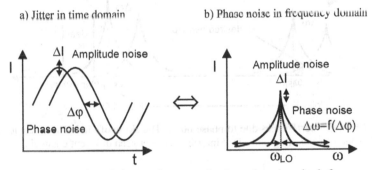

Fig. 11.6a,b. Amplitude and phase noise appearing in: **a** time domain; **b** frequency domain

Assuming sinusoidal characteristics, in time domain, the output current of an oscillator can be described by

$$i(t) = (I + \Delta I) \cdot \cos(\omega_{LO} t + \Delta\varphi), \qquad (11.23)$$

where ΔI is the amplitude noise and $\Delta\varphi$ is the phase noise contribution. Figure 11.6 illustrates the effect of these noise types in both the time and frequency domain. Time and frequency representations of periodic signals are correlated by means of Fourier relations. It is straightforward that amplitude variations in the time domain convert to amplitude variations in the frequency domain and vice versa.

Phase variations in the time domain appear as jitter and exhibit phase noise in the frequency domain. Since phase is the integral of frequency with respect to time, phase variations correspond to frequency variations. Thus, if we talk about phase noise, we may also talk about frequency noise, however, this expression is rarely used in literature. Since oscillators operate in saturation, amplitude variations caused by noise are compressed, lowering their impact, whereas phase noise does not benefit from such a compression effect. This is one of the reasons why the level of phase noise power is typically more than 20 dB higher than the amplitude noise power.

Phase noise can significantly degrade the performance of wireless communication systems as illustrated in Fig. 11.7. A detailed tutorial about this subject has been published in [Hau05]. Consider that the adjacent channels are close together to maximise the possible number of potential users. For mobile phones, the difference between adjacent channels is as small as a few tens of kHz. Due to the noise in the local oscillator used for mixing of the RF frequencies, the IF spectrum is widened. Thus, signal parts of two adjacent channels can overlap leading to significant performance degradations.

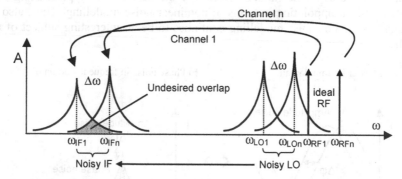

Fig. 11.7. Channel interferences due to phase noise. The skirts of the LO frequencies influence the IF spectra generating undesired interferences between different channels

11.4.1 Types of Phase Noise

Noise in oscillators can have several interrelated origins, e.g. the wideband thermal noise, the thermal noise shaped in a narrowband resonator/feedback loop and the latter one impacted by 1/f noise.

11.4.1.1 Noise Floor

Referring back to Sect. 4.3.1, the noise of a resistor representing the resistive losses in the resonator and the active device may be described by a noise current source yielding a mean-squared spectral noise current density of

$$\overline{I_{th}^2} = 4 \cdot F \cdot kT \cdot \frac{1}{R_R} \cdot \Delta f . \tag{11.24}$$

Compared to a purely passive element, the noise factor F>1 has been added to consider the active properties of the incorporated transistors. Since this spectral density has no dependency on frequency, it exhibits a constant response and is also called noise floor or white noise. Following the equipartition theorem of thermodynamics, the levels of thermal amplitude and phase noise are equal. Assuming that the impact of amplitude noise can be neglected with respect to phase noise, the total spectral noise power amounts to

$$S_n = \frac{\overline{I_{th}^2} \cdot R_R}{\Delta f} = 2 \cdot F \cdot kT. \tag{11.25}$$

As shown in the next section, Eq. (11.25) holds only for large $\Delta\omega$ with respect to ω_{LO}; or in other words for frequencies outside the 3 dB resonator bandwidth $\omega_R = \frac{\omega_{LO}}{2Q}$. The latter relation is based on $Q = \frac{\omega_{LO}}{2 \cdot \omega_R}$ derived from a simple LC tank employing the general Q factor definition $Q = \frac{1}{2\pi} \cdot \frac{\text{stored energy}}{\text{loss energy per cycle}}$. The equation describes the slope of the power spectrum around ω_{LO}. The higher the slope and Q, the lower the energy in the noise bands. A factor $\frac{1}{2}$ arises from the fact that we have two sidebands each carrying loss energy thereby reducing the Q accordingly.

11.4.1.2 Noise Within Resonator Bandwidth

The impedance of a parallel LCR tank is given by

$$Z_R = \frac{1}{\dfrac{1}{R_R} + \dfrac{1}{j\omega L_R} + j\omega C_R}. \tag{11.26}$$

We investigate the characteristics inside the resonator bandwidth for $\Delta\omega = \omega - \omega_{LO} < \frac{\omega_{LO}}{2Q} \ll \omega_{LO}$. Linearisation around $\Delta\omega$ yields[3]

$\frac{\omega}{\omega_{LO}} - \frac{\omega_{LO}}{\omega} \approx \frac{2\Delta\omega}{\omega_{LO}}$. With $Q = \omega_{LO} \cdot C_R \cdot R_R = \sqrt{\frac{C_R}{L_R}} \cdot R_R$ we can approximate Z_R for frequencies close to the resonance frequency:

---

[3] The reader may verify this relation by means of a simple numerical example

$$\underline{Z}_R\left(\omega_{LO}+\Delta\omega\right)\approx\frac{R_R}{1+j\cdot\dfrac{\Delta\omega}{2Q\cdot\omega_{LO}}}.\qquad(11.27)$$

In high-Q resonators, the imaginary part of $\underline{Z}_R$ dominates. This claim is corroborated by the fact that in oscillators the real part of the impedance must be compensated by the negative resistance of the active device. Thus, we can write

$$\underline{Z}_R\left(\omega_{LO}+\Delta\omega\right)\approx -j\cdot 2Q\cdot R_R\cdot\frac{\omega_{LO}}{\Delta\omega}.\qquad(11.28)$$

$\underline{Z}_R$ describes the transfer function of the resonator $\underline{H}(j\omega)=\dfrac{\underline{V}_{th}}{\underline{I}_{th}}$. Together with Eq. (11.25) we can model the noise properties by the following identity:

$$\frac{\overline{V_{th}^2}}{\Delta f}=\frac{\overline{I_{th}^2}}{\Delta f}\cdot Z_R^2=2\cdot F\cdot kT\cdot R_R\cdot\left(\frac{\omega_{LO}}{2\cdot Q\cdot\Delta\omega}\right)^2.\qquad(11.29)$$

The latter relation is in accordance with the often-cited equation of Leeson [Lee66]. Within the resonator bandwidth, the equation reveals the following properties:

- The noise power is proportional to $\dfrac{1}{Q^2}$, implicating the need for maximum Q.

- Attributed to the factor $\left(\dfrac{\omega_{LO}}{\Delta\omega}\right)^2$, the noise decreases with frequency offset from the carrier resulting in a skirt-like shape.

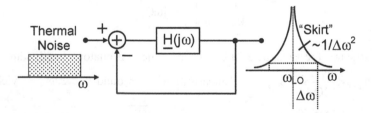

Fig. 11.8. Shaping of thermal noise in feedback loop

Until now, we have considered the oscillator as a resonator based circuit. However, we have seen that there exist different descriptions of oscillators leading to similar results. Optionally, we can review the characteristics on the basis of a simple feedback system fed by the thermal noise as illustrated in Fig. 11.8 [Raz96]. The transfer function around ω_{LO} yields

$$\frac{V_{out}}{V_{in}}(j\omega_{LO}) = \frac{H(j\omega_{LO})}{1+\underline{H}(j\omega_{LO})}. \tag{11.30}$$

Taylor approximation reveals that

$$\underline{H}(j(\omega_{LO}+\Delta\omega)) \approx \underline{H}(j\omega_{LO})+\Delta\omega\frac{d\underline{H}(j\omega_{LO})}{d\omega}. \tag{11.31}$$

With

$$\left|\Delta\omega\frac{d\underline{H}(j\omega_{LO})}{d\omega}\right| << \left|\underline{H}(j\omega_{LO})\right|, \tag{11.32}$$

and assuming an inverting amplifier with $\underline{H}(j\omega_{LO}) = -1$ (which may be provided by a simple common source stage) we obtain

$$\frac{V_{out}}{V_{in}}(j(\omega_{LO}+\Delta\omega)) = \frac{H(j\omega_{LO})+\Delta\omega\frac{d\underline{H}(j\omega_{LO})}{d\omega}}{1+\underline{H}(j\omega_{LO})+\Delta\omega\frac{d\underline{H}(j\omega_{LO})}{d\omega}} \approx \frac{-1}{\Delta\omega\frac{d\underline{H}(j\omega_{LO})}{d\omega}}. \tag{11.33}$$

The slope around $H(j\omega_{LO})$ increases with raised Q and is inversely related to ω_{LO}. Thus, we may approximate the slope around $H(j\omega_{LO})$ by

$$\frac{d\underline{H}(j\omega_{LO})}{d\omega} = \frac{2Q}{\omega_{L0}}. \tag{11.34}$$

Again, the factor 2 considers the two-sided spectrum. With Eqs. (11.33) and (11.34) we get a squared voltage transfer function of

$$\left|\frac{V_{out}}{V_{in}}(j(\omega_{LO}+\Delta\omega))\right|^2 = \left(\frac{\omega_{LO}}{2\cdot Q\cdot\Delta\omega}\right)^2. \tag{11.35}$$

Compared to the resonator based model, the dependencies regarding the Q-factor and the impact of the $\frac{\omega_{LO}}{\Delta\omega}$ ratio are equal. This is a remarkable result when considering the different point of views and assumptions. By taking into account Eq. (11.25), we get the same expression for the feedback shaped noise as in Eq. (11.29).

11.4.1.3 Flicker Noise

As treated in Sect. 4.3.3, the $1/f$ noise, which is also known as flicker noise, dominates the noise properties towards low frequencies. The reader may argue – who cares about the noise close to DC – the operation frequency ω_{LO} is far away from DC. We have to keep in mind that an oscillator is a highly nonlinear circuit also acting as a mixer, e.g. with mixing product of $\omega_m = \omega_{LO} \pm \omega_{1/f}$ leading to a conversion of $1/f$ noise into the ω_{LO} signal since $\omega_{1/f}$ is small. Thus, the mixed signal appears within the resonator bandwidth, where the $1/f$ noise is shaped. To account for this effect, the frequency response according to Eqs. (11.29) and (11.35) must be multiplied by the factor $\dfrac{\omega_{1/f}}{\Delta\omega}$.

11.4.1.4 Overall Frequency Response

A reasonable figure of merit is the ratio between parasitic noise power and desired local oscillator power P_{LO}. In practice, we frequently find the following measure:

$$L(\Delta\omega) = 10\log\left[\frac{S_n(\omega_{LO} + \Delta\omega)}{P_{LO}(\omega = \omega_{LO})}\right], \tag{11.36}$$

where $S_n(\omega_{LO} + \Delta\omega)$ represents the spectral noise power density at frequency offset $\Delta\omega$. Denoted in logarithmic scale[4], the unit of $L(\Delta\omega)$ is dBc/Hz, representing dB per Hz with respect to the carrier (LO) power at a certain offset $\Delta\omega$. S_n is associated with a noise power determined at a bandwidth of 1 Hz. A typical oscillator exhibits an $L(\Delta\omega)$ of around -100 dBc/Hz at 1 MHz offset. For wireless applications, it makes sense to refer the frequency offsets to the closest channel. Dependent on the standard, the offsets range between tens of kHz up to several MHz.

Based on the insights gained in the preceding sub-sections, we can now assemble the behaviour of each frequency segment to an overall relation yielding

$$L(\Delta\omega) = 10\log\left[\underbrace{\frac{2FkT}{P_{LO}}}_{\substack{\text{Noise} \\ \text{floor}}}\underbrace{\left(1 + \left(\frac{\omega_{LO}}{2Q\Delta\omega}\right)^2\right)}_{\substack{\text{Resonator/} \\ \text{feedback} \\ \text{Noise}}}\cdot\underbrace{\left(1 + \frac{\omega_{1/f}}{\Delta\omega}\right)}_{\substack{\text{Flicker} \\ \text{Noise}}}\right]. \tag{11.37}$$

---

[4] From pure mathematical point of view, performing a logarithmic operation on basis of values not having the same unit is not correct. However, since this definition has found wide exception in the community, we report it as it is.

The latter equation is known as the extended Leeson equation. In a first order approximation, a linearised large signal noise factor is considered by means of the small signal noise factor F. However, in practice, we may observe significant deviations between small and large signal noise properties. Hence, for enhanced modelling, F may be treated as fitting parameter. In Fig. 11.9, the characteristics of an ideal and more or less realistic spectrum according to Eq. (11.37) are illustrated.

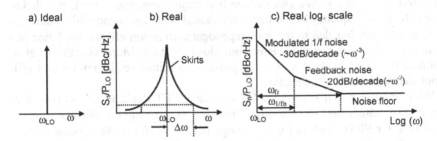

Fig. 11.9a-c. Phase noise: **a** ideal; **b** real; **c** real in logarithmic scale

11.4.1.5 Minimisation of Phase Noise

Based on Eq. (11.37), we are now able to derive strategies for minimisation of $L(\Delta\omega)$:

a. Maximisation of LO Power
Proper choice of the bias and the loadline is important to maximise the LO power. Corresponding design issues are very similar to those of power amplifiers. To increase the signal swing, complementary p- and n-MOS structures are efficient in topologies capable of adding the individual voltage swings of the devices [Fon04]. However, one has to consider that p-MOS transistors suffer from higher parasitics at given g_m.

b. Optimum Gain
To ensure start-up, the negative resistance and consequently the gain provided by the active device must be high to compensate the losses in the system within a wide frequency band, temperature range and process variations. In steady-state, the LO signal is subject to power compression, whereas the phase noise is not compressed. That means that at a certain gain level the noise is amplified but not the LO signal leading to a degradation of L. Thus, depending on the system requirements a reasonable tradeoff has to be found for the gain.

c. Maximisation of Resonator Q
The Q of the resonator has a significant impact on the noise and must be kept as high as possible. At low to moderate frequencies, the Q factor of the resonator is clearly determined by the inductor in the resonator requiring lots of turns and large area. Thus, the inductor exhibits significant resistive losses associated with the large series resistance of the lines and the coupling to the lossy substrate. Corresponding design and optimisation strategies have been addressed in Sect. 6.2.

Towards very high frequencies, the varactor becomes more and more the limiting element since the varactor Q is more or less inversely proportional to frequency.

d. Choice of Transistor

The noise floor is impacted by the noise properties of the device. A low average noise figure along the whole signal swing is important but more or less determined by the technology used.

Close to the LO frequency the noise is significantly increased by the up-converted flicker noise. We can deduce two major constraints. First, the flicker noise activation frequency of the transistor should be as low as possible leading to a low $\omega_{1/fn}$. Note that due to the interdependent conversion effects, the latter two frequencies are not necessarily equal but close together. Moreover, the level of nonlinearities must be kept as low as possible to minimise the conversion efficiency demanding a low loop gain.

What about the transistor type? On one hand, FET devices are less nonlinear than bipolar transistors. On the other hand, the $\omega_{1/fn}$ of bipolar devices is much lower than for FETs. Values for $\omega_{1/fn}$ range from 1 to 10 kHz for bipolar transistors and 0.1 to 1 MHz for FETs. Proved by experiments, the impact of the $\omega_{1/fn}$ properties is more significant than the level of mixing. Thus, typically, bipolar devices provide lower $1/f$ noise, which can dominate the total VCO noise performance.

Why is the $\omega_{1/fn}$ of bipolar transistors lower than that of FETs? This question is still subject to research. However, let us outline some indications. Recall that the flicker noise is generated by statistical charge effects on the basis of traps, which are impurities in the fabricated materials. The highest impurities occur at the surface of the devices, where different materials are put together. Reviewing the current flow in the transistors with respect to these surface traps gives us useful insights. By inspection of Figs. 4.8, 5.7, 5.32 and 5.38 we observe that in bipolar transistors the current flows mainly vertically through the device, whereas in FETs the current flows laterally and in parallel to the surface traps. That means that in FETs the current passes the traps during a much longer time than in BJTs. Thus, the probability for charge actions is much higher for FETs than for BJTs. This may explain why the $\omega_{1/fn}$ of bipolar transistors is lower than for FETs leading to the superior performance observed.

11.4.1.6 Phase Noise Figure of Merit

Referring to the equation of Leeson and reviewing the shape of the phase noise spectrum reveals that the noise performance depends on the frequency offset. Higher $\Delta\omega$ implies lower phase noise. Low phase noise comes at the expense of high power consumption. The figure of merit

$$\text{FOM} = L(\Delta\omega) - 20\log\left(\frac{\Delta\omega}{\omega_{LO}}\right) - 10\log\frac{[\text{mW}]}{P_{dc}[\text{mW}]} \qquad (11.38)$$

considers corresponding tradeoffs and is frequently applied for VCO comparisons.

11.5 Topologies

The proper choice of the oscillator topology is mandatory to achieve optimum performance with respect to maximum oscillation frequency, bandwidth, output power, phase noise and power consumption. The most frequently used topologies in the area of high-speed electronics are the gate resonator, Colpitts, cross-coupled and ring oscillator. They are studied in the following sections.

11.5.1 LC Gate Resonator Oscillator

The single ended FET oscillator with a LC resonator between gate and ground is successfully used for both integrated and hybrid-type oscillators [War89]. A detailed analysis has been presented in [Len04]. The simplified circuit schematic is illustrated in Fig. 11.10a for a circuit employing a resonator with the elements L_R and C_R connected in parallel. A part of C_R is given by the parasitic capacitance of L_R. The losses of the resonator and the active device are incorporated into R_R. As discussed in Sect. 6.2, the equivalent circuit of an inductor is composed of series and parallel networks. Rearrangement of the network elements allows an equivalent representation in terms of parallel elements only, simplifying the analyses. Optionally, an equivalent circuit only consisting of series components would also be feasible. We have to keep in mind that these modified networks are only exactly equivalent at one single frequency. However, within a reasonable bandwidth of around 10%, the equivalent circuits typically provide good accuracy.

a)   b)

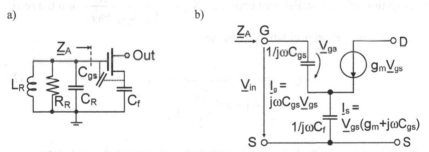

Fig. 11.10a,b. Oscillator with gate resonator: **a** topology; **b** simplified equivalent circuit of FET including feedback capacitance

The feedback capacitance C_f is instrumental in making the device unstable. Referring to Fig. 11.10b and Sect. 7.1.2, the transistor input impedance can be calculated by

$$Z_A = \frac{V_{in}}{I_g} = r_A + jX_A \qquad (11.39)$$

$$= - \frac{g_m}{\omega_{LO}^2 \cdot C_f \cdot C_{gs}} - j \frac{C_f + C_{gs}}{\omega_{LO} \cdot C_f \cdot C_{gs}} . \qquad (11.40)$$

For simplicity we have assumed that the input capacitance of the transistor is given by C_{gs}. It is clear that better accuracy can be obtained by considering the increased input capacitance associated with the Miller effect. According to the first term, a negative resistance of

$$r_A = - \frac{\omega_t}{\omega_{LO}^2 \cdot C_f} \qquad (11.41)$$

is generated, which has to compensate the losses of the passive resonator[5]. We can conclude that:

- The magnitude of r_A can be raised by increasing ω_t. A result, which may not be very surprising since g_m and gain at given frequency become larger.

- Since $r_A \sim \frac{1}{\omega_{LO}^2}$, the magnitude of the negative resistance significantly decreases with raised frequency, hence limiting oscillation towards high frequencies.

- Lowering C_f results in an increased r_A. However, decreased C_f adversely reduces the maximum output power since the raised impedance drops the RF source current. Obviously, a tradeoff exists.

By investigation of the parallel connection of $X_A = - \dfrac{C_f + C_{gs}}{C_f \cdot C_{gs} \cdot \omega_{LO}}$ and the resonator elements, we get an oscillation frequency of

$$\omega_{LO} = \frac{1}{\sqrt{L_R \left(C_R + \dfrac{C_{gs} \cdot C_f}{C_{gs} + C_f} \right)}} . \qquad (11.42)$$

As usual in the area of circuit design, the single ended gate resonator approach offers both advantages and drawbacks. A high efficiency (output power/DC power) can be obtained since no output buffer is necessary for impedance matching of the

---

[5] If we want to check the oscillation conditions, we have to take care that the negative resistance and the losses of the resonator are both modelled in either Z or Y representation. Mixing of the representations would lead to wrong calculations. If we consider $\underline{Z}_A$, the resonator hast to be modelled in Z-representation as well. Hence, we have to transform the parallel (Y) model in Fig. 11.10a into a network consisting of series elements only. Optionally, we could also apply the depicted Y model of the resonator. However, in this case, we have to compare the associated Y values with the Y representation of the negative resistance.

output. Moreover, compared to differential approaches, no DC current source has to be included.

Especially on silicon-based technologies, single ended circuits exhibit significant disadvantages. They are susceptible to noise pick-up from the substrate and the DC supply. Furthermore, undesired coupling can impact other components of the system. Consequently, single ended topologies are mainly suitable for III/V technologies providing high substrate isolation [Ell28]. However, due to the good isolation properties of advanced RF silicon-based technologies in terms of substrate isolation, good results have been achieved in 0.25 μm BiCMOS technology [Bür33].

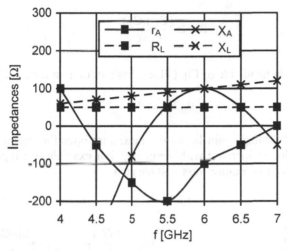

Fig. 11.11. Small signal impedances of typical LC gate resonator MESFET oscillator with R_L=50 Ω and series output bond inductor L=0.5 nH

The small signal impedances in the plane cut between the active device and a 50 Ω load is illustrated in Fig. 11.11. The load exhibits inductive properties since it includes the bond wire connection between load and device. According to the impedances, the circuit is capable to oscillate at 5 GHz since $X_A = -X_L$ and $r_A < -R_L$. Moreover, within a certain frequency range, we achieve $r_A \approx -3R_L$. Referring to Sect. 11.3.3, high output power can be expected in this range.

11.5.2 Typical Colpitts Oscillator

Figure 11.12 sketches the typical schematic and equivalent circuit of a single-ended Colpitts oscillator. Please note that further topologies are possible. We focus on that most frequently used for integrated circuits. Similarly to the gate oscillator, the oscillation frequency can be estimated by

$$\omega_{LO} \approx \frac{1}{\sqrt{L_R\left(C_R + \dfrac{C_1 \cdot C_2}{C_1 + C_2}\right)}}.$$ (11.43)

Compared to the gate resonator based topology, the resonator is located at the drain. To isolate the resonator from further circuits connected to the output, an output buffer is required. The amount and phase of the constructive feedback can be set by C_1 and C_2. It can be shown that the voltage gain required to compensate the resonator losses R_p is given by [Raz03]

$$g_m \cdot R_R \geq \frac{(C_1 + C_2)^2}{C_1 \cdot C_2}.$$ (11.44)

The term on the right hand side of Eq. (11.44) has a minimum for $C_1 = C_2$. We find for this case that

$$g_m R_R \geq 4.$$ (11.45)

This relatively high voltage gain limits the maximum operation frequency when assuming that the losses increase with frequency. For example, the topology presented in the next section requires less voltage gain.

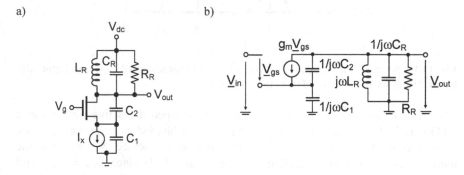

Fig. 11.12a,b. Colpitts oscillator: **a** topology; **b** simplified equivalent circuit

11.5.3 Cross-Coupled Oscillator

The cross-coupled oscillator is a differential topology mainly consisting of two equal active RF devices acting as both signal amplifiers and active feedback elements. A rigorous analysis and fruitful design strategies have been presented in [Ham01]. According to Fig. 11.13 we can recall the Barkhausen criteria for start-up with the transistor transfer functions:

$$|\underline{H}_1(j\omega_{LO})| \cdot |\underline{H}_2(j\omega_{LO})| \geq 1 \tag{11.46}$$

and

$$\angle\underline{H}_1(j\omega_{LO}) + \angle\underline{H}_2(j\omega_{LO}) = n \cdot 360°. \tag{11.47}$$

If the latter two equations hold, the signal increases at every cycle until it is saturated by means of the IV boundaries. Consider now the schematic in Fig. 11.13b, where the voltage transfer functions of the common source stages can be approximated by

$$\underline{H}(j\omega_{LO}) = -g_m \cdot \underline{Z}(j\omega_{LO}). \tag{11.48}$$

In this context, one should refer to Eq. (7.4). To get a high voltage gain, we can either raise g_m by increasing the transistor width and bias current, or by making $|\underline{Z}(j\omega_{LO})|$ large. The latter approach is advantageous since the power consumption can be kept low.

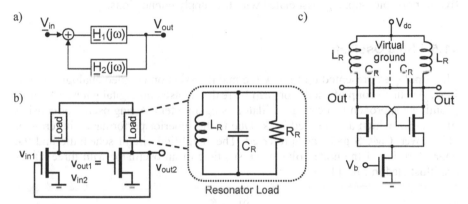

Fig. 11.13a–c. Cross-coupled oscillator: **a** feedback representation; **b** simplified circuit schematic with coupled transistors; **c** schematic including current source required for differential operation

A high load impedance can be realised by a parallel LC resonator. When neglecting any resistive parasitics, the impedance and thus the loop gain approach infinity at the resonance frequency. Consequently, oscillation occurs at the resonance frequency

$$\omega_{LO} = \frac{1}{\sqrt{L_R C_R}} \tag{11.49}$$

with L_R and C_R as the effective resonator components. Note that all other parasitics including the ones of the transistors and interconnects are incorporated in these elements. The maximum impedance is limited by the equivalent parasitic resistance R_R, which can be estimated in accordance with Sect. 6.2. We can deduce the following requirement:

$$\left|\underline{H}_1(\omega)\right| = \left|\underline{H}_2(\omega)\right| = g_m \cdot R_R \geq 1. \tag{11.50}$$

With typical values of R_R, a voltage gain of above unity can be achieved up to high frequencies of around 50% of f_t. Compared to the Colpitts oscillator, much lower voltage gain is required making higher operation frequencies feasible. Thus, the cross-coupled topology is an excellent candidate for oscillators covering highest frequencies. At resonance, the phase of the resonator equals zero. Both common source stages have a phase transfer function of 180° resulting in a total loop phase of 360° satisfying the mandatory condition postulated in Eq. (11.47).

So far, we have treated the cross-coupled oscillator as two single-ended cascaded stages with symmetrical feedback. In practice, a current source is added as illustrated in Fig. 11.13c to allow differential operation with common mode rejection. An example of a 60-GHz cross-coupled oscillator will be presented in Sect. 11.6.1. Due to the differential structure, cross-coupled VCOs are less sensitive to noise pick-up, e.g. associated with the supply connections.

11.5.4 Ring Oscillator

The topologies presented up to now are mainly based on coupled analogue amplifiers. Another group of oscillators can be realised based on digital gates. The most prominent approach is the ring-oscillator using inverters. A ring oscillator consists of n stages, which are coupled back. Due to the superior performance in terms of DC power, CMOS gates are preferred. The simplified circuit schematic and the characteristics of the node voltages versus time of an oscillator with three stages are illustrated in Fig. 11.14.

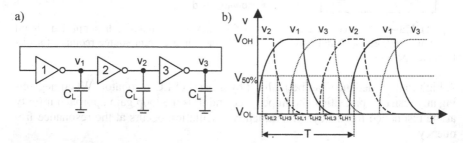

Fig. 11.14a,b. Ring oscillator with three inverter stages, C_L illustrates the input capacitance of next stage: **a** simplified circuit schematic; **b** voltage characteristics versus time

An odd number of stages is required for periodic variation of the stage voltages. This is explained in Fig. 11.15 for even and odd conditions. We can also observe that no periodic variation is generated for n=1. Consequently, the lowest number of stages equals three if single-ended inverter gates are used. Given an odd number of stages, we can deduce from Fig. 11.14b that the circuit oscillates with a fixed period of

$$T = \tau_{HL2} + \tau_{LH3} + \tau_{HL1} + \tau_{LH2} + \tau_{HL3} + \tau_{LH1} = 2 \cdot n \cdot \tau_p, \quad (11.51)$$

where τ_{LHn} and τ_{HLn} are the rise and fall times related to 50% of the full voltage magnitude, and $\tau_p = \frac{1}{2} \cdot (\tau_{LHn} + \tau_{HLn})$ is the average propagation delay. One significant advantage of the ring oscillator topology is that it needs only compact transistors and resistors and no bulky inductors as required for the analogue amplifier based approaches. However, due to the minimum number of three stages, the corresponding maximum oscillation frequency f is bounded by

$$f_{L0} = \frac{1}{T} \leq \frac{1}{6 \cdot \tau_p}, \quad (11.52)$$

implying that the speed is determined by the parasitics of at least three transistors. Thus, the maximum oscillation frequency of ring oscillators is typically lower than the frequency achievable with the analogue amplifier based topologies.

An oscillation frequency of 13 GHz and an ultra fast gate delay of 4.7 ps have been achieved using a ring oscillator with nine stages in 90-nm SOI CMOS technology [Rod9]. This circuit indicates that an oscillation frequency of up to 39 GHz may be possible with three stages. The need of adequate frequency tuning elements such as varactors decreases the oscillation frequency. Further disadvantages of ring oscillators are the relatively high phase noise and power consumption. Ring-oscillators are well suited for applications demanding multiple phase outputs, e.g. for quadrature-type receivers, which can be realised from of a four stage ring oscillator providing phase paths of 0°, 90°, 180° and 270°. This approach has been demonstrated in a 10-GHz circuit integrated together with a complete phase locked loop [Kos04].

Even (e.g. n = 2)				Odd (e.g. n =3)			
Period	V_1	V_2		Period	V_1	V_2	V_3
P1	1	0		P1	1	0	1
P2	1	0		P2	0	1	0
P3	1	0		P3	1	0	1

⇒ static, no oscillation   ⇒ time varying signal, oscillation

Fig. 11.15. Example of wave amplitudes vs periods for even and odd number of stages demonstrating why only odd stages allow oscillation, single-ended CMOS gates are applied

11.5.5 Frequency Tuning

For several reasons, oscillators need frequency tuning. First, for course tuning associated with the compensation of process variations, aging and temperature effects, which can change the reactive values of the circuit elements. For example typical yield specifications define C_{gs} values in the order of ±15%. By inspection of Eq. (11.54), we can determine the corresponding frequency variations. Second, the channel selection demands for fine-tuning of the oscillator. Note that most systems employ a fixed IF frequency, which is superior concerning filtering, and an LO matched in accordance to the desired RF frequency. The trend for multifunctional transceivers demands for large frequency tuning as well. One example is a system capable of WLAN operation around 2.4 and 5.5 GHz.

Both voltage and current controlled oscillators are feasible. This section is devoted to the former ones, which are referred to as VCOs (Voltage Controlled Oscillators). How can we vary the frequency? Generally, by changing the reactive values of the elements. We know that the capacitances of active devices can be changed by varying the gate and/or drain voltage. Unfortunately, this bias tuning may degrade the large signal properties of the device, which is not desired. Furthermore, powerless or low-power control is mandatory for many systems, which would not be possible by tuning the drain node drawing high current.

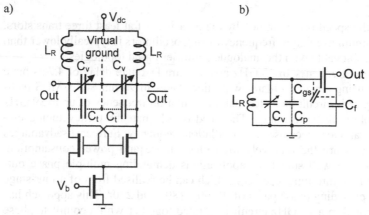

Fig. 11.16a,b. VCOs: a cross-coupled; b LC gate resonator

What about the resonator elements? Active integrated inductors capable of tuning can be realised in compact form [Har91, Cho97]. Regrettably, such active inductors are problematic, since they generate a significant amount of noise, which is unacceptable. A better solution is to employ varactors. Recall from Sect. 6.5 that the capacitive tuning ratio can be defined by

$$t_v = \frac{C_{vmax}}{C_{vmin}}, \tag{11.53}$$

where the maximum and minimum values of C_v are denoted by C_{vmax} and C_{vmin}, respectively. The tuning range can be estimated based on the resonance frequency of the resonator loop

$$\omega_{LO} = \frac{1}{\sqrt{L_R \cdot C_v}}. \tag{11.54}$$

In the ideal case, where we assume that all parasitic capacitances are determined by the varactor, the tuning range can be calculated by

$$\Delta\omega_{LO} = \frac{1}{\sqrt{L_R \cdot C_{v\,min}}} - \frac{1}{\sqrt{L_R \cdot C_{v\,max}}}. \tag{11.55}$$

A ΔC of 3 would yield a maximum $\Delta\omega_{LO}$ of around 30% with respect to centre frequency. However, in practice we have to consider additional parasitic capacitances of the active devices, inductors, varactors, capacitors, interconnects, pads, etc., which can have a remarkable impact towards high frequencies. Consider that active devices exhibit a large capacitance at the input and a moderate capacitance at the drain. Generally, we can identify parasitic capacitances C_t with fixed value connected either in parallel or in series with the varactor resulting in

$$\Delta\omega_{LO} = \frac{1}{\sqrt{L_R \cdot (C_{v\,min} + C_t)}} - \frac{1}{\sqrt{L_R \cdot (C_{v\,max} + C_t)}} \tag{11.56}$$

and

$$\Delta\omega_{LO} = \frac{1}{\sqrt{L_R \cdot \dfrac{C_{v\,min} \cdot C_t}{C_{v\,min} + C_t}}} - \frac{1}{\sqrt{L_R \cdot \dfrac{C_{v\,min} \cdot C_t}{C_{v\,min} + C_t}}}, \tag{11.57}$$

respectively. From these equations we can easily deduce that for high tuning range, large varactor capacitances are required if C_t is connected in parallel, whereas small varactor capacitances are superior when C_t appears in series.

For the cross-coupled topology depicted in Fig. 11.16a, where all major parasitics are more or less parallel, the first equation holds. For many topologies, e.g. the LC gate oscillator according to Fig. 11.16b, we find both series and parallel elements in the resonator loop, which can be taken into account in the expanded equation

$$\Delta\omega = \frac{1}{\sqrt{L_R \cdot \left(C_{v\,min} + C_p + \dfrac{C_{gs} \cdot C_f}{C_{gs} + C_f}\right)}} - \frac{1}{\sqrt{L_R \cdot \left(C_{v\,max} + C_p + \dfrac{C_{gs} \cdot C_f}{C_{gs} + C_f}\right)}}, \tag{11.58}$$

with C_p as the total capacitance at the gate node including the parasitic capacitance of the inductor and the fixed part of the varactor capacitance to ground. With all these parasitic capacitances in mind, it is not surprising that with typical FET varactors providing $\Delta C = 3$, in practice, we achieve not more than $\Delta\omega \approx 5\text{–}15\%$.

It has been demonstrated that improved tuning range can be obtained for LC gate resonator VCOs by implementing an additional varactor in the feedback [And88]. By doing so, it is possible to adjust the negative resistance during frequency tuning thereby allowing for wider bandwidth, where the oscillation conditions are meet. However, since the Q-factor of varactors is lower than that of MIM capacitors, a varactor in the source can increase the noise, which may not be acceptable.

11.5.6 Comparison of Oscillator Types

The key performance parameters of different VCO topologies are compared and discussed in Table 11.2.

Table 11.2. Comparison of oscillator topologies

Parameter	Ring oscillator	Colpitts	LC gate resonator	Cross-coupled
Max. oscillation frequency up to which parasitics can be compensated	Low/moderate, parasitics of at least three gates determine speed in CMOS technology	Moderate, high voltage gain of approximately 4 required	High	Very high since minimum voltage gain as low as 1 is required
Output power	Depends on output buffer	Depends on output buffer	Moderate to high. Output buffer is not necessarily required	Depends on output buffer, advantage of differential operation with double voltage swing but disadvantage since additional voltage headroom required for current source
Efficiency	Low, several gates have to be fed, capacitive load of following stage has to be driven, multistage buffer required for impedance matching	Moderate to low since output buffer required	High since no output power or impedance transformer necessarily required	Moderate to low depends on buffer
Noise (L)	High since no high-Q frequency stabilisation	Moderate to low	Moderate, noise generated in gate resonator is amplified	Low given weak loading of core by means of high impedance buffer, high voltage swing, differential inductor has higher Q
Immunity against external noise/ common mode rejection	High using differential gates, low with single-ended CMOS gates	Low if single-ended	Low since single-ended	High since differential
Circuit size	Very compact since no inductors required	Moderate, at least one inductor required	Moderate, at least one inductor required	Moderate, at least one differential inductor required
Suitability for integrated systems	High, suited to generate multiphase outputs	Up to low/moderate frequencies, high in III/V technologies	Up to moderate frequencies, high in III/V technologies	Very high

11.6 Design Examples

In the following sections, we present three typical oscillator examples: a 60-GHz CMOS cross-coupled VCO, a 40-GHz CMOS quadrature cross-coupled VCO and a power efficient 4-GHz Class-E LC gate resonator VCO.

11.6.1 60-GHz CMOS Cross-Coupled VCO

The simplified schematic of the circuit realised in 90-nm IBM VLSI SOI technology is shown in Fig. 11.17 [Ell36]. A common drain output buffer with high input and low output impedance is used, which minimises the loading of the oscillator core and simplifies the 50 Ω output matching, respectively. The minimum width of the buffer transistor w_b is determined by the output power of the oscillator core.

Thick oxide MOSFET capacitors with a capacitance control range $t_v = \dfrac{C_{v\,max}}{C_{v\,min}}$ of

approximately 2 are used for frequency tuning.

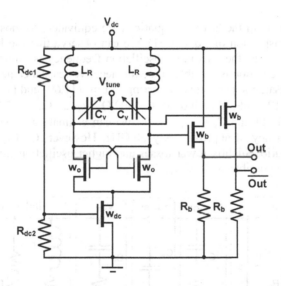

Fig. 11.17. Cross-coupled VCO schematic, C_v=[25–50 fF], L_R=90 pH, w_o=w_b=16 μm, w_{dc}=32 μm, R_b=75 Ω, R_{dc1}=1.4 kΩ, R_{dc2}=2 kΩ, compare [Ell36] © IEEE 2004

The major design goal was to reach an operation frequency as high as possible. Moreover, a frequency tuning range of more than 10% should be achieved to allow compensation of process tolerances. To gain first insights, the maximum oscillation frequency is estimated based on the equivalent circuit depicted in Fig. 11.18. The losses of the tank represented by the parasitic conductance g_{par} must be compensated by the g_m of the active devices. Under large signal conditions, g_m changes from 0 in the threshold region to a maximum of $g_{m,max}$ at gate voltage $V_{gs,max}$. For small signal calculations, we may take the mean value of g_m over V_{gs} at n values between V_{th} and $V_{gs,max}$ and equidistant V_{gs} steps:

$$g_{ma} \approx \frac{1}{n} \sum_{1}^{n} g_{m,n}(V_{gs,n}) \tag{11.59}$$

yielding approximately 12.5 mS for a 90-nm n-channel SOI FET with w_g=16 μm. This value corresponds to 60% of $g_{m,max}$. Written in admittance and transconductance form, the mandatory relation for oscillation is given by

$$g_{ma} \geq g_{par} \tag{11.60}$$

with

$$g_{par} \approx \frac{1}{R_{ds}} + \frac{1}{R_p} + \frac{R_i}{R_i^2 + \frac{1}{\omega^2 C_i^2}} + \frac{R_v}{R_v^2 + \frac{1}{\omega^2 C_v^2}} + \frac{R_s}{R_s^2 + \omega^2 L_R^2} . \tag{11.61}$$

The relations made in the last two equations are equivalent to those presented in Sect. 11.3. By inspection of Eq. (11.61), we can observe that the losses increase with frequency. Thus, the maximum oscillation frequency is limited. Above this frequency, no oscillation is feasible since the negative resistance/gain is not sufficient to compensate the parasitics. Referring to Eqs. (11.60) and (11.61), and with values of g_{ma}=12.5 mS, R_{ds}=300 Ω, R_p=500 Ω, R_s=2 Ω, C_i≈20 fF, R_i=60 Ω, L_R=90 pH, R_v=15 Ω and C_v=C_{vmax}=50 fF, we can estimate a maximum possible oscillation frequency of approximately 60 GHz. However, C_v=C_{vmax} considers a worst-case scenario. Slightly lower losses and higher oscillation frequencies of at least 65 GHz are possible at C_v=C_{vmin}=25 fF.

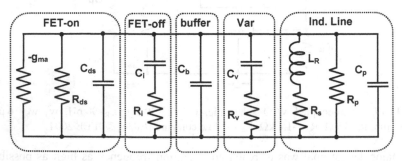

Fig. 11.18. Equivalent small signal circuit of VCO, FET-on: FET switched on, FET-off: contribution of FET when switched off, Var: varactor, ind. line: inductive line

We are now able to identify the main source of the losses, where we assume that C_v=C_{vmax}. The quantitative contributions at 60 GHz are

$$g_{par} \approx \frac{1}{300\Omega} + \overset{\text{Inductor}}{\frac{1}{500\Omega} + \frac{1}{353\Omega} + \frac{1}{202\Omega}} + \frac{1}{802\Omega} \approx 12.5\text{mS}. \qquad (11.62)$$

FET-on   FET-off Varactor

From the last relation, we can get the following interesting insights: at these high frequencies, the losses are dominated by the varactor and not by the inductor as usually claimed for circuits operating at lower frequencies. To verify this fact we have to take a look at the corresponding Q-factors. Due to the thin metal thickness, the Q-factor of the compact inductive line is mainly determined by the series resistance. According to Sect. 6.2.2, the Q-factor may be approximated by $Q_L \approx \frac{\omega \cdot L_s}{R_s}$

yielding a high value of 17 at 60 GHz since Q_L increases with ω. This is not the case for the varactor Q, which may be estimated by $Q_v \approx \frac{1}{\omega \cdot C_v \cdot R_v}$. A low value of approximately 3.5 is obtained at these high frequencies since Q_v is proportional to $1/\omega$.

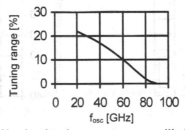

Fig. 11.19. Simulated tuning range versus oscillation frequency

A frequency tuning range of approximately 55 to 65 GHz can be estimated from

$$\Delta\omega_{LO} \approx \frac{1}{\sqrt{L_R\left(C_{v\,min} + C_{par}\right)}} - \frac{1}{\sqrt{L_R\left(C_{v\,max} + C_{par}\right)}} \qquad (11.63)$$

with $C_{par} \approx C_{ds} + C_b + C_p + C_i$, where $C_b \approx C_{ds} \approx 3.75$ fF and $C_p \approx 2$ fF. By means of Eq. (11.63), we can deduce that high operation frequencies demand for small varactor sizes with respect to the transistor sizes leading to small tuning range. Thus, maximum operation frequency and tuning range are conflicting goals. The simulated tuning range versus the maximum centre oscillation frequency is depicted in Fig. 11.19. A photograph of the MMIC is shown in Fig. 11.20. With an overall chip size of 0.3×0.25 mm^2, the chip is very compact.

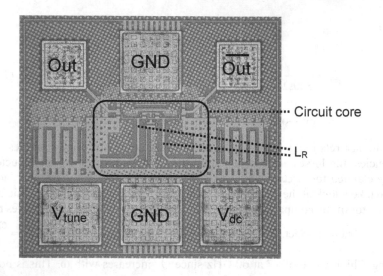

Fig. 11.20. Photograph of IC with compact size of 0.3×0.25 mm^2, compare [Ell36] ©
IEEE 2004

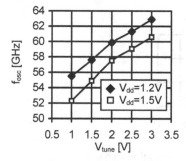

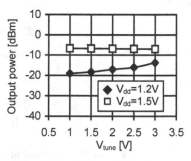

Fig. 11.21. Measured oscillation frequency vs tuning voltage, compare [Ell36]
© IEEE 2004

Fig. 11.22. Measured output power
versus tuning voltage, compare
[Ell36] © IEEE 2004

In Fig. 11.21, the measured tuning range versus varactor voltage is illustrated. A
high tuning range of 55.5–62.9 GHz and 52.3–60.6 GHz at supply voltages of
V_{dc}=1.2 V and V_{dc}=1.5 V, respectively are measured. The corresponding supply
currents are 8 mA and 12 mA, respectively. Despite the assumptions and simplifications, the measured tuning range agrees well with the estimations made before.
The measured output power versus tuning voltage is illustrated in Fig. 11.22. Over
the full tuning range, an output power of –6.8±0.2 dBm is achieved at V_{dc}=1.5 V.
A typical measured output spectrum of the oscillator is depicted in Fig. 11.23. At
V_{dc}=1.2 V, a frequency of approximately 60 GHz and a frequency offset of 1 MHz
from the carrier, a phase noise of –94 dBc/Hz is achieved. The measured phase

noise for a frequency offset of 1 MHz is shown in Fig. 11.24 versus tuning range. Within the full tuning range, the phase noise is lower than -85 dBc/Hz. A comparison with state-of-the-art VCOs at millimetre wave frequencies is given in Table 11.3. All circuits listed are cross-coupled oscillators.

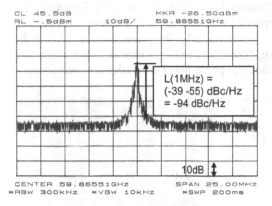

Fig. 11.23. Measured output spectrum at 59.9 GHz, L: phase noise V_{dc}=1.2 V and V_{tune}=2 V, losses of cables not deembedded, compare [Ell36] © IEEE 2004

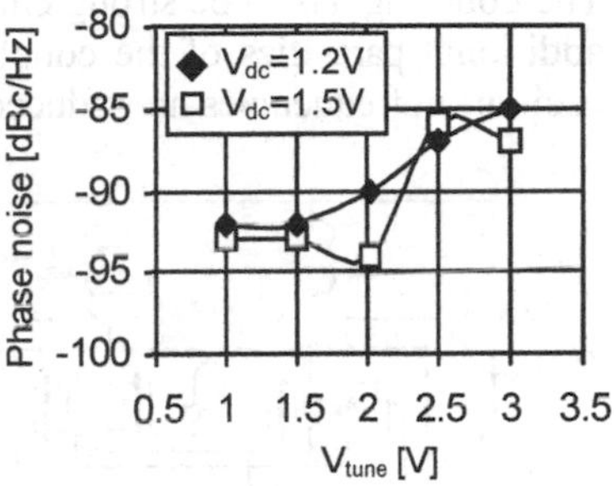

Fig. 11.24. Measured phase noise at 1 MHz offset versus tuning voltage, compare [Ell36] © IEEE 2004

Table 11.3. Comparison of state-of-the-art cross-coupled VCOs at millimetre wave frequencies

Technology/speed	Centre frequency	Tuning range	Phase noise @1MHz	Supply power	Ref.
Compound					
InP HBT, f_{max}=210 GHz	67 GHz	3%	−90 dBc/Hz	4.5 V× 70–170 mA	[Sch99]
SiGe HBT/f_t=120 GHz	43 GHz	11.8%	−96 dBc/Hz	3 V×121 mA	[Sha03]
CMOS					
0.12-μm CMOS/n.a.	51 GHz	2%	−85 dBc/Hz	1.5 V× 6.5 mA	[Tie02]
0.13-μm SOI CMOS/f_{max}=168 GHz	40 GHz	15%	−90 dBc/Hz	1.5 V× 7.5 mA[a]	[Fon04]
90-nm SOI CMOS/f_{max}=160 GHz	57 GHz	16%	−90 dBc/Hz	1.5 V× 14 mA	This section
	60 GHz	14%	−94 dBc/Hz	1.2 V×8 mA	

Phase noise performances specified at centre frequencies, [a]without buffer

11.6.2 40-GHz Cross-Coupled Quadrature VCO

For many applications such as image reject receivers, half rate transceivers and phase rotator based phase locked loops, quadrature VCOs (QVCOs) with phase outputs of 0°, 90°, 180° and 270° are required. The phase offsets can be generated by symmetrical coupling of two differential LC VCOs [Wan03]. This coupling introduces additional challenges compared to the conventional differential approach. The coupling has to be strong enough to ensure the 90° phase locking. Due to the additional parasitics of the coupling transistors and interconnects, the maximum oscillation frequencies are reduced and the phase noise properties are deteriorated.

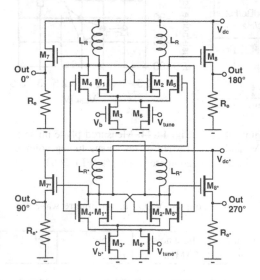
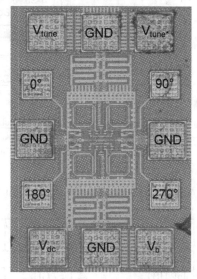

Fig. 11.25. Circuit schematic, L_R=110 pH, w_g of M1, M2, M4, M5 = 16 μm, w_g of M3, M6 = 32 μm, R_e=75 Ω, V_{dc}=V_{dc*}, V_b=V_{b*}, V_{tune}=V_{tune*}, compare [Ell30] © IEEE 2004

Fig. 11.26. Photograph of compact chip, size: 0.4×0.3 mm², compare [Ell30] © IEEE 2004

The presented circuit has been fabricated in 90 nm IBM VLSI SOI CMOS technology [Ell30]. In Fig. 11.25, the simplified circuit schematic is shown. Based on two cross-coupled VCOs, the circuit consists of the core transistors M1-M2, the inductors L_r with Q of approximately 18 at 40 GHz, and a current source M3 biased at V_b=0.8 V. The VCOs are coupled by the transistors M4-M5, which can be used for frequency tuning since the bias dependent coupling determines the oscillation frequency. According to the equations presented in [Liu99], the frequency tuning can be derived. No varactors are required. This is a significant advantage since most VLSI technologies do not feature high-Q varactors. At 40 GHz, a Q of approximately 5 is measured for the MOSFET based varactors. This Q is significantly lower than the one achievable for inductors. Careful choice of the coupling transistors and the bias is important to achieve sufficient coupling for 90° phase locking within the full tuning range. The coupling increases with gate width w_g and bias of M4-M5. On the other hand, increased coupling raises the parasitics, decreases the maximum possible oscillation frequency and deteriorates the phase noise. FETs with relative large w_g equal to the ones of M1-M2 are chosen. Down to the threshold voltage, they provide sufficient coupling within a large tuning range and allow the compensation of process variations. For the reasons mentioned in the preceding section, source followers are used as output buffers. To maintain the desired phase offsets, symmetrical layout is mandatory.

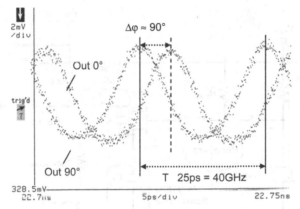

Fig. 11.27. Measured waveforms of channels Out $0°$ and Out $90°$, V_c=0.5 V, V_b=0.8 V, V_{DD}=1.5 V, compare [Ell30] © IEEE 2004

A photograph of the fabricated chip with compact size of 0.4×0.3 mm^2 is shown in Fig. 11.26. Together with results of other circuits, the measured performance of the circuit is summarised in Table 11.4. In Fig. 11.27, the output waveforms of the coupled VCOs are illustrated verifying the phase offset of approximately $90°$. With respect to multiphase generation, up to moderate frequencies, ring oscillator based circuits provide a good alternative to the cross-coupled approach [Kos04].

Table 11.4. Quadrature VCOs

Technology	f_{osc}	P_{out}	Phase noise @offset	V_{dc}/I_{dc}	Chip size	Ref.
0.18-µm CMOS	5.8–6 GHz	n.a.	−105 dBc/Hz @5 MHz	1.8 V/ 10 mA	1.1×0.9 mm^2	[Wan03]
SiGe HBT	24.8–28.9 GHz	−14.7 dBm[a]	−85 dBc/Hz @1 MHz	5 V/ 25.8 mA	0.55×0.45 mm^2	[Hac03]
90-nm SOI CMOS	38–44 GHz	−4.7 to −7.5 dBm[b]	−80to −87 dBc/Hz @3 MHz	1.5 V/ <54 mA	0.4×0.3 mm^2	This circuit

[a]No 50 Ω output buffer [b]Per channel

11.6.3 4-GHz MESFET LC Gate Resonator Oscillator

A class-E VCO is presented in this section, which has been fabricated in commercial 0.6-μm MESFET technology [Ell28]. It has been optimised for high efficiency at an ultra-low supply voltage of 1 V. Figure 11.28 shows the simplified equivalent circuit of the VCO. The key blocks of the circuit are the resonator, feedback in the source and class-E output matching. In Table 11.5, the tasks of the elements are outlined.

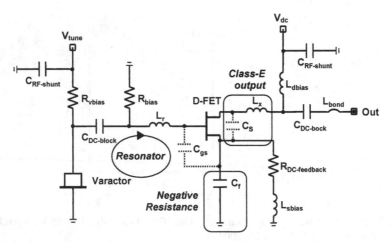

Fig. 11.28. Class-E tuned MESFET VCO, C_s is the output capacitance of the FET, compare [Ell28] © IEEE 2001

Figure 11.29 shows the equivalent class-E output network of the VCO. The theory of class-E oscillators is very similar to the one of amplifiers discussed in Sect. 9.3.2. Thus, the design equations can be mapped. V_s and I_s are the effective drain source voltage and current of the FET acting as switch. R_{eff} is the effective load resistance being the sum of the load resistance and the resistive parasitics of the output network. $V_{dc,eff}$ is the effective DC supply voltage, determined by the source voltage drop of the bias network, the knee (drain source saturation) voltage and the resistive losses of the FET. L_{eff} is the class-E inductance L_x including the output bondwire L_{bond} ($L_{eff}=L_x+L_{bond}$). Referring to Eq. (9.54), the required class-E DC current can be approximately calculated as follows:

$$I_{dc} \approx \frac{2V_{dc,eff}}{g^2 R_{eff}} \tag{11.64}$$

with g=1.862. According to Eq. (9.57), the class-E capacitance C_s can be estimated by

$$C_s \approx \frac{2}{g^2 \omega_{LO} R_{eff} \pi} \tag{11.65}$$

with ω_{LO}, the oscillation frequency. Using Eq. (9.62), the required class-E inductance L_x can be calculated as follows:

$$L_x \approx L_{eff} - L_{bond} \approx \frac{1.153 R_{eff}}{\omega_{L0}} - L_{bond} \qquad (11.66)$$

Figure 11.30 plots the simulated class-E voltage and current waveforms of the FET. The peaks of V_s and I_s have the typical class-E phase offset, which reduces the total DC power consumption P_s through the transistor.

Table 11.5. Circuit discussion

Common Source	High efficiency, high internal gain
D-FET	Good output power, higher than for E-FET
w_g=300 μm	Necessary to create required class-E capacitance C_s=C_{ds}. Note that $C_{ds} \propto w$
D-FET Bias	Class-E bias optimised for $V_{dc} \approx 1.2$ V. Self-biasing, single supply mode
Varactor	Resonator C tuneable with V_{Tune}. Deep depletion FET is used since it has good Q. The varactor diode is realised using G-FET with drain and source connected together. Tuning voltage positive because gate connected to ground
w_g=1000 μm	Tradeoff between tuning range (make w small) and parasitic series resistance (make w large). The diode characteristics of this varactor have been shown in Fig. 6.29
L_x=3 nH	Class-E inductance. Together with C_s, it creates reactive power in the switching FET and real power in the load thus maximizing the efficiency
C_s=150 fF	Class-E capacitance. In this case it is the output capacitance C_{ds} of the FET
L_r=5.5 nH	Resonator L
C_f=1.3 pF	Feedback capacitance makes circuit unstable to guarantee oscillation
L_{dbias}=7 nH	RF isolating drain biasing
R_{bias}=5 kΩ	RF isolating gate bias. Note that the gate current is very low
R_{vbias}=5 kΩ	Varactor bias. Note that the varactor control current is very low
$C_{DC-block}$=7 pF	DC block
$C_{RF-shunt}$=7 pF	RF block
$R_{DC-feedback}$ =55 Ω	Adjustment and stabilization (threshold voltage variations, temperature changes and aging) of D-FET current
L_{sbias}=7 nH	RF isolating DC bias at source
L_{bond}=0.5 nH	External bond wire has to be included since it influences the class-E characteristics

To generate the negative resistance, the feedback capacitance C_f can be estimated according to Eq. (11.41). With respect to the 50 Ω output, a negative resistance of –200 Ω is generated to allow oscillation within the whole band of interest and compensation of potential process variations. The oscillation frequency and band-width have been optimised on basis of Eqs. (11.64, 11.65). Figure 11.31 depicts a photograph of the MMIC bonded into a test package. The oscillation frequency and the frequency tuning range are illustrated in Fig. 11.32 versus supply voltage. The tuning voltage V_{tune} is varied from 0 to V_{dc}. At a V_{dc} of 1.8 V, the measured oscillation centre frequency is 4.4 GHz. A frequency tuning range of 150 MHz was measured. Applying a V_{dc} of 0.9 V yields a centre frequency of 3.6 GHz and a tuning range of 80 MHz.

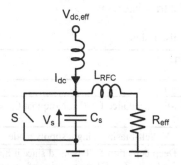

Fig. 11.29. Equivalent class-E output circuit of VCO including bias, compare [Ell28] © IEEE 2001

Fig. 11.30. Simulated voltage (V_S) and current (I_S) waveforms of class-E tuned FET, V_{dc}=1.2 V, V_{tune}=0.6 V, compare [Ell28] © IEEE 2001

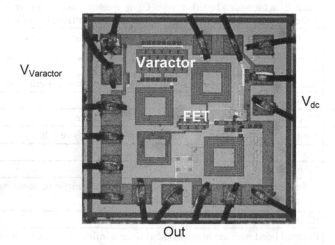

Fig. 11.31. Photograph of oscillator with chip size 0.9×0.9 mm^2, compare [Ell28] © IEEE 2001

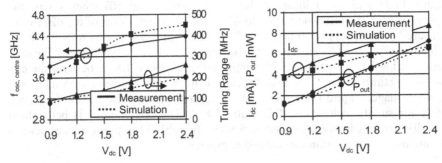

Fig. 11.32. Oscillation centre frequency and frequency tuning range, $V_{tune, max} \leq V_{dc}$ compare [Ell28] © IEEE 2001

Fig. 11.33. Output power and supply current versus supply voltage, compare [Ell28] © IEEE 2001

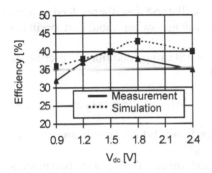

Fig. 11.34. Efficiency versus supply voltage, compare [Ell28] © IEEE 2001

Fig. 11.35. Phase noise versus frequency offset, compare [Ell28] © IEEE 2001

Table 11.6. Data of C-Band VCOs

Techn.	Supply voltage	Class	f_{osc}/ tuning	P_{out}	η	Phase noise @ Δf=1 MHz	Ref.
0.6-μm GaAs MESFET, f_{max}=18 GHz	0.9 V	E	3.6 GHz 2.5%	1.5 mW	36%	−112 dBc/Hz	This section
PMOS, f_{max}=20 GHz	2.5 V	A	5.4 GHz 5.6%	-	-	−119 dBc/Hz	[Hun01]
SiGe HBT, f_{max}=75 GHz	3 V	A	6 GHz 15%	0.1 mW	7%	−110 dBc/Hz	[Kle01]

Figures 11.33 and 11.34 illustrate the supply current and efficiency versus the supply voltage. An efficiency of up to 43% and an output power of 6.5 dBm were measured using a supply voltage of 1.8 V. At a supply voltage of only 0.9 V, an efficiency of up to 36% and an output power of 1.1 dBm were achieved. The efficiency of the VCO decreases with supply voltages above 1.8 V, because the circuit was optimised for 1.2 V and the class-E mismatch increases with increased supply voltage. Figure 11.35 shows the measured phase noise versus frequency offset for supply voltages of 0.9 V and 1.8 V. The phase noise is better than -132 dBc/Hz at a frequency offset of 5 MHz. In Table 11.6, the results of several C-band oscillators are summarised and compared.

11.7 Tutorials

1. Why do we need oscillators? Sketch a typical transceiver architecture including the oscillator.
2. What are the general design goals for oscillators? Oscillators feed mixers. Consequently, how much output power/voltage swing should an oscillator provide?
3. Outline three methods suitable to explain the start-up conditions in oscillators. For which perspectives are they favourable?
4. Are purely passive oscillators possible?
5. In start-up, the amplitude grows and grows. Which effects prevent transistors from destruction?
6. A start-up signal is required as initial condition. Which signals are suitable? Which one is already inherent in transistors?
7. Explain oscillation by means of the feedback model. What are the two mandatory criteria for oscillation? Who invented them? From which great institution was this genius? When did he live? Compare the impact of inverting and non-inverting amplifiers. Which kind of basic transistor circuits belong to the first and second group, respectively?
8. Explain the model of negative resistance. Which start-up conditions are required in terms of impedances and reflection coefficients? Prove that the representation by impedances and reflection parameters are equivalent. What is the equivalent of negative resistance at the transistor level? What determines the oscillation frequency? Why? What limits the maximum oscillation frequency of oscillators?
9. How can we generate negative resistance in a gate LC oscillator? Calculate the series feedback capacitance required to compensate the losses of a resonator at 5 GHz with $R_R = 5\ \Omega$ using a simple equivalent circuit given by $g_m = 50$ mS and $C_{gs} = 100$ fF. What are the pros and cons of choosing a smaller feedback capacitance? Choose a suitable varactor size, and calculate L_R and $\Delta\omega_{LO}$ assuming $\Delta C = 3$. Compare the properties of a series and parallel resonator. Generally, when is which one superior?

10. In steady-state, how do the conditions change? Generally, what are the limitations of oscillator small signal investigations? Assume a typical class-A and a class-B small signal bias with a certain value for g_m. How do the average values change in the large signal case?

11. How can you estimate the output power of oscillators? Why can we map the theories for the IV characteristics of amplifiers to oscillators? What is the major difference?

12. How is efficiency defined for oscillators? Does the PAE measure make sense for oscillators?

13. Explain the functional principle of the cross-coupled oscillator. Derive the Barkhausen criteria for this circuit. Explain each element and the functionality of the oscillator treated in Sect. 11.6.1.

14. Generally, can highest operation frequency and tuning range be achieved at the same time? What are the tradeoffs?

15. Explain each element and the functionality of the oscillator treated in Sect. 11.6.3. Recall the basic idea of class-E. How can we map the insights of the amplifier theory? Why can we achieve high efficiency? The circuit operates with very low supply voltages. What is the challenge in terms of output power and efficiency? In this context, review the losses appearing in the IV curves.

16. Explain the functional principle of the ring oscillator. How many stages are at least required for oscillation based on a) CMOS gates and b) differential gates? How does this impact the maximum frequency of oscillation?

17. Review the capability of maximum oscillation frequency for all oscillator principles you know.

18. What are the advantages of differential oscillators concerning system implementations? What are the pros and cons of the single-ended counterparts?

19. Oscillators exhibit both amplitude and phase noise. Which one is more significant? Why?

20. Why is phase noise a problem for wireless systems?

21. How does LO noise impact RF noise?

22. Which kind of noise figure of merit is defined for oscillators? Comment on the tradeoffs. What are typical values for the figure of merit? What does the offset frequency mean? What are meaningful offset specifications for GSM and WLAN?

23. Discuss the noise sources and mechanisms for oscillators. Which one plays the most important role? Why? Discuss the extended formula of Leeson (the one which considers the overall frequency response) and the corresponding frequency characteristics.

24. How is 1/f noise generated? At which frequencies does it appear? Why? Why is 1/f noise observed at high RF frequencies? What are the two major measures determining the 1/f noise at RF? Compare these measures for both the FET and the BJT. Which transistor type provides lower 1/f noise? Why is the 1/f corner frequency of FETs higher than for BJTs?

25. How can we minimise the phase noise? Which parameters are fixed by technology and which ones can be optimised by the designer?

26. For which kind of applications do we need quadrature oscillators? Explain at least two different quadrature oscillator topologies.
27. What are the performances typically achieved by leading-edge oscillators?
28. One of the major impacts of phase noise is the frequency drift. Which kind of circuit technique can be employed to improve the frequency stability? You may answer this question after studying the next chapter.

References

[And88] J. E. Andrews, T. J. Holden, K. W. Lee, A. F. Podell, "2-5-6 GHz broadband GaAs MMIC VCO", IEEE Microwave Theory and Techniques Symposium, pp. 491–494, 1988.

[Bür33] G. von Büren, F. Ellinger, L. Rodoni, H. Jäckel, "Low power consuming single-ended VCO for wireless LAN at C-band", IEEE Bipolar Circuits and Technology Meeting, pp. 249–252, Sept. 2004.

[Cho97] Yong-Ho Cho, Song-Cheol Hong, Young-Se Kwon, "A novel active inductor and its application to inductance-controlled oscillator", IEEE Transactions on Microwave Theory and Techniques, Vol. 45, No. 8, pp. 1208–1213, Aug. 1997.

[Ell28] F. Ellinger, U. Lott and W. Bächtold, "Design of a low supply voltage, high efficiency class–E voltage controlled MMIC oscillator at C-band", IEEE Transactions on Microwave Theory and Techniques, Vol. 49, No. 1, pp. 203–206, Jan. 2001.

[Ell30] F. Ellinger and H. Jäckel, "38-43 GHz quadrature VCO on 90 nm VLSI SOI CMOS with feedback frequency tuning", IEEE MTT-S International Microwave Symposium, June 2005.

[Ell36] F. Ellinger, T. Morf, G. von Büren, C. Kromer, G. Sialm, L. Rodoni, M. Schmatz and H. Jäckel, "60 GHz VCO with high tuning range fabricated on VLSI SOI CMOS technology", IEEE Microwave Theory and Techniques Symposium, Vol. 3, pp. 1329–1332, June 2004.

[Fon04] N. Fong, J.-O. Plouchart, N. Zamdmer, D. Liu, L. Wagner, C. Plett, G. Tarr, "A low-voltage 40 GHz complementary VCO with 15% tuning range in SOI CMOS technology", IEEE Journal of Solid-State Circuits, Vol. 39, No. 5, pp. 841–846, May 2004.

[Gon97] G. Gonzales, Microwave Transistor Amplifiers, Prentice Hall, 1997.

[Hac03] S. Hackl, J. Bock, G. Ritzberger et al. "A 28 GHz monolithic integrated quadrature oscillator in SiGe bipolar technology", IEEE JSSC, pp. 135–137, Jan. 2003.

[Ham01] D. Ham, A. Hajimiri, "Concepts and methods in optimization of integrated LC VCOs", IEEE Journal of Solid-State Circuits, Vol. 36, No. 6, pp. 896–908, June 2001.

[Har91] S. Hara, T. Tokumitsu, "Monolithic microwave active inductors and their applications", IEEE International Symposium on Circuits and Systems, Vol. 3, pp. 1857–1860, June 1991.

[Hau05] H. Hausmann, "The effect of high stability reference oscillators on system phase noise", Microwave Journal, Feb. 2005.

[Hun01] C. Hung, B. A. Floyd, N. Park, K. K. O., "Fully integrated 5.35 GHz CMOS VCOs and prescalers", IEEE Transactions on Microwave Theory and Techniques, Vol. 49, No. 1, pp. 17–22, Jan. 2001.

[Kle01] B. U. Klepser, M. Scholz and J. J. Kucera, "A 5.7 GHz Hiperlan SiGe BiCMOS voltage-controlled oscillator and phase-locked-loop frequency synthesizer", Radio Frequency Integrated Circuits Symposium, pp. 61–64, 2001.

[Kos04] M. Kossel, T. Morf, W. Baumberger et al., "A multiphase PLL for 10 Gb/s links in SOI CMOS technology", IEEE MTT-S, pp. 207–210, June 2004.

[Kur69] K. Kurokawa, "Some basic characteristics of broadband negative resistance oscillator circuits", The Bell System Technical Journal, July 1969.

[Lee66] D. B. Leeson, "A simple model of feedback oscillator noise spectrum", Proc. of IEEE, Vol. 54, No. 2, pp. 329–330, 1966.

[Lee00] T. H. Lee, A. Hajimiri, "Oscillator phase noise: a tutorial", IEEE Journal of Solid-State Circuits, Vol. 35, No. 3, pp. 326–336, March 2000.

[Len04] F. Lenk, M. Schott, J. Hilsenbeck, W. Heinrich, "A new approach for low phase noise reflection-type MMIC oscillators", IEEE Transactions on Microwave Theory and Techniques, Vol. 52, No. 12, pp. 2725–2731, Dec. 2004.

[Liu99] T.P. Liu, "A 6.5 GHz monolithic CMOS voltage-controlled oscillator", IEEE ISSCC, pp. 404–405, Feb. 1999.

[Ngu92] N. M. Nguyen, R.G. Meyer, "Start-up and frequency stability in high frequency oscillators", IEEE Journal of Solid-State Circuits, Vol. 27, pp. 810–820, May 1992.

[Raz03] R. Razavi, Design of integrated circuits for optical communication, McGraw-Hill, 2003.

[Raz96] B. Razavi, "A study of phase noise in CMOS oscillators", IEEE Journal of Solid State Circuits, Vol. 31, No. 3, pp. 331–343, March 1996.

[Raz98] B. Razavi, RF Microelectronics, Prentice Hall, Upper Saddle River, 1998.

[Rod9]   L. C. Rodoni, F. Ellinger, H. Jäckel, "CMOS inverter with 4.7 ps gate delay fabricated on 90 nm SOI technology", Electronics Letters 2004, Vol. 40, No. 20, pp. 1251–1252, Sept. 2004.

[Sch99] V. Schwarz, T. Morf, A. Huber, H.-R. Benedickter and H. Jäckel, "Differential InP-HBT Current Controlled LC-Oscillators with Center Frequencies of 43 GHz and 67 GHz", Electronics Letters, vol. 35, No. 14, pp. 1197–1198, July 1999.

[Sha03] D. K. Shaeffer and S. Kudszus, "Performance-optimized microstrip coupled VCOs for 40-GHz and 43-GHz OC-768 optical transmission", IEEE Journal of Solid-State Circuits, Vol. 38, No. 7, July 2003.

[Tie02]   M. Tiebout, H.-D. Wohlmuth and W. Simbürger, "A 1 V 51 GHz fully-
          integrated VCO in 0.12 um CMOS", IEEE International Solid-State
          Circuits Conference, pp. 300–301, Feb. 2002.

[Wan03]   Wang H. et. al, "Some design aspects on 5 GHz CMOS quadrature
          VCO with fully integrated LC tank, Conference on ASIC, pp. 1010–
          1013, Oct. 2003.

[War89]   A. Warren, J. M. Golio and W. L. Seely, "Large and small signal
          oscillator analysis", Microwave Journal, pp. 229–246, May 1989.

12 Phase Locked Loops and Synthesisers

Life is a face locked loop.
Unknown

In the preceding chapter we have observed that the frequency stability of oscillators is significantly deteriorated by noise. Moreover, we have to consider the impact of process variations, temperature changes and aging. These effects make it impossible to reproduce a certain oscillation frequency at given varactor voltage. To meet the challenging requirements in synthesisers, tracking filters, clock recovery circuits and PLL (Phase Locked Loop) techniques are commonly applied. Referring to Sect. 11.4, phase noise and jitter appearing in the frequency and time domain, respectively, are useful measures for the signal purity. PLLs can be realised in analogue, digital or mixed analogue/digital form. In this chapter, we focus on analogue PLLs, which have lower complexity than their digital counterparts.

PLL based systems are nonlinear. Due to the interdependencies of the inherent components analytical and numerical analyses are complex. As for other circuits, we can make life much easier by linearisation of the dynamic operation around a static operation point exhibiting average characteristics. Usually, this simplification provides good accuracy as long as the elongations around the static operation point are small. These simplified analyses are sufficient to understand the fundamental relations. Further optimisations are usually made on basis of CAD tools and large signal simulations.

For detailed information about PLLs, the reader is referred to the specific literature [Bes93, Raz96, Rat01].

12.1 Phase Locked Loop Basics

A PLL synchronises the output by means of a reference input. Feedback from the output to the input is employed for this task. The parameter to be locked in communication systems is the frequency. In the locked case, the input and output frequencies ω_{in} and ω_{out} are equal. The phase is the integral of the frequency implying that in locked state the phase difference $\Delta\varphi(t)$ between the input and output phase $\varphi_{in}(t)$ and $\varphi_{out}(t)$ must be constant demanding for $\dfrac{d\varphi_{out}}{dt} - \dfrac{d\varphi_{in}}{dt} = 0$ and subsequently $\omega_{out} = \omega_{in}$. The latter property will serve as an important control goal.

We can conclude that in the locked state, the input and output signals are synchro-
nised in frequency and phase. An unlocked loop yields $\omega_{in} \neq \omega_{out}$ and exhibits a $\Delta\varphi$
varying with time. Referring to the block diagram in Fig. 12.1, the PLL mainly
consists of a VCO (Voltage controlled Oscillator), a PD (Phase Detector) and a LP
(Lowpass) filter. These components will be treated in the following sections.
There are three key parameters that are of particular importance in practical
systems:

- Capture range: input/reference frequency range for which the loop can achieve
  lock.
- Lock range: input/reference frequency range over which the loop will remain
  locked – usually larger than capture range.
- Settling time: required time for the loop to lock onto a new frequency.

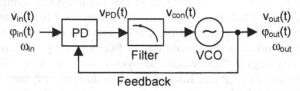

Fig. 12.1. Block diagram of basic PLL in time domain

12.1.1 Phase Detector

The main task of the PD is the detection of the phase difference between two signals. Hence, the PD may be described as an error amplifier. In the PLL, the phase
difference or phase error is given by $\Delta\varphi(t) = \varphi_{out}(t) - \varphi_{in}(t)$. Dependent on this
phase difference, an output voltage $v_{PD}(t)$ is generated in the phase detector. For
analysis and further processing it is convenient if this relation is linear. Referring
to Fig. 12.2a, we may write

$$v_{PD}(t) = K_{PD} \cdot \Delta\varphi(t) \tag{12.1}$$

where the slope K_{PD} has a unit of V/rad. Let's review these idealised characteristics with real word circuits. Mixer based phase detectors are frequently employed
as phase detectors. One example is the Gilbert cell mixer treated in Sect. 10.3.3.
The circuit acts as signal multiplier. Corresponding theoretical background can
also be found in Sect. 4.4.3. After adequate filtering, multiplication of the input
signals $v_{in}(t) = V_{in} \cos(\omega_1 t)$ and $v_{out}(t) = V_{out} \cos(\omega_2 t + \Delta\varphi)$ yields the following
dependency:

$$v_{PD}(t) = c \cdot \cos\left[(\omega_2 - \omega_1)t - \Delta\varphi\right] \tag{12.2}$$

with $c \sim V_{in} \cdot V_{out}$ as the device specific conversion coefficient. Assuming
$\omega_{in} \approx \omega_{out}$ yields

$$v_{PD}(t) = c \cdot \cos\Delta\varphi . \tag{12.3}$$

The characteristics of v_{PD} are plotted in Fig. 12.2b vs $\Delta\varphi$ revealing that we can accomplish a linear approximation around the zero crossings, thereby verifying the claim made in Eq. (12.1).

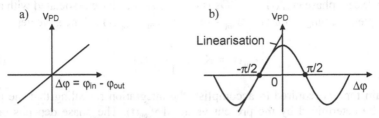

Fig. 12.2a,b. Characteristics of PD with phase difference as input and voltage as output: **a** ideal characteristics; **b** linearisation of multiplier/mixer based PDs

12.1.2 Voltage Controlled Oscillator

VCOs have been elaborated in Chap. 11. According to Fig. 12.3, the input is represented by the DC control voltage $v_{con}(t)$. The output is characterised by the frequency ω_{out} and the excess phase $\varphi_{out}(t)$. By assuming linear dependency between ω_{out} and $v_{con}(t)$ as illustrated in Fig. 12.4, we can specify

$$\omega_{out}(t) = \omega_{fr} + K_{VCO} \cdot v_{con}(t), \tag{12.4}$$

where ω_{fr} serves as the reference frequency, which may be the centre frequency of the VCO, and K_{VCO} with unit of rad/Vs determines how much voltage is required for a specific frequency change $\Delta\omega(t) = \omega_{out}(t) - \omega_{fr} = K_{VCO} \cdot v_{con}(t)$. The parameter K_{VCO} may be considered as the VCO gain.

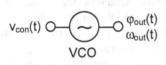

Fig. 12.3. VCO input and output parameters

Fig. 12.4. Linear approximation of VCO output frequency vs control voltage

However, in which extent is the assumption of a linear voltage to frequency dependency feasible? Neglecting the parasitic capacitances, the oscillation frequency of many VCOs may be approximated by $\omega_{out} = \dfrac{1}{\sqrt{L_R \cdot C_{var}(v_{con})}}$ with L_R as the resonator inductance and C_{var} the varactor capacitance. Given that $C_{var} \sim v_{con}^2$, ω_{out} would depend linearly on v_{con} as assumed. In reality and depending on the applied varactor, the accuracy of this assumption is indeed limited but may serve as basis for qualitative investigations. The maximum frequency tuning range $\omega_{min} - \omega_{max}$ is defined by the capacitance control range of the varactor.

The excess phase $\varphi_{out}(t)$ in VCOs is defined as the phase associated with a certain frequency change $\Delta\omega(t) = \omega_{out} - \omega_{fr} = K_{VCO} \cdot v_{con}(t)$. Hence, we get

$$\varphi_{out}(t) = K_{VCO} \int v_{con}(t) \cdot dt. \tag{12.5}$$

A certain time is required to accomplish the integration revealing that the phase cannot be determined by the present value of $v_{con}(t)$. The phase depends on the history of $v_{con}(t)$ being a state variable. We have to change $v_{con}(t)$ and subsequently the frequency to adjust the phase. This is instrumental to compensate the phase-error in a PLL.

For PLL analyses, it is convenient to define the input to output transfer function given by the ratio between φ_{out} and v_{con} in the frequency domain. The conversion between time and frequency domain is performed by the Laplace transformation. An integration in the time domain generates a factor $1/s$ in the frequency domain. We obtain for the transfer function in the frequency domain:

$$H_{VCO}(s) = \frac{\varphi_{out}(s)}{V_{con}(s)} = \frac{K_{VCO}}{s}. \tag{12.6}$$

Due to the pole, the VCO exhibits low-pass characteristics providing some level of attenuation for undesired harmonics.

12.1.3 Filter

We have to bear in mind that the mixer-based phase detectors need a hard input drive to allow operation. Thus, in addition to the desired DC control voltage undesired harmonics and intermodulation products are generated. To suppress these products, V_{PD} is low-pass filtered yielding the DC control voltage V_{con} for the VCO. Different types and orders of low-pass filters are possible. A simple RC lowpass filter as illustrated in Fig. 12.5a can be applied yielding a transfer function of

$$K_{LPF}(s) = \frac{1}{1 + \dfrac{s}{\omega_{LPF}}}, \tag{12.7}$$

where $\omega_{LPF} = 1/RC$. As treated later in Sect. 12.1.5, the filter characteristics have a strong impact on the bandwidth, gain, stability and settling time.

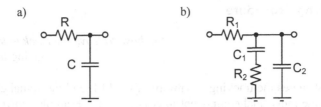

Fig. 12.5a,b. Example of RC lowpass filters: **a** first order; **b** adding of zeros and poles to increase flexibility

Since the optimum filter parameters are subject to tradeoff, the adding of further zeros and poles (see Fig. 12.5b) can be advantageous. The increased flexibility at different frequency bands simplifies the optimisation.

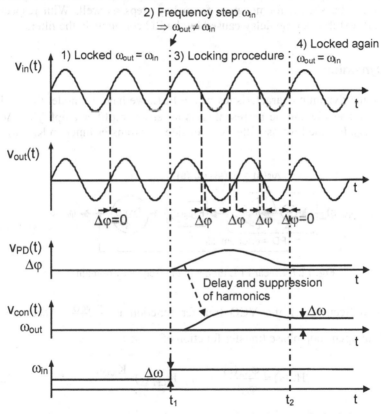

Fig. 12.6. Tracking procedure of PLL, $t<t_1$: system locked, $t=t_1$ unlocked due to step in input frequency, $t_1<t<t_2$ locking procedure, $t=t_2$ lock achieved again

12.1.4 Locking Procedure

You have to leave the track to learn more.
Saying from Africa

Figure 12.6 sketches the tracking capability of the PLL and the signal characteristics between the individual component interfaces. In the example shown, the locking is lost due to a step in the input frequency. After the locking procedure, the input frequency is tracked again. How does this looking procedure work? The frequency step generates a phase error, which is measured in the phase detector. Accordingly, the phase detector yields an increased $v_{PD}(t)$, which is converted to $v_{con}(t)$ after filtering. The latter control voltage increases the output frequency until the phase error vanishes to zero. Consequently, after a certain locking time, the input and output frequencies are equal. Hence, locking is achieved. Corresponding properties can be shown for negative frequency steps as well. With respect to $v_{PD}(t)$, $v_{con}(t)$ exhibits a time delay caused by the RC constants in the filter.

12.1.5 Dynamics

To deduce the dynamic characteristics of the PLL, we have to model the applied components. The PD can be represented as an error amplifier amplifying $\Delta\varphi(s)$ with gain K_{PD}. For the lowpass filter, we consider the transfer function $K_{LPF}(s)$.

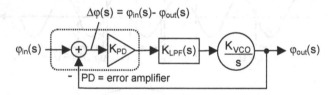

Fig. 12.7. Linear PLL model in the frequency domain

Referring to Sect. 12.1.1 the VCO transfer function is $\dfrac{K_{VCO}}{s}$. According to Fig. 12.7, an open-loop phase transfer function of

$$H_o(s) = \frac{\varphi_{out}(s)}{\varphi_{in}(s)} = K_{PD} \cdot K_{LPF}(s) \cdot \frac{K_{VCO}}{s} \qquad (12.8)$$

can be calculated. We suppose a passive filter, where the amplitude of $K_{LPF}(s)$ does not exceed unity. Consequently, the open loop gain is mainly determined by $K = K_{PD} \cdot K_{VCO}$. The close-loop phase transfer function yields

$$H(s) = \frac{H_o(s)}{1 + H_o(s)} = \frac{K \cdot K_{LPF}(s)}{s + K \cdot K_{LPF}(s)}. \tag{12.9}$$

By inspection of Eq. (12.9) we observe lowpass filter characteristics for $H(s)$. Due the lowpass properties of the inherent VCO this is not surprising. However, the suppression of harmonics is usually not sufficient. Thus, we have to implement additional filtering by means of the lowpass filter. By including the lowpass filter characteristics according to Eq. (12.7), Eq. (12.9) can be written as

$$H(s) = \frac{K}{\dfrac{s^2}{\omega_{LPF}} + s + K}. \tag{12.10}$$

In control theory it is common practice to represent the denominator in the normalised form: $s^2 + 2\zeta\omega_n s + \omega_n^2$ with ζ as the damping factor and ω_n as the normalised frequency. Hence, we can rearrange $H(s)$ to

$$H(s) = \frac{\omega_n^2}{s^2 + 2\zeta\omega_n s + \omega_n^2} \tag{12.11}$$

with

$$\omega_n = \sqrt{\omega_{LPF} K} \tag{12.12}$$

and

$$\zeta = \frac{1}{2}\sqrt{\frac{\omega_{LPF}}{K}}. \tag{12.13}$$

Note that ω_n does not have any direct relations with ω_{in} or ω_{out}. Assuming that the bandwidth is determined by the lowpass filter, ω_n may be seen as a measure for the gain-bandwidth product of the loop. To achieve a high dynamic with capability for small settling times and quick frequency changes, a large ω_n, and subsequently a high ω_{LPF} and K are desired. However, we have to keep in mind that K impacts the damping and thus the shape of the signal response. A damping of $\zeta = \frac{1}{\sqrt{2}}$ provides an optimally flat response [Bes93]. According to Eqs. (12.12) and (12.13), a large ω_n and a strong damping demand for high ω_{LPF}. On the other hand, to suppress undesired harmonics and to provide a DC control signal as pure as possible, ω_{LPF} should be small. We can conclude that both K and ω_{LPF} have to be carefully chosen to achieve an optimum tradeoff. Moreover, the impact on noise has to be taken into account. Increased flexibility can be obtained by using filters with higher order as sketched in Fig. 12.5b [Bes93, Raz96].

12.1.6 Noise

In extension to the conclusions made in the last section, we can derive some insights with respect to phase noise. PLLs are suited to impact and minimise the noise. We can distinguish between two types of noise:

1. Noise generated by the VCO, which is directly transferred to the output. In PLLs, the associated frequency shifts can be compensated to a certain level as long as the speed of the correction is high enough favouring a large gain-bandwidth product. Only the VCO noise appearing within the bandwidth of the feedback loop can be suppressed. The noise being outside the bandwidth is not filtered.
2. The conditions are different if noise is injected at the input of the loop, e.g. generated by the input source and the detector. To minimise these contributions, the gain-bandwidth product should be as low as possible.

Depending on the significance of 1 or 2, the designer should strive for a corresponding tradeoff regarding the gain-bandwidth.

12.2 Integer-N Synthesiser

A wise guy considers advice even from idiots.
French saying

Synthesisers apply PLL techniques to improve the frequency stability of VCOs. They are key components in wireless systems communicating at different channels allocated at specific frequencies. Refer to a typical wireless architecture shown in Fig. 12.8. By tuning the LO frequency in correspondence with the RF frequency, the desired channels 1, 2 .. n can be selected while keeping the IF frequency constant, thereby relaxing the IF filter requirements. To maximise the number of channels, the RF channel distance must be minimised accordingly. To avoid undesired interferences between the different channels, the signal purity and frequency accuracy must be as high as possible.

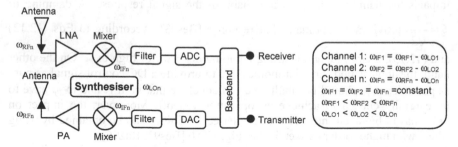

Fig. 12.8. Simplified architecture of synthesiser applied in wireless systems

Figure 12.9 illustrates the schematic of an integer N-synthesiser as frequently used in mobile systems. We can map many insights from the preceding PLL discussions. The basic idea is to perform the synchronisation of the RF output with respect to a low frequency input signal. At low frequencies, a very stable quartz oscillator can be employed as reference, which provides excellent frequency stability. Typically, the frequency of these quartz oscillators is fixed and allocated in the range of around 1–100 MHz. No comparable reference performances are possible at RF frequencies and on basis of ICs. For detection of the undesired frequency offset between input and output, the RF output frequency ω_{out} is downconverted. Frequency dividers are employed for this task yielding a frequency of

$$\omega_{div} = \frac{\omega_{out}}{N},$$
(12.14)

where the divider ratio N is an integer value. Figure 12.10a shows the architecture of a commonly used differential frequency-divider with N=2 consisting of two coupled latches. According to Fig. 12.10b, the input period of T is converted into an output period of 2T leading to the division of frequency. With this approach, frequencies up to 28 GHz have been achieved in 80-nm CMOS technology [Bür06]. In many cases, the system speed is limited by the divider properties.

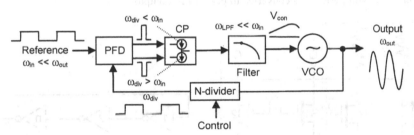

Fig. 12.9. Schematic of integer-N PLL synthesiser, PFD: Phase Frequency Detector, CP: Charge Pump

To enhance the understanding, we review the principle of clocked latches, e.g. that of a SCFL (Source Coupled FET Logic) circuit sketched in Fig. 12.11. Suppose that the input signal is high within the first half of the period. Hence, the data is read by the transistors R1 and R2 of the first differential pair. Meanwhile the current source of the second differential cell is switched off since the input signal is inverted. Thus, in this time frame, the two differential pairs are isolated. In the second half of the period, the signal previously read by the first stage is saved in the second stage.

Because of the constructive coupling of S1 and S2 the signal is amplified until saturation occurs due to the voltage drop associated with the load. We can conclude that the first differential pair acts as read stage, whereas the second serves as regenerative memory for an additional period. Two of those latches are cascaded and coupled back. Consequently, the total output period is doubled with respect to the input, thereby dividing the input frequency by two. We may find similarities to a ring oscillator with two stages. Higher division rations can be achieved by increasing the number of cascaded latches. Optionally, we can simply cascade divide-by-two circuits as shown in Fig. 12.12 yielding division ratios of $N=2^n$.

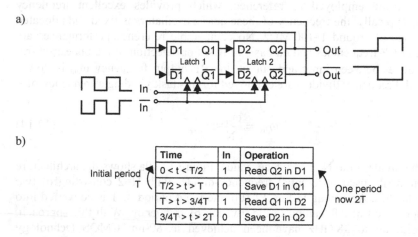

a)

b)

	Time	In	Operation
Initial period T	0 < t < T/2	1	Read Q2 in D1
	T/2 > t > T	0	Save D1 in Q1
	T > t > 3/4T	1	Read Q1 in D2
	3/4T > t > 2T	0	Save D2 in Q2

One period now 2T

Fig. 12.10a,b. Frequency-divider with N=2: **a** schematic; **b** operations vs time demonstrating that input with period T is converted to period 2T at output

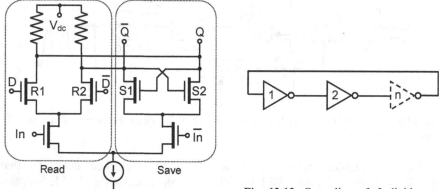

Fig. 12.11. Source coupled FET logic latch

Fig. 12.12. Cascading of :2 dividers to obtain division ratios of $N=2^n$

Figure 12.13a shows a common architecture of a phase frequency detector based on two clocked edge-triggered set-reset flip-flops (DFFs). The two frequencies ω_{in} and ω_{div} are fed into a phase frequency detector, which acts similar as the phase detector discussed in Sect. 12.1.2. However, it performs both phase and frequency detection at the same time. Both inputs of the DFFs are permanently fed with a high state (1). The signal with frequencies ω_{in} and ω_{div} are applied as clocks. As long as one of the clocks is high as well, the state 1 is transferred to the associated

DFF output. If both outputs are high at the same time, the DFFs are reset by means of an AND gate. The corresponding frequency difference tracking capability is demonstrated in Fig. 12.13b for the example of $\omega_{in}>\omega_{div}$. On the condition that no positive edge appears at the divider DFF, the DFF of the reference keeps transferring its high state to the output. This is exactly the case where the divider signal is delayed with respect to the reference or in other words $\omega_{in}>\omega_{div}$. In the connected charge pump the produced up-signal is converted into a current and accumulated in the filter capacitances generating the control voltage for the VCO. Since the control voltage is raised, ω_{div} is increased until $\omega_{in}=\omega_{div}$. In this case a further increase of the control voltage is prevented since a down signal with opposite magnitude would be generated if $\omega_{in}<\omega_{div}$ leading to a incremental decrease of the VCO control voltage. Consequently, frequency tracking is possible independent if $\omega_{in}>\omega_{div}$ or $\omega_{in}<\omega_{div}$. The polarities of the two current sources in the charge pump are as follows: current flows in the filter capacitances as long as the up-signal is high, whereas current flows out of the filter capacitances if the down-signal is high. By inspection of the signals, we can also show that the input and output phases are tracked.

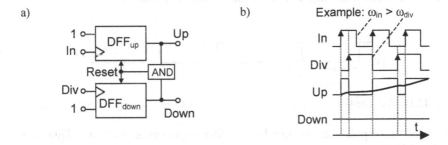

Fig. 12.13a,b. Phase frequency detector, In: input with reference frequency ω_{in}, Div: input with divided frequency ω_{div}: **a** architecture; **b** example for locking procedure if divider frequency is smaller than reference frequency

Suppose that tracking has been successful and $\omega_{in}=\omega_{div}$. Depending on N, which can be adjusted in the divider by means of an external control unit, the exact output frequency $\omega_{out}=\omega_{div}\cdot N$ can be set. The minimum resolution between two frequency settings is defined by $(N+1-N)\cdot\omega_{in}=\omega_{in}$. To get maximum frequency resolution, ω_{in} must be as low as possible. On the other hand, this imposes several drawbacks. To guarantee sufficient suppression of the undesired harmonics, the corner frequency of the filter ω_{LPF} must be well below ω_{in} – a rule of thumb specifies $\omega_{LPF}<10\ \omega_{in}$. Hence, at high N and consequently low ω_{in}, very large filter capacitors would be required. With respect to circuit area and costs this may not be acceptable. Moreover, according to Sect. 12.1.5 a high ω_{LPF} is preferred regarding the dynamic performance. A large filter capacitance would need lots of time to change currents. Last but not least, $N = \dfrac{\omega_{out}}{\omega_{in}}$ and the resulting ω_{LPF} have a significant impact on the noise as outlined in Sect. 12.1.6. Recall that only the part of the VCO noise within the bandwidth of the feedback loop is filtered. As higher N, as lower the loop bandwidth and the noise suppression of the VCO.

12.3 Fractional-N Synthesisers

To maximum the number of potential users allocated within a certain bandwidth, the distance between adjacent channels must be as low as possible. In case of integer-N synthesisers, small channel spacing mandates a low ω_{in} and even lower ω_{LPF}. The drawbacks in terms of system dynamics and noise have already been mentioned. A smart idea mitigating these drawbacks is the fractional-N synthesisers, which is capable of generating fractions of N, which are below unity. Consequently, at given channel spacing and determined by the reciprocal value of the fraction, ω_{in} can be much higher than in the integer-N approach.

Fig. 12.14. Weighted time average of N and N+1 values yield fractional divider values

The up and down pulses are summed in the filter capacitance over time. Thus, the associated VCO control voltage is the time average of these pulses. By periodically change between two divider values, e.g. between N and N+1 and time-wise weighting, fractions between these two values can be generated. Consider the illustration in Fig. 12.14. With T_N being the duration the divider value N is applied, and T_{N+1} representing the time the divider value N+1 is active, the VCO frequency is given by

$$\omega_{out} = \omega_{in} \cdot \frac{\left[T_N \cdot N + T_{N+1}(N+1) \right]}{T_N + T_{N+1}} \tag{12.15}$$

$$= \omega_{in} \cdot \left[N + \text{fraction} \right] \tag{12.16}$$

with

$$\text{fraction} = \frac{T_{N+1}}{T_N + T_{N+1}}. \tag{12.17}$$

This means that the divider fraction can be adjusted by setting of T_{N+1} with respect to the overall period given by the sum of T_N and T_{N+1}. This idea can be realised in a circuit shown in Fig. 12.15. In addition to the circuit blocks already known for the integer-N synthesiser, a dual modulus divider providing divider values of both N and N+1, and a clocked accumulator is required. The desired fraction is applied at the accumulator input. During each clock cycle, one faction is added until the defined accumulator capacity is reached. As long as the accumulator capacity is not approached, the divider value N is set. If the divider capacity is reached the divider value is changed to N+1 followed by a memory reset. Consequently, by proper choice of T_N, T_{N+1} and the accumulator capacity, the desired fractional output frequency is achieved. Example: you would like to have a divider value of N.1 corresponding to a faction of 0.1. The accumulator capacitance is given by 1. That means that we have totally $T_{N+1}+T_N=10$ cycles until the accumulator is reset. From Eq. (12.16), we get $\dfrac{T_{N+1}}{10}=0.1$ leading to $T_{N+1}=1$ and $T_N=9$.

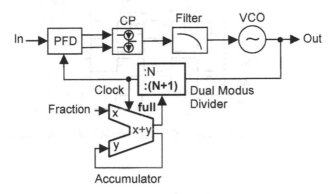

Fig. 12.15. Fractional-N architecture

Multi modus divider functionality can, for example, be realised by the phase selection method [Mue03]. Based on a common :2 divider, the signal transitions can be delayed by switching to the opposite phase after a cycle is finished. The divider is differential featuring outputs with 180° offset. A feasible approach is sketched in Fig. 12.16a. Switching is accomplished by changing the value of the parameter S from high to low or vice versa. This parameter controls a simple selector circuit consisting of two AND gates with outputs connected together. An example demonstrating the functionality is illustrated in Fig. 12.16b. The divider value changes from :2 to :3.

Most approaches have advantages as well as disadvantages. This is also the case for the fractional-N approach. The resulting fractional frequency is based on the defined repetition of pulses with duration of T_N and T_{N+1}. Unfortunately, these pulses generate spurs due to their periodic character. Moreover, the required switching operations are nonlinear, leading to harmonics. Both the spurs and harmonics are undesired since they may modulate the VCO. Since the filtering of those signals may be challenging, these products must be avoided as much as possible. One solution is to randomise the divider modulus while preserving the average division factor and thus the output frequency.

a)

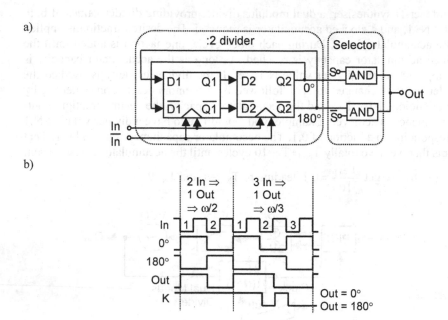

b)

Fig. 12.16a,b. Example of fractional-N frequency divider: **a** schematic; **b** generation of fraction

12.4 Tutorials

1. How do free-running VCOs perform regarding frequency stability? Which are the impact parameters? How can we improve the VCO frequency stability?

2. Explain how PLLs can be employed to stabilise the frequency of VCOs. Discuss each circuit component required.

3. Why do we need a loop filter? With regard to optimum dynamic and purity of the VCO control voltage, do we prefer a large or a low filter bandwidth?

4. Can PLLs be used for noise shaping? Why? Which kind of noise types can we identify in a PLL loop? Concerning noise, is a large or a small filter bandwidth advantageous?

5. Illustrate the schematics of a transceiver and discuss the functionality of synthesisers. How does frequency stability effect communications systems with high number of users?

6. Explain the functionality of an integer-N synthesiser. What is the advantage of the approach concerning communication systems? What is the tradeoff concerning the divider value N?

7. Explain how a divide-by-two circuit operates.

8. How does a frequency phase detector work and contribute to frequency tracking? What are the tasks of the charge pump and the filters? Illustrate the input and output signals of all components in a synthesiser with initial conditions of $\omega_{in} < \omega_{div}$.

9. Explain the idea behind the fractional-N synthesiser. What are the enhancements with respect to the integer-N synthesiser? Which additional components are required? How do they work?

10. Why is an accumulator needed? You would like to realise a fractional divider ratio of N,01. How do you have to set the fraction input and the capacity? What are the relations between T_N and T_{N+1}?

11. In general, how can we realise fractional-N and multi-mode frequency dividers? Discuss the phase selector method. How could we generate a divider ratio of 3/8?

12. Which disadvantages are imposed by the fractional-N approach? How could we mitigate the problems?

References

[Bes93] R. E. Best, Phase-locked Loops, Theory, Design and Applications, McGraw Hill, 1993.

[Bür06] G. von Büren, C. Kromer, F. Ellinger, A. Huber, M. Schmatz and H. Jäckel, "A combined dynamic and static frequency divider for a 40 GHz PLL in 80 nm CMOS", IEEE International Solid-State Circuits Conference, pp. 598–599, Feb. 2006.

[Mue03] B. de Muer, CMOS Fractional-N Synthesizers, Kluwer, 2003.

[Rat01] H. R. Rategh, T. H. Lee, Multi-GHz Frequency Synthesis & Division, Kluwer Academic, 2001.

[Raz96] B. Razavi, Monolithic Phase-locked Loops and Clock Recovery Circuits – a tutorial, Wiley, April 1996.

13 Amplitude Control and Switches

To realise the possible, the impossible has to be tried over and over again.
Herman Hesse

Amplitude control is required for various applications. Passive attenuators or VGAs (Variable Gain Amplifiers) are used for this task. A typical application is sketched in Fig. 13.1a, where the attenuators prevent the system from saturation in case that the input power at the receiver is too high.

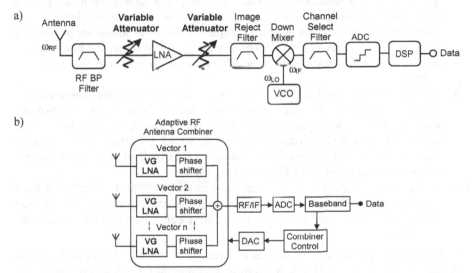

Fig. 13.1a,b. Application scenarios for amplitude control: **a** attenuator in receiver prevents from saturation in case of high input power; **b** adaptive antenna receiver with n active antenna paths. VGLNAs (Variable Gain LNAs) are used to weight the amplitude of each antenna path

Amplitude control components are also used in adaptive antenna receivers [Ell29]. An application scenario employing variable gain LNAs is illustrated in Fig. 13.1b. Moreover, amplitude control is, e.g. needed for measurement setups, signal generators and feed-forward systems. Typical requirements for amplitude control components are:

- High gain control range = maximum gain + maximum attenuation (for passive circuits: attenuation control range = maximum attenuation − minimum attenuation). Since active circuits are capable of providing gain, active circuits usually yield higher amplitude control ranges than passive circuits.

- Minimum control complexity on the basis of one single DC control signal. The relation between gain/attenuation and control signal should be as linear as possible.
- In many applications, the input and output impedances should be preserved within the full gain control range to prevent any impact on the circuits in front and behind the amplitude control component. Reflections, which may lead to stability problems, should be avoided.
- For some applications, e.g. for adaptive antenna combining, the transmission phase should stay constant vs gain control. The variations of the transmission phase may be compensated by phase shifters. However, in this case, amplitude and phase can't be controlled independently. A feedback control loop would be required, which would increase the control complexity. The design of an amplitude control circuit with constant phase is challenging since the parameters of both active and passive control devices change with bias leading to a variation of the inherent RC time constants, and subsequently to a non-constant phase. Active circuits are more sensitive than the passive counterparts. This is attributed to the strong impact of the drain source supply current flowing in active devices, which varies significantly with control bias. For example, the reader may consider the Miller capacitance seen at the input of a transistor.
- Power handling and linearity should be sufficient within the whole control range. To achieve good large signal performance, the bias current of active circuits must be high to prevent from saturation. Class-A bias points are well suited. The high power consumption is a drawback for active circuits. Due to their low nonlinearities, passive attenuators yield excellent large signal properties without consuming any DC power supply.
- Noise is an important issue for amplitude control components located in front of receivers. Recall that noise in active circuits is strongly impacted by the bias current. In passive attenuators, the noise figure equals approximately the resistive losses.

From the listed points, we can deduce that passive attenuators have advantages in terms of large signal performance, power consumption and transmission phase variations. However, they are not capable of providing any gain and generate significant noise.

It is straightforward that amplitude control components can be used for switches and signal selectors. In this case, we change only between maximum and minimum gain.

13.1 Attenuators and Switches

The channel conductivity of FETs can be controlled by means of the gate voltage. This property is exploited for attenuators and switches. Since the principle of these two circuits is similar, we treat both of them together in this section. Due to their passive nature, attenuators and switches do not consume any DC supply power. Moreover, as already mentioned, they have superior properties in terms of power handling and linearity. To yield similar large signal performances, a very high DC power would be required for amplifiers. To guarantee passive operation, DC-wise, the drain and the source are connected to ground. Consequently, the drain source voltage is zero and the channel resistance is determined by R_{ds}. FETs biased accordingly are also referred to as cold FETs. Two possible bias networks are shown in Fig. 13.2. Since the DC current is very low, high-ohmic bias resistors R_{bias} can be applied to feed the bias. The higher the impedance, the lower the associated RF losses. Due to the compact size, this network is frequently used. However, the power handling of this approach is limited by the voltage drops generated across the resistors, which may drive the device into active operation. In turn this may lead to strong nonlinearities. Consequently, the value of R_{bias} must be traded off for RF loss and linearity. The resistors can be replaced by inductors, which mitigate the DC voltage drop. Significant drawback of inductors is the large size. An RF shunt capacitance is used to attenuate any undesired signals appearing at the gate bias terminal. For the sake of simplicity, the DC bias networks are not shown for the circuits presented in the next sections.

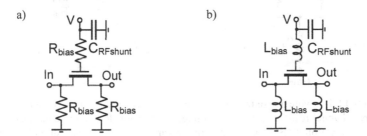

Fig. 13.2a,b. Typical bias networks for cold FETs

13.1.1 Modelling

Important performance parameters of switches and attenuators are the on- and off-resistances of the transistors. They impact the undesired insertion loss and the wanted attenuation. An ideal switch has zero on-resistance and infinity off-resistance. Figures 13.3a and 13.4a depict the simplified equivalent circuit of a resistive FET and the corresponding values in on- and off-mode for a transistor in 90-nm CMOS technology. Since the device is biased in the resistive region (V_{ds}=0 V), the channel resistance is determined by R_{ds}. If switched on, the attenuation of the FET is mainly defined by R_{ds}. Neglecting R_d and R_s results in an on-resistance of

$$Z_{on} \approx R_{ds,on} \approx \frac{l_g}{w_g \cdot \sigma} , \qquad (13.1)$$

where σ is the channel conductivity being a function of V_{gs}. In the off-mode, where R_{ds} is very large, the isolation is particularly determined by the parasitic capacitance C_{par}, which is a function of C_{gs}, C_{gd} and C_{ds}. Note that $C_{gs}=C_{gd}>C_{ds}$. Thus, the impedance in off-mode may be approximated by

$$Z_{off} \approx \frac{1}{\omega C_{par,off}} \approx \frac{1}{\omega \cdot l_g \cdot w_g \cdot C_{par}^{"}} , \qquad (13.2)$$

where $C_{par}^{"}$ is the parasitic capacitance per area. A first order model of a FET attenuator is plotted in Fig. 13.3b.

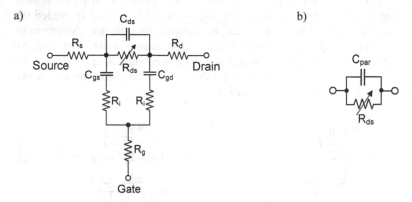

Fig. 13.3a,b. FET as voltage controlled resistor at zero DC drain source voltage, $R_{ds}=f(V_g)$: drain source channel resistance, R_g: gate connection resistor, R_d: drain connection resistor, R_s: source connection resistor, C_{gs}: gate source capacitance, C_{gd}: gate drain capacitance, $R_i=f(R_{gs}, R_g)$: gate resistor: **a** equivalent circuit, substrate capacitances are neglected; **b** simple model, C_{par}: considers all capacitances

From Eqs. (13.1) and (13.2) we can conclude that:

- $|Z_{off}|$ decreases with raised operation frequency. Thus, the performance is getting worse towards high frequencies.
- A small l_g lowers $|Z_{on}|$ and increases $|Z_{off}|$. Consequently, the relation of $|Z_{off}|/|Z_{on}|$ which may be seen as a figure of merit for switches can be enhanced by lowering l_g. Needless to mention that the decrease of l_g is limited by the available technology.

- A large w_g lowers $|Z_{on}|$, but unfortunately at the same time decreases $|Z_{off}|$. We can identify a general tradeoff between low insertion loss and high attenuation. Consequently, w_g has to be carefully chosen.

- High σ and low C_{par}'' are advantageous but are defined as well by the used technology. In this context, we can deduce that n-channel devices are superior with respect to the p-channel counterparts. At given w_g and C_{par}'', we achieve a lower $|Z_{on}|$.

- By the way, we have not yet considered the losses of the extrinsic interconnects. They may have a non-negligible impact on the overall $|Z_{on}|$.

The variation of the channel resistance vs gate voltage is shown in Fig. 13.4b for a 90-nm n-channel MOSFET. With this device an R_{ds} range of 6–2000 Ω can be covered.

a)

Parameter	On	Off
V_{gs}	0.9 V	0 V
R_{ds}	6 Ω	2000 Ω
C_{gs}	55 fF	30 fF
C_{gd}	55 fF	30 fF
R_i	17 Ω	20 Ω
C_{ds}	12 fF	
R_g	~2Ω	
R_s, R_d	~1 Ω	

b)

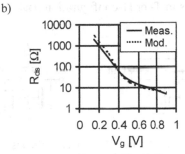

Fig. 13.4a,b. A 90-nm n-channel SOI FET with gate width of 64 μm at V_{ds}=0 V: **a** model parameters; **b** drain source resistance vs DC bias

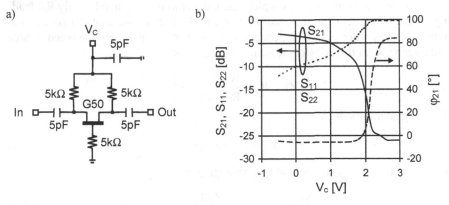

Fig. 13.5a,b. Series attenuator: **a** circuit schematics including bias network; **b** measured attenuation, transmission phase and return losses at 5.2 GHz

13.1.2 Transmission Phase Variations

Attenuators exhibit phase variations vs amplitude control due to the associated changes of the RC time constants. The circuit schematic, attenuation control performance, corresponding transmission phase variation and the return losses of a series-type attenuator at an operation frequency of 5.2 GHz are shown in Fig. 13.5. The 0.6-μm GaAs MESFET has a gate width of 50 μm. A minimum loss of 3 dB and a maximum attenuation range of 23 dB were measured. Within an attenuation control range of 10 dB and 20 dB, the transmission phase variations are 6° and 38°, respectively. Up to 10 dB amplitude control, the generated phase variations are moderate and lower than those of a typical active amplifier.

13.1.3 Impedance Matching

For many applications, the impedance matching must be preserved within the full attenuation range. This requirement can be met by combining of series- and shunt-FETs in T- or Π-configuration [Bah05] as illustrated in Fig. 13.6.

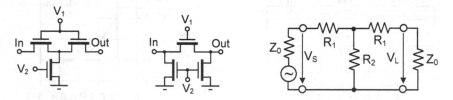

Fig. 13.6. T- and Π-attenuator circuits, for sake of simplicity, the DC supply networks are not included

Fig. 13.7. Equivalent T-circuit

We evaluate the T-circuit. To simplify the calculations, we consider only R_{ds} leading to the equivalent circuit according to Fig. 13.7. Suppose that the source and load impedance are equal. Impedance matching at the junctions between source and attenuator, and load and attenuator demands for

$$Z_0 = R_1 + \frac{R_2(R_1 + Z_0)}{R_2 + R_1 + Z_0} .\tag{13.3}$$

Rearrangement results in

$$Z_0^2 = R_1^2 + 2R_1R_2 .\tag{13.4}$$

The relation between source and load voltage is given by

$$\frac{V_L}{V_S} = \frac{Z_0 R_2}{R_1^2 + 2R_1R_2 + R_1Z_0 + R_2Z_0} .\tag{13.5}$$

With Eq. (13.4) we can write for the voltage attenuation

$$L_V = \frac{R_1 + R_2 + Z_0}{R_2}. \tag{13.6}$$

The power attenuation can be calculated by

$$L_P = 20 \log L_V. \tag{13.7}$$

By means of Eqs. (13.4) and (13.6), the resistances for a certain desired voltage attenuation can be written as

$$R_1 = Z_0 \cdot \frac{L_V - 1}{L_V + 1} \tag{13.8}$$

and

$$R_2 = Z_0 \cdot \frac{2}{L_V - \frac{1}{L_V}}. \tag{13.9}$$

Similarly, we can calculate the relations for the Π-circuit. In Table 13.1, the key equations and the element values required for common power attenuations are summarised.

Table 13.1. Key relations and element values for attenuators

Topology	T-type		Π-type	
Schematics				
Conditions for impedance matching	$Z_0^2 = R_1^2 + 2R_1 R_2$		$Z_0^2 = \dfrac{R_2^2}{1 + \dfrac{2R_2}{R_1}}$	
Voltage attenuation	$L_V = \dfrac{R_1 + R_2 + Z_0}{R_2}$		$L_V = \dfrac{R_2 + Z_0}{R_2 - Z_0}$	
Element relations	$R_1 = Z_0 \dfrac{L_V - 1}{L_V + 1}$	$R_2 = Z_0 \dfrac{2}{L_V - \dfrac{1}{L_V}}$	$R_1 = \dfrac{Z_0}{2}(L_V - \dfrac{1}{L_V})$	$R_2 = Z_0 \dfrac{L_V + 1}{L_V - 1}$
Power attenuation				
$Z_0 = 50\ \Omega$	R_1	R_2	R_1	R_2
0 dB	$0\ \Omega$	∞	0	∞
3 dB	$9\ \Omega$	$142\ \Omega$	$18\ \Omega$	$292\ \Omega$
6 dB	$17\ \Omega$	$67\ \Omega$	$37\ \Omega$	$150\ \Omega$
10 dB	$26\ \Omega$	$35\ \Omega$	$71\ \Omega$	$96\ \Omega$
20 dB	$41\ \Omega$	$10\ \Omega$	$248\ \Omega$	$61\ \Omega$

13.1.4 Antenna Switch

Figure 13.8a,b shows the schematics of a typical antenna switch, which is also known as SPDT (Single Pole Double Throw) switch. It is suited for the transceiver architecture treated in Sect. 2.3. Two control voltages are required, which are inversely fed to the transistor pairs. One of these voltages is high and the other is low. As a consequence, in transmit mode, the shunt transistor in the transmit path is off, whereas the series FET is on. Thus, the transmit signal is transferred to the antenna. At the same time, the shunt transistor in the receiver branch is on, while the series device is off thereby blocking the transmit signal in the receiver. The opposite scenario is achieved in the receive mode, where the control voltages are inverted. As mentioned before, the gate widths of the transistors have to be carefully chosen to achieve an optimum tradeoff between insertion loss in on-mode and isolation in off-mode. Low loss demands for large w_1 and small w_2. Opposite constraints hold for maximum isolation.

The performances of SPDT switches in 0.18-μm CMOS technology are shown in Fig. 13.8c. Losses of below 1.6 dB and isolations of beyond 30 dB are simulated from DC up to 10 GHz. Slightly better performance is feasible in III/V technology. For further information, the reader is referred to literature [Jin05, Hua00].

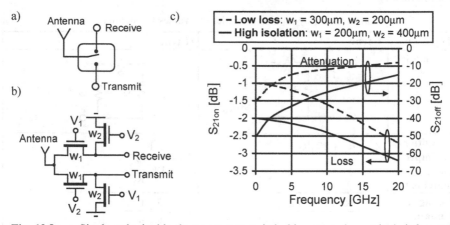

Fig. 13.8a–c. Single pole double throw antenna switch, bias networks not included: **a** generic schematic; **b** circuit schematic; **c** insertion loss in on-mode (S_{21on}) and attenuation in off-mode (S_{21off}) for typical circuit in 0.18-μm CMOS technology

13.1.5 Digitally Adjustable Attenuators

Continuously adjustable amplitude control is possible by sweeping the analogue control voltage of an attenuator. The major advantage is the high amplitude resolution attributed to the fact that theoretically every individual gain/attenuation can be set within the feasible control range. Unfortunately, this approach requires power consuming and costly DACs with high resolution. Similar to the digitally adjustable phase shifters which will be treated in Chap. 14, the DACs can be saved by using circuits requiring only "digital" control voltages with high or low values.

The idea is to switch between different paths with individual fixed attenuators by means of SPDT switches. An example of a 4 bit attenuator with 15 dB amplitude control range is illustrated in Fig. 13.9. The amplitude resolution of the circuit is determined by the least significant bit, which in this case is 1 dB. Eight (2 times bit-amount) digital DC control voltages are required for this task. We have to consider the complexity of the wiring and the chip size increase due to the large contact pads. Fortunately, these connections are uncritical DC paths. For further information concerning the implementation and results of such digital attenuators, the reader is referred to the literature [And94, Din97, Bed91]. By the way, digital control of amplitude-weighted amplifiers is also possible [Hwa84].

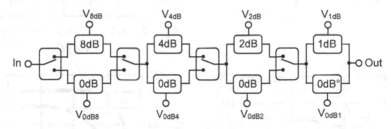

Fig. 13.9. Digital adjustable attenuator with 4 bit resolution

13.2 Variable Gain Amplifiers

According to Chap. 5, the gain of transistors and amplifiers can be controlled by varying one of the supply voltages. Two design examples are presented now. Further information can be found in [Ell43] and [Ell51].

13.2.1 GaAs MESFET Variable Gain Common Source Amplifier

We focus on a combination of common source amplifiers as depicted in Fig. 13.10a. The circuit is optimised for 5.2 GHz and 50 Ω terminations. A 0.6-μm GaAs MESFET technology is applied for the design providing maximum gain of 11 dB at V_{ds}=1.5 V and V_{gs}=0.5 V. The V_{th} of the used enhancement FET is 0.14 V. Recall that the power gain is related with the voltage gain given by $a_v = -g_m \cdot Z_L$. Decreasing of V_{gs} lowers g_m, and subsequently the gain. Figure 13.10b shows the simulated phase and amplitude characteristics of the variable gain amplifier when V_{gs} is varied and V_{ds} is fixed. Within a V_{gs} ranging from 0 V to 0.5 V and an associated amplitude control of 20 dB, a phase deviation of ±24° is generated, which is not acceptable for phase critical applications. Is there a way to compensate this phase error without external correction? Let us first review the characteristics when decreasing V_{ds} while keeping V_{gs}=0.25 V constant. The corresponding results are plotted in Fig. 13.10c. Within a certain amplitude control range, the phase increases with lowered values of V_{gs}, whereas the phase falls with lowered values for V_{ds}. We can conclude that the phases change in opposite directions. This implies the combination of the two bias control methods to compensate the phase errors, e.g. by cascading two amplifiers as shown in Fig. 13.10e.

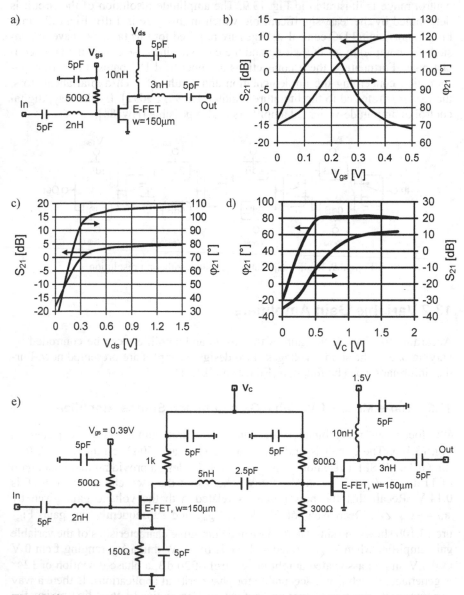

Fig. 13.10a–e. Amplitude control of common source amplifier at 5.2 GHz: **a** circuit schematic of simple stage; **b** gain and transmission phase versus V_{gs} at V_{ds}=1.5 V; **c** gain and transmission phase versus V_{ds}, V_{gs}=0.25 V; **d** control of both V_{gs} and V_{ds}; **e** circuit schematic of two stage amplifier, compare [Ell27] © IEEE 2001

The control voltage V_C connects the V_{ds} of the first stage with the V_{gs} of the second stage. A voltage divider is applied to set the optimum V_{gs} of the second amplifier. To lower the impact of V_{th} variations with respect to process variations, a DC feedback is implemented in the source of the first stage. The simulated characteristics of the proposed amplifier are shown in Fig. 13.10d. Within a control voltage from 0.6 V to 1.75 V and a corresponding gain control range of 20 dB, the phase variation can be reduced to $\pm 3°$, which is by a factor of 8 lower than in the uncompensated version.

For details concerning this method and the implemented circuit, the reader is referred to [Ell27, Ell59]. A further method for transmission phase error compensation of a cascode amplifier will be presented in the next section.

13.2.2 SiGe Cascode Variable Gain LNA at 5–5.5 GHz

A 4.8–5.7-GHz cascode VGLNA (Variable Gain LNA) fabricated in commercial IBM 6HP SiGe BiCMOS technology [Ell34] is presented. The same technology as for the wideband LNA treated in Sect. 8.3.1 is applied. A cascode is used since it provides both higher maximum gain and larger attenuation compared to single transistor approaches. The latter property is possible because of the lower parasitic input to output capacitance, which may be approximated by series connection of the feedback capacitances of the common emitter stage and the common base stage. The attenuation is also limited by the substrate coupling, which can be significant for silicon-based circuits. Ground shields are used for inductors and pads to decrease the impact. Reactive impedance matching to 50 Ω is accomplished by reusing the LC bias elements. Figures 13.11 and 13.12 show the circuit schematic and photo of the compact IC. Equal transistors are employed. If not otherwise mentioned, the bias is as follows: V_{cc}=1.2 V, V_{b1}=0.6 V and I_{cc}=1 mA. At this bias, 12.7 dB gain, 2.4 dB NF, 0 dBm OIP3 and input and output return losses of 7 dB and 9.6 dB, respectively, are measured at 5.2 GHz. All results will be referred to the latter frequency.

13.2.2.1 Gain Control

The gain of a cascode circuit can be decreased by lowering of one of the bias voltages (V_{b1}, V_{b2}, V_{cc}), while keeping the remaining two voltages constant:

Mode 1: In this mode, V_{b1} is decreased. Due to the falling current, g_m and subsequently the gain are lowered.

Mode 2: Decreasing of V_{b2} lowers the gain, since the collector emitter voltage of the common emitter stage $V_{ce,CE}$=V_{b2}–V_{b1} is driven into the resistive region. Due to the raised resistive parasitics and the falling supply current, the gain is decreased.

Mode 3: Lowering of the supply voltage V_{cc} leads to a continuously growing gain drop since the collector emitter voltage of the common base stage $V_{ce,CB}$=V_{cc}–V_{b2}+V_{b1} is driven into the resistive region.

For comparison of the different bias modes, the key characteristics were measured vs gain and are discussed in following sub-sections.

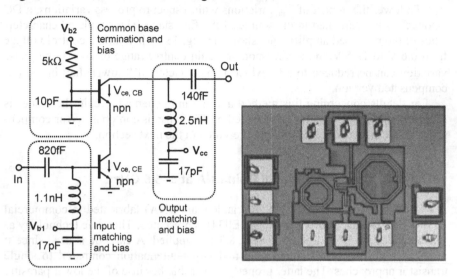

Fig. 13.11. Simplified schematic of cascade VGLNA, compare [Ell34] © IEEE 2004

Fig. 13.12. Photograph, overall chip size is 0.7×0.9 mm², compare [Ell34] © IEEE 2004

a. Control Linearity

High control linearity minimises the resolution required for the D/A converter and the control components. As shown in Fig. 13.13, the best linearity is reached for bias mode 1. For all modes, an amplitude control range of more than 35 dB is achieved.

b. Current Consumption

According to Fig. 13.14, the supply currents decrease with lowered bias voltages. As expected, the highest decrease of the supply current is observed with bias mode 1. Recall that the supply current and V_{b1} are related by an exponential function. Drawback of bias mode 3 is that a relatively high control current has to be provided. Bias modes 1 and 2 have the advantage that their control current (base current) is very low.

c. Noise Figure

At maximum control voltages, the optimum current with respect to noise is provided. Thus, it is clear that the noise figure raises with decreased bias. The properties are illustrated in Fig. 13.15. In bias mode 1, the noise is significantly degraded since a decreased V_{b1} quickly moves the bias current away from the bias current for minimum noise. A medium noise increase was measured for bias mode 2. Noise is increased since the common emitter stage is driven into the resistive region. The lowest increase of the noise figure was obtained for bias mode 3. The noise of the common base stage (output stage) is increased because it is driven into the resistive region, whereas the noise increase and gain decrease of the common emitter (input stage) is weak. According to the formula of Friis, the system noise is dominated by the input stage. Thus, the total noise increase vs gain is small for bias mode 3.

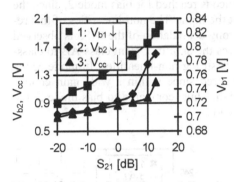

Fig. 13.13. Measured bias control linearity vs gain, compare [Ell34] © IEEE 2004

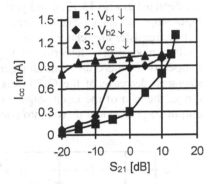

Fig. 13.14. Measured supply current vs gain, compare [Ell34] © IEEE 2004

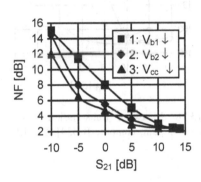

Fig. 13.15. Measured noise characteristics versus gain, compare [Ell34] © IEEE 2004

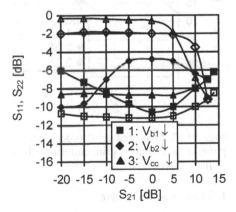

Fig. 13.16. Measured input return loss (filled symbols) and output return loss (symbols not filled) versus gain, compare [Ell34] © IEEE 2004

d. Return Losses

As illustrated in Fig. 13.16, the best results with relatively high return losses within a wide gain control range were obtained for bias mode 1. A strong degradation of the output return losses is observed for bias mode 3. The output impedance gets very high-ohmic since the output stage is driven in the resistive region.

e. Signal Linearity

As depicted in Fig. 13.17, good performance is reached for bias mode 2, since the input stage acts as a variable attenuator at the input. The non-linearities of this resistive transistor are relatively weak. A strong degradation of the OIP3 is observed for bias mode 1. The lowering of V_{b1} drives the amplifier from class-AB to class-C operation, where the transistor generates strong non-linearities. In mode 3, the strongest decrease of the OIP3 was measured since within a wide gain control range, the gain of the input stage stays high. Thus, compared to bias mode 2, the output stage is sooner driven into compression.

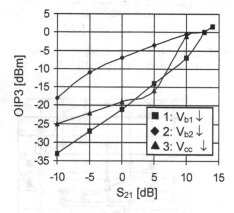 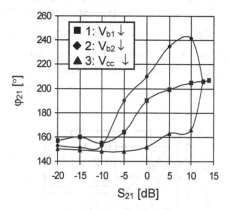

Fig. 13.17. Measured third order intercept point at the output vs gain, compare [Ell34] © IEEE 2004

Fig. 13.18. Measured transmission phase variations vs gain, compare [Ell34] © IEEE 2004

f. Transmission Phase

Figure 13.18 shows the measured performances of the three control modes. The lowest phase variation was measured in bias mode 1. High phase variations were observed with bias mode 2 and 3. These phase variations are generated by the strong resistive and capacitive variations appearing at the transition from the forward active to the resistive region.

13.2.2.2 Bias Technique for Constant Transmission Phase

A smart biasing method is presented now, which significantly decreases the transmission phase variations. The basic idea is that within a given gain range, bias mode 1 and bias mode 2 have opposite phase characteristics. Decreasing of V_{b1} lowers the phase, whereas reducing of V_{b2} increases the phase. Thus, similar to the circuit presented in the preceding section, within a certain amplitude control range, the phase variations can compensate each other. The measured results are shown in Fig. 13.19 demonstrating that the phase can be kept constant for an amplitude range of approximately 25 dB.

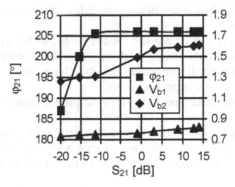

Fig. 13.19. Bias method with constant transmission phase vs gain, compare [Ell34] © IEEE 2004

13.3 Tutorials

1. What are common applications for amplitude control components?
2. What are the typical performance parameters of amplitude control components? Review the importance of these parameters with respect to the application. In which case would you use a passive or active approach?
3. What are the pros and cons for placing an attenuator in front or behind the LNA as depicted in Fig. 13.1a?
4. Illustrate an equivalent circuit of an attenuator FET including the bias network. What are the major differences with respect to a FET in active operation? How can we estimate the behaviour in on- and off-mode? Outline the impact on the high frequency behaviour. Which parameters determine the on- and off-resistance?
5. Review qualitatively the amplitude control range with respect to the gate width of the applied transistors for both passive and active approaches.
6. Why do amplitude control components exhibit phase variations vs amplitude control? How could we compensate them?
7. Why do cascode amplifiers provide higher amplitude control range than single stage transistors?
8. Design an antenna switch with high isolation between the receive and transmit path. How do you have to choose the gate width of the transistors? Discuss the tradeoff if you would like to optimise the switch for minimum insertion loss.
9. Suppose that the PA transmits 1 W. The maximum power the LNA can handle is 5 dBm. What is the minimum required isolation of the switch if the antenna should be used for both the receive- and the transmit- path?
10. Depict the circuit schematic of an attenuator circuit with 7 dB control range and 50 Ω terminations, which should provide attenuation with 3 bit resolution on the basis of resistors and cold FETs. Is it important to incorporate the parasitics of the FETs? How could we compensate them? What is the drawback?
11. Explain the gain, noise, linearity and DC characteristics of the possible bias methods of a cascode variable gain amplifier.
12. Which amplitude control circuits are more sensitive with respect to process variations? The active or passive ones?

References

[And94] A. Anderson, J. Joshi, "Wideband constant phase digital attenuators for space applications", Microwave Engineering Europe, pp. 25–30, Feb. 1994.

[Bah05] Inder Bahl, Lumped Elements for RF and Microwave Circuits, Artech House, 2005.

[Bed91] B. Bedard, B. Maoz, "Fast GaAs MMIC attenuators has 5 bit resolution", Microwave & RF, pp. 71–76, Oct. 1991.

[Din97] S. Dingo, R. Meierer and R. North, "Computer-aided design of MMIC variable attenuators", Microwave Journal, pp. 134–140, Nov. 1997.

[Ell27] F. Ellinger, U. Lott and W. Bächtold, "An antenna diversity MMIC vector modulator for HIPERLAN with low power consumption and calibration capability", IEEE Transactions on Microwave Theory and Techniques, Vol. 49, No. 5, pp. 964–969, May 2001.

[Ell29] F. Ellinger, R. Vogt and W. Bächtold, "Calibratable adaptive antenna combiner at 5.2 GHz with high yield for PCMCIA card integration", IEEE Transactions on Microwave Theory and Techniques, Special Issue, Vol. 48, No. 12, pp. 2714–2720, Dec. 2000.

[Ell34] F. Ellinger, C. Carta, L. Rodoni, G. von Büren, D. Barras, M. Schmatz and H. Jäckel, "BiCMOS variable gain LNA at C-band with ultra low power consumption for WLAN", International Telecommunication Conference, CD ROM, Aug. 2004.

[Ell43] F. Ellinger, R. Vogt, W. Bächtold, "Comparison of variable gain LNA MMICs at C-band using GaAs enhancement, depletion and deep-depletion MESFETs", IEEE/SBMO International Microwave and Optoelectronics Conference, Conference CDROM, Belem, Brasil, Aug. 2001.

[Ell51] F. Ellinger, U. Lott and W. Bächtold, "A 5.2 GHz variable gain LNA MMIC for adaptive antenna combining", IEEE Radio Frequency Integrated Circuit Symposium, Anaheim, pp. 197–200, June 1999.

[Ell59] F. Ellinger, "Method to compensate transmission phase errors in variable gain amplifiers", European patent application, Jan. 2001.

[Hua00] F.J. Huang, K. O., "A 0.5 um T/R switch for 900 MHz wireless applications", IEEE Journal on Solid-State Circuits, Vol. 36, No. 3, pp. 486–492, May 2000.

[Hwa84] Y. Hwang, Y. Chen, R. Naster, "A microwave phase and gain controller with segmented dual gate MESFETs in GaAs MMIC", IEEE Microwave and Millimeter Wave Circuits Symposium Digest, San Francisco, pp. 1–5, 1984.

[Jin05] Y. Jin and C. Nguyen, "A 0.25 μm CMOS T/R switch for UWB wireless communication," IEEE Microwave and Wireless Component Letters, 2005.

14 Phase Shifters

Phase shifters [Kou91] are applied for many RF circuits, e.g. in adaptive antenna combiners (refer to Fig. 2.16 and [Ell21, Ell29]), measurement equipment, phase locked loops and PA linearisation systems. On the wish list for phase shifters we can find the following performance constraints: 360° phase shift, low loss or even gain, adequate bandwidth, RF power handling and linearity, low power consumption, and small size. In most cases, phase shifters require inductors consuming lots of circuit area. However, small circuits are favourable concerning costs. The control complexity should be as low as possible. One control signal is preferred. Moreover, the amplitude variations versus phase control should be small. All those constraints make the design of integrated phase shifters to a challenging task. As for other circuits, reasonable tradeoffs have to be found. Both active and passive approaches have been invented to meet the requirements for specific applications. The most common phase shifter approaches including varactor tuned transmission lines, reflective type phase shifters and vector modulators will be elaborated in the next sections.

In advance and from system perspective, the designer should carefully take into account stability. Recall that phase has a significant impact on stability. Phase conditions with 180° offset may lead to opposite stability properties. Consequently, stability investigations versus the full phase range must be performed. Non-ideal grounds play an important role.

14.1 Varactor Tuned Transmission Lines

The phase can be controlled by tuning of transmission lines [Nag99]. Lumped element equivalent structures are preferred since they require much smaller circuit area. Distributed lowpass- or highpass-structures in T- or Π-configuration are used for this purpose. Common FET varactors are applied to vary the capacitance. Since the DC currents through the varactors are very small, the control voltage may be fed via compact high-ohmic resistors. Optionally, tuneable active inductors could be considered as reported in [Hay96]. Unfortunately, active inductors consume high DC power and increase the circuit and control complexity, thereby limiting the suitability for mobile applications. Therefore, we focus on varactor tuned transmission lines comprising spiral inductors with fixed impedances.

Lowpass structures in Π-configuration, as shown in Fig. 14.1, are the best choice, since they require a minimum number of area consuming inductors. Neighbouring varactor capacitances can be combined. Furthermore, the distributed phase shifter with Π lowpass structures allows minimum bias complexity. The anodes of the varactors are grounded without requiring additional bias circuitries. All cathodes of the varactors are connected by the inductors. Therefore, a positive control voltage can be applied, which is fed into only one inductor node of the whole distributed phase shifter. N-channel D-FETs are favourable concerning the large signal properties. The drain and source form the cathode. By varying the cathode potential, a positive voltage can be applied within the main control range. The gate is grounded.

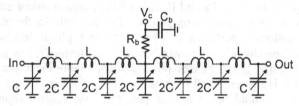

Fig. 14.1. Varactor tuned transmission line phase shifter using distributed lowpass structures realised by lumped elements. Neighbouring varactor capacitances of Π-sections can be combined, R_b: bias network, C_b: RF shunt

14.1.1 Characteristics of One Tuned Lowpass Section

According to Sect. 6.9 and [Gar74, Ell19], the design equations for the inductance and the capacitor of one Π-lowpass segment with characteristic length φ are given by

$$C = -\frac{\tan\left(\dfrac{\varphi}{2}\right)}{\omega \cdot Z_0} \tag{14.1}$$

and

$$L = -\frac{Z_0 \sin(\varphi)}{\omega}. \tag{14.2}$$

With $y_C = \omega C \cdot Z_0$ and $x_L = \dfrac{\omega L}{Z_0}$, the corresponding transmission properties can be characterised by

$$\underline{S}_{21} = \frac{2}{2(1 - y_C x_L) + j(x_L + 2y_C - y_C^2 x_L)} \tag{14.3}$$

and

$$\varphi_{21} = \varphi = \tan^{-1}\left[\frac{y_C^2 x_L - 2y_C - x_L}{2(1 - y_C x_L)}\right]. \tag{14.4}$$

Note that the latter equation is only valid within $0 \geq \varphi \geq -90$. Now, we investigate the phase control based on the variation of the varactor capacitance and the resulting characteristic length of the virtual line. The impedance $x_{L,0}$ is fixed. With the minimum and maximum varactor capacitances C_{min} and C_{max}, y_C can be varied from y_{Cmin} to y_{Cmax} yielding a capacitance control ratio of

$$t_v = \frac{C_{vmax}}{C_{vmin}} = \frac{y_{Cmax}}{y_{Cmin}}. \tag{14.5}$$

To obtain symmetrical variation of y_C around its mean value $y_{C,0}$ we define

$$y_{Cmax} = y_{C,0}\sqrt{t_v} \tag{14.6}$$

and

$$y_{Cmin} = \frac{y_{C,0}}{\sqrt{t_v}}, \tag{14.7}$$

which is a good compromise between low transmission loss associated with the reactive mismatch and high transmission phase control range. On the basis of Eqs. (14.4), (14.6) and (14.7), the phase control range of one ideal Π lowpass section can be calculated by

$$\Delta\varphi = |\varphi_{21}(y_{Cmax}) - \varphi_{21}(y_{Cmin})| \tag{14.8}$$

$$= \left|\tan^{-1}\left[\frac{y_{C,0}^2 \cdot t_v \cdot x_{L,0} - 2y_{C,0}\sqrt{t_v} - x_{L,0}}{2(1 - y_{C,0}\sqrt{t_v} \cdot x_{L,0})}\right] - \tan^{-1}\left[\frac{\dfrac{y_{C,0}^2 x_{L,0}}{t_v} - \dfrac{2y_{C,0}}{\sqrt{t_v}} - x_{L,0}}{2\left(1 - \dfrac{y_{C,0}x_{L,0}}{\sqrt{t_v}}\right)}\right]\right|. \tag{14.9}$$

Figure 14.2a demonstrates the transmission phase control range vs capacitance control ratio and centre transmission phase. By raising the characteristic length, the phase control range can be improved. In turn, this demands for increased values of the lumped capacitances and inductances as emanated from Eqs. (14.1) and (14.2). In this case, the resonance frequency of the structure

$$\omega_r = \frac{1}{\sqrt{LC}}, \tag{14.10}$$

around which maximum phase variations are obtained, decreases. Thus, the distance between the operation frequency and the resonance frequency is lowered as well. Unfortunately, close to or above the resonance frequency, the signal is attenuated due to the lowpass characteristics. Therefore, a compromise is aimed for between maximum transmission phase control range and minimum transmission loss at given capacitance control range.

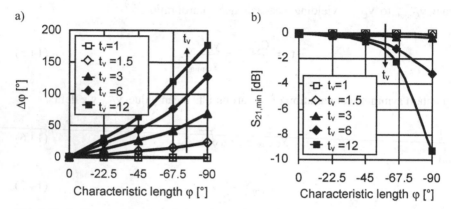

Fig. 14.2a,b. Versus characteristic length φ and capacitance control range ΔC at centre operation frequency of one lowpass section assuming ideal elements are plotted: **a** transmission phase control range $\Delta\varphi$; **b** maximum transmission loss $S_{21,min}$ due to reactive mismatch

On the condition that ideal elements are used, the minimum transmission loss $S_{21,max}$ obtained at centre frequency ω_0 and φ equals unity. Due to the reactive mismatch transmission loss is generated if the phase is varied. The maximum transmission loss is obtained at y_{Cmax} since the lowpass effect of the structure is most significant at maximum capacitance C_{max}, hence:

$$S_{21,min} = S_{21}(y_{Cmax}). \tag{14.11}$$

Deduced from Eq. (14.3) we obtain

$$S_{21,min} = \frac{2}{\sqrt{4(1 - y_{Cmax}x_{L,0})^2 + (x_{L,0} + 2y_{Cmax} - y_{Cmax}^2 x_{L,0})^2}}. \tag{14.12}$$

Equations (14.6) and (14.7) allow the rearrangement to

$$S_{21,min} = \frac{2}{\sqrt{4\left(1-y_{C,0}\cdot\sqrt{t_v}\cdot x_{L,0}\right)^2 + \left(x_{L,0}+2y_{C,0}\sqrt{t_v}-y_{C,0}^2\cdot t_v\cdot x_{L,0}\right)^2}}. \quad (14.13)$$

Finally, we can write for the maximum possible transmission loss variation:

$$\Delta S_{21} = S_{21,min}. \quad (14.14)$$

Figure 14.2b shows the maximum transmission loss versus characteristic length and capacitance control range. The phase control range can be increased by cascading of lowpass sections. Accordingly, the losses of the virtual transmission line rise as well.

14.1.2 GaAs MESFET Varactor Tuned Transmission Line Phase Shifter at 4.5–6.5 GHz

As design example, a C-band circuit is presented in this section [Ell19, Ell25]. The circuit was fabricated on the commercial Triquint TQTRx GaAs MMIC process featuring MESFETs with gate lengths of 0.6 μm and transit frequencies around 18 GHz. A characteristic length of approximately −50° at the centre frequency of 5.5 GHz was chosen for each lowpass element. This value is a good compromise between low insertion loss and high phase control for capacitance control ranges typically provided by common MESFET varactors. Deep depletion FETs (G-FETs) with capacitance control ratio of approximately 3 are used as varactors. According to Eqs. (14.9) and (14.13), the theoretical transmission phase control range and the transmission loss of one element are approximately 30° and 0.15 dB, respectively. The required inductor and centre capacitance values for a characteristic wave impedance of 50 Ω are approximately 1.2 nH and 0.3 pF, respectively. In practice, the capacitance control ratio and phase control are slightly decreased by the capacitive parasitics of the inductors. In addition to the lowpass characteristics of the lumped element equivalent of the transmission line, the transmission loss is determined by the Q-factors of the inductors and varactors. A relatively high number of 16 cascaded lowpass elements is used to guarantee a phase shift of 360° within a sufficient frequency bandwidth. Via a high-ohmic resistor with value of 5 kΩ, the DC control voltage V_c is fed into one inductor node connecting all varactor cathodes. The anodes of the varactors are grounded. At 5.5 GHz and control voltages of 0 V and 5 V, the Q-factors are around 14 and 32, respectively. The inductor Q is approximately 23. A photograph of the IC is depicted in Fig. 14.3. The phase control and transmission losses are shown in Fig. 14.4. Within a measured phase control range of 360°, the measured transmission losses at 4.5 GHz, 5.5 GHz and 6.5 GHz are 3.6±1.8 dB (V_c=−0.4 .. 5 V), 3.9±.5 dB (V_c=-0.25 .. 5 V) and 4.5±1.6 dB (V_c=0.1 .. 5 V). The transmission loss and the phase control range raise with increased frequency, since the distance between the operation frequency and the resonance frequency decreases. Thus, the lowpass effect becomes more significant. A similar effect can be observed for decreased V_c associated with an increase of the varactor capacitance. Within 4.5–6.5 GHz, the return losses are higher than 10 dB. A 1 dB input compression point of higher than 12 dBm was measured. The control linearity (function of phase versus control voltage) is good, keeping the resolution requirements and costs for the DAC low.

In

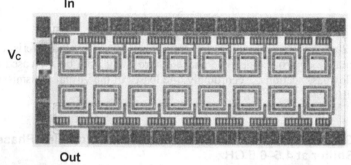

Out

Fig. 14.3. Photograph of varactor tuned transmission line phase shifter consisting of 16 lumped lowpass elements in Π configuration. All non-marked pads are ground pads. Chip size is 1.4×0.6 mm^2, compare [Ell19] © IEEE 2003

a)   b)

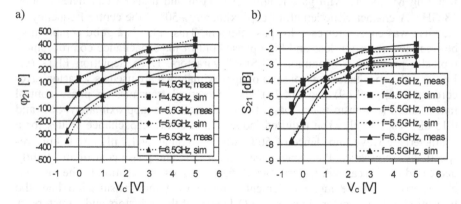

Fig. 14.4a,b. Versus control voltage and operation frequency are shown the measured and simulated transmission: **a** phase; **b** loss, compare [Ell19] © IEEE 2003

14.2 Reflective Type Phase Shifters

Figure 14.5 sketches the architecture of a reflective type phase shifter consisting of a branch-line coupler and two reflective loads with equal impedances Z_r. Branch-line couplers have been discussed in Sect. 6.10.2. If not mentioned otherwise, ideal conditions with lossless elements are assumed in the analysis. The input signal is divided into two parts. Each part is reflected at one of the reflective loads. Given that the reflective loads are purely reactive, no power is lost in the reflective loads and all signal power is coupled into the output. One significant benefit of this configuration is the fact that the input and output impedance matching is preserved, although the coupler is terminated with reactive loads. In other words, the coupler serves as an impedance isolator.

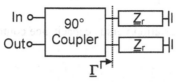

Fig. 14.5. Circuit topology of the RTPS using a branch-line coupler with lumped elements. Phase control is obtained by varying the phase of the reflection factor Γ

Following [Ell26], we derive now the analytical relations for the circuit. With the characteristic coupler impedance Z_0, the reflection coefficient observed between Z_r and coupler ports is given by

$$\underline{\Gamma} = \frac{\underline{Z}_r - Z_0}{\underline{Z}_r + Z_0}. \tag{14.15}$$

The phase of this reflection coefficient can be changed by varying $\underline{Z}_r$. Variable active inductances and capacitances can be applied for this task. Due to advantages in terms of loss, size and power consumption, varactors are employed in most cases. The phase variation of Γ determines the transmission phase of the circuit and can be calculated on basis of Eq. (14.15) yielding

$$\Delta\varphi = 2\left[\arctan\left(\frac{Z_{max}}{Z_0}\right) - \arctan\left(\frac{Z_{min}}{Z_0}\right)\right]. \tag{14.16}$$

With minimum and maximum varactor capacitances C_{min} and C_{max}, we obtain maximum and minimum reflection impedance magnitudes of $Z_{max}=1/\omega C_{min}$ and $Z_{min}=1/\omega C_{max}$, respectively. Theoretically, a maximum phase control range of 180° can be achieved if $\Delta C \rightarrow \infty$. The centre value C_0 of the varactor capacitance has to be optimised to achieve maximum phase control range with limited ΔC. Deduced from Eq. (14.16) we obtain

$$C_0 = \frac{1}{j\omega Z_0} \quad \text{or equivalently} \quad \left|\frac{\underline{Z}_{C0}}{Z_0}\right| = 1 \tag{14.17}$$

with

$$C_{max} = C_0\sqrt{t_v} \quad \text{and} \quad C_{min} = \frac{C_{vo}}{\sqrt{t_v}}. \tag{14.18}$$

These relations lead to a maximum phase control range of

$$\Delta\varphi = 2\left[\arctan\left(\sqrt{t_v}\right) - \arctan\left(\frac{1}{\sqrt{t_v}}\right)\right].$$ (14.19)

The corresponding phase variation versus capacitance control ratio is illustrated in Fig. 14.6a for $\left|\dfrac{Z_{C0}}{Z_0}\right| = 1$.

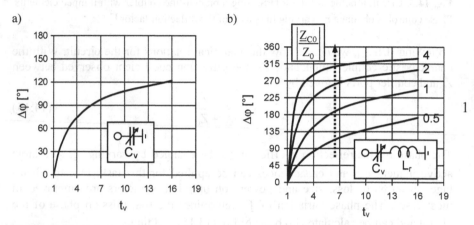

Fig. 14.6a,b. Calculated phase variation of the reflection coefficient versus capacitance control range, parasitics not taken into account: **a** only varactor, $\left|\dfrac{Z_{C0}}{Z_0}\right| = 1$; **b** resonated with inductor, compare [Ell26] © IEEE 2001

By resonating the capacitance of the varactor with an inductor L_r, the phase control range can be significantly increased. The reflector impedance magnitudes are now given by $Z_{max} = \omega L_r - 1/\omega C_{min}$ and $Z_{min} = \omega L_r - 1/\omega C_{max}$. Calculating the corresponding maximum of Eq. (14.16) modifies the maximum phase control range to

$$\Delta\varphi = 4\arctan\left(\left|\frac{Z_{C0}}{Z_o}\right| \cdot \frac{1}{2}\left(\sqrt{t_v} - \frac{1}{\sqrt{t_v}}\right)\right)$$ (14.20)

with the following relation between load inductor L_r and Z_{C0}:

$$L_r = \frac{|Z_{C0}|}{2\omega_o}\left(\sqrt{t_v} + \frac{1}{\sqrt{t_v}}\right).$$ (14.21)

In Fig. 14.6b, the phase variations are illustrated versus t_v and $\left|\dfrac{Z_{C0}}{Z_0}\right|$. To raise $\Delta\varphi$,

either t_v or $\left|\dfrac{Z_{C0}}{Z_0}\right|$ can be increased. For $t_v\to\infty$, or $\left|\dfrac{Z_{C0}}{Z_0}\right|\to\infty$, a phase control range

of 360° can be achieved. The increase of $\left|\dfrac{Z_{C0}}{Z_0}\right|$ demands for raised L_r and lowered

C_0 leading to an increase of the series resistance of both devices. This circumstance slightly worsens the phase control range and significantly increases the signal loss given by

$$|\underline{S}_{21}| = 20\log\left(|\underline{\Gamma}|\right).\tag{14.22}$$

An optimum tradeoff between high phase control range (high L_r and low C_0 resulting in high resistive parasitics) and low loss (low resistive parasitics demanding low L_r and high C_0) has to be found for the design.

14.2.1 GaAs MESFET Reflective Type Phase Shifter at 6.1–6.3 GHz with Resonated Load

A reflective type phase shifter with centre frequency of 6.2 GHz is treated in this section, which has been realised with the same technology as the circuit discussed in the previous section [Ell26]. Referring to Sect. 6.10.2, the element values for the 90° coupler are L=0.91 nH, C_1=0.51 pF and C_2=0.21 pF. The varactor control voltage is applied by means of high-ohmic resistors with value of 5 kΩ.

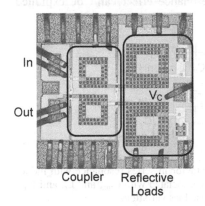

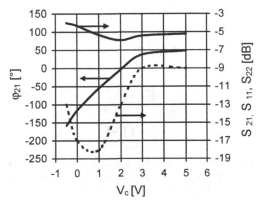

Fig. 14.7. Photograph, overall chip size is 0.85 × 1.1 mm², compare [Ell26] © IEEE 2001

Fig. 14.8. Measured phase, signal loss and return losses (dotted) versus control voltage, f=6.2 GHz, compare [Ell26] © IEEE 2001

An LC series load with C_0=235 fF and L_l=3.5 nH is chosen corresponding to $\left|\dfrac{Z_{C0}}{Z_0}\right|$=2.16. According to our deduced equations and with t_v=4 provided by the optimised MESFET varactors, a good trade-off between high phase control range of 205° and moderate insertion loss in the order of 3.1±0.4 dB is predicted in the ideal case. In practice, we can expect higher attenuation since coupler and mounting losses have not been taken into account yet. Full 360° phase control range can be obtained by series connection of two of these RTPS. Figure 14.7 shows a photograph of the IC mounted and bonded on a test substrate. A signal loss of -4.9 dB±0.9 dB and return losses of higher than 8 dB are measured within a maximum phase control range of 210° at 6.2 GHz (see Fig. 14.8). As expected, the losses are slightly higher compared to the ideal calculations. The 1 dB compression point is above 5 dBm. Mainly limited by the strong resonance of the reflective loads, the bandwidth amounting to 300 MHz is only moderate.

This circuit approach has also been successfully implemented in SiGe technology for operation at 17 GHz [Ell32] and in GaAs at L-band [Ell44].

14.2.2 GaAs MESFET Reflective Type Phase Shifter at 5.1–5.7 GHz with Parallel Resonated Loads

The next circuit to be treated uses the same technology as the latter two circuits and is optimised for a centre frequency of 5.2 GHz [Ell23]. Enhanced reflective loads are applied providing higher phase control than the preceding topology. Recall that the varactor load according to Fig. 14.9a can be resonated with an inductor as shown in Fig. 14.9b, thereby raising the phase control from 60° to 190°. It is straightforward that the highest phase variation and control sensitivity are obtained at resonance. However, it is clear that the resonance effect can't be exploited within the full ΔC range.

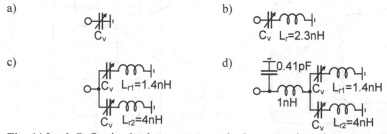

a) C_v

b) C_v L_r=2.3nH

c) C_v L_{r1}=1.4nH
     C_v L_{r2}=4nH

d)   0.41pF   1nH C_v L_{r1}=1.4nH
                     C_v L_{r2}=4nH

Fig. 14.9a–d. Reflective loads: **a** varactor only; **b** varactor in series resonance with inductor; **c** two parallel loads with overlapping resonances associated with C_{max} and L_{r1} and C_{min} and L_{r2}; **d** RL3 plus transformation network to decrease loss variations

What about using networks with multiple resonances? Indeed, further advancements are possible by connecting two different loads in parallel with overlapping resonances as sketched in Fig. 14.9c. The inductor values are chosen to enable a resonance at low bias voltage associated with C_{max}=0.7 pF and L_{r1}=1.4 nH, and a resonance at high bias voltage with C_{min}=0.18 pF and L_{r2}=4 nH. A promising phase control range of approximately $410°$ was simulated for this configuration. Unfortunately, when considering real elements and their inherent losses, the magnitude of the reflection factor shows significant variations for this configuration, resulting in relatively large loss variations.

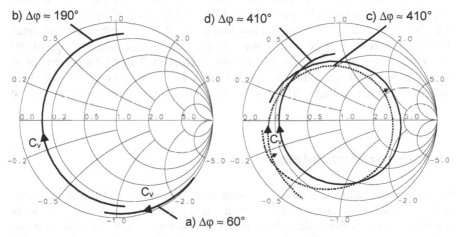

Fig. 14.10. Simulated reflection coefficient of reflective loads including parasitics at 5.2 GHz, V_C varied from -0.5 V to 5 V corresponding to capacitance variation from C_{min}=0.18 pF to C_{max}=0.7 pF, RL: reflective load, compare [Ell23] © IEEE 2002

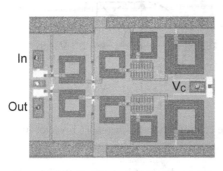

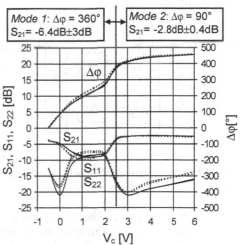

Fig. 14.11. Photograph of IC, chip size is 0.85 × 1.1 mm², compare [Ell23] © IEEE 2002

Fig. 14.12. Measured phases, signal losses and return losses of different samples versus control voltage at 5.2 GHz, mode 1: large phase control, mode 2: low loss variations, compare [Ell23] © IEEE 2002

Referring to Figs. 14.9d and 14.10, a simple LC transformation network can be added to decrease the variation of the reflection factor and the resulting loss variations. This load was used for circuit implementation.

A photograph of the fabricated IC and the measured results are shown in Figs. 14.11 and 14.12, respectively. A phase control range of over 360°, a signal loss of approximately -6.4±3 dB and return losses of higher than 7.2 dB were measured within a control voltage range from -0.5 V to 2.5 V (mode 1). Optionally, the circuit can be operated in a low loss mode. Within a voltage control range from 2.5 V to 6 V, a signal loss of only -2.8±0.4 dB, a phase control range of 90° and return losses of higher than 13.5 dB were achieved. The 1 dB input compression point is 2 dBm at a control voltage of -0.5 V and increases up to 15 dBm with increased control voltage. The bandwidth of the circuit ranges from 5.1 to 5.7 GHz.

14.3 Vector Modulators

A further phase shifter circuit is the vector modulator. Figure 14.13a sketches the architecture of a vector modulator with a phase control range of 90°. The input signal is divided in two paths, which are amplitude-weighted by variable gain amplifiers or attenuators. After generation of a phase offset between these two paths, the signals are combined. Various modifications are possible, e.g. the phase offset may also be implemented in front of the amplifiers, the amplifiers may act as divider or combiner, the combiner and the phase offset can be accomplished by means of a 90° hybrid, etc.

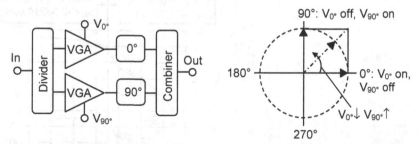

Fig. 14.13a,b. Vector modulator with 90° phase control range: **a** schematic; **b** vector diagram

The resulting phases can be inspected by means of the vector diagram depicted in Fig. 14.13b. If the amplitude of the 0° path is maximum and the one of the 90° branch minimum, the resulting phase is 0°. In the opposite case we obtain 90°. On condition that both branches have equal amplitude the resulting phase is 45°. Every phase between 0° and 90° can be achieved by weighting of the path amplitudes. To obtain constant output amplitude, the amplitudes of the branches have to be set properly. E.g. if full amplitudes are fed in the 45° case, the resulting amplitude would be by a factor of $\sqrt{2}$ larger than in the 0° and 90° states. The magnitudes of the other phases would be between 1 and $\sqrt{2}$. The impact raises when it comes to the power relations, which are proportional to the voltage amplitude by the power of two. Consequently, at full amplitudes, the 45° signal magnitude would be 3 dB higher as in the 0° and 90° modes, which is not acceptable. However, if we vary the relative amplitude between unity in the 90° and 0° cases, and $\dfrac{1}{\sqrt{2}}$ for the 45° phase, it is possible to get constant output magnitudes.

Full 360° phase control requires four paths each exhibiting phase offsets of 90°. Usually, these four paths need four independent DC control signals. Other phase offsets are feasible as well. High phase control ranges demand for large phase offsets. On the other hand, by inspections of the vector diagrams, we observe that the resulting amplitudes decrease for increasing phase offsets. They vanish for phase differences towards 180°. A phase offset of 120° may be a good compromise between phase control range and amplitude loss. Since only three paths and control voltages are required, we can reduce the circuit size and control complexity.

Amplitude control components are required for vector modulators. In this context, the reader is referred to Sect. 13. For these components we can deduce some important requirements:

- In real circuits, amplitude control is always associated with a certain phase variation, which should be as low as possible and definitely much lower than the phase control range of the phase shifter. If significant, the phase variations of the amplifiers must be compensated when setting the overall phase.
- The amplitude control range should be high enough that each path can more or less be switched off with respect to the others. If this condition is not met, the phase of the associated path being switched on can't be approached exactly. Typically, the amplitude control change should be above 20 dB.
- Moreover, the input and output impedances of the circuits should be constant vs amplitude control. Otherwise, undesired amplitude and or phase variations can appear in the circuits connected in front or behind these components.

By the way, if the amplitude control elements provide sufficient gain, vector modulators can be employed for both phase and amplitude control. In this case, not only the relative magnitudes are set. In addition, the overall amplitudes are controlled. Complex look-up tables are required for this approach.

14.3.1 GaAs MESFET Active Vector Modulator at 5–5.5 GHz

Figure 14.14a shows an example of a vector modulator, which provides a phase shift of 360° [Ell27]. Optimised for operation at 5–5.5 GHz and 50 Ω terminations, the circuit is fabricated with the same GaAs technology as the previous phase shifters. Low power consuming enhancement MESFETs with gate widths of 50 μm and gate lengths of 0.6 μm are used to divide, respectively to combine the signal paths. The 120° phase offsets of the signal paths are generated by highpass

(HP) and lowpass (LP) structures with an impedance of approximately 200 Ω. This impedance simplifies the interstage matching. The values of these filters can be calculated in accordance to Sect. 6.9.

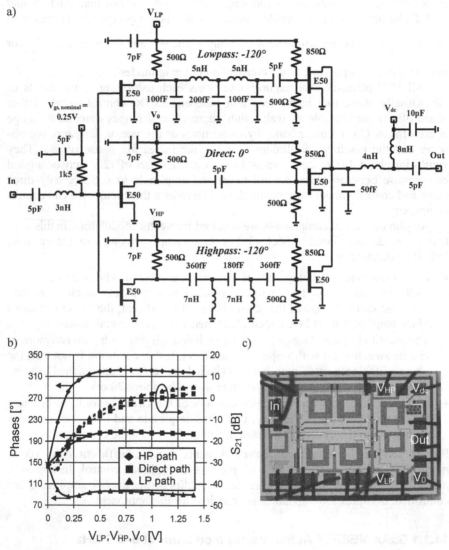

Fig. 14.14a–c. Active vector modulator with 360° phase control: **a** circuit schematic; **b** gain and phase characteristics of the three paths; **c** chip photo, size 1.3×1 mm², compare [Ell27] © IEEE 2001

By means of the voltages V_{HP}, V_0 and V_{LP} ranging from 0 V and 1.4 V, the amplitudes of the signal paths can be controlled. The amplitude control principle presented in Sect. 13.2.1 is used to compensate the phase errors of the amplifiers versus gain control. A low power consuming circuit with cascode variable gain amplifiers and Wilkinson combiner capable of operating between 4.7–5.7 GHz has been published in [Ell48]. Figure 14.14b shows the measured phase and amplitude characteristics of each of the three signal paths, when the other two paths are switched off. At 1.5 V voltage supply, lower than 7 mA current supply and a frequency of 5.2 GHz, the gain of the paths is above 2 dB. Within amplitude control ranges of 15 dB, the maximum phase errors of the cascaded amplifiers are only ±4°. Versus full phase control of 360°, an overall gain of 0.6 dB was measured. The −1 dB input compression point is higher than −9 dBm and the NF is better than 7 dB. A photograph of the vector modulator chip bonded on a measurement substrate is shown in Fig. 14.14c.

14.3.2 GaAs MESFET Passive Vector Modulator at 4.7–5.7 GHz

A simple, passive vector modulator based phase shifter is described in this section fabricated with the same 0.6-µm GaAs MESFET IC technology as the previously presented circuits [Ell45]. The schematic and photo of the circuit are shown in Fig. 14.15a,b. Input and output are matched to 50 Ω. Deep depletion (G) FETs with negative V_{th} and w_g=200 µm are used to control the attenuation and to switch between the signal paths. These cold FETs are biased via resistors with high impedances. The 120° phase offsets of the signal paths are generated by LC highpass and lowpass filters. The attenuations of the different signal paths can be continuously controlled by the voltages V_{HP}, V_0 and V_{LP} within a voltage range from 0 V (min. attenuation) to −3 V (max. attenuation). The insertion loss and maximum attenuation per single path are around 3 dB and 15 dB, respectively. To obtain a phase control range of 360°, the attenuations of two paths are weighted, while the third path is switched off. The highest insertion loss measured within the 360° phase range was 9 dB. This loss is in particular determined by dividing/combining losses and the non-zero Z_{on} of the FETs. These losses help to improve the return losses. The phase of the FETs slightly varies with attenuation, thereby creating phase changes of up to 15°. For further information concerning the FEF attenuator characteristics, the reader is referred back to Sect. 13.1.2. A nominal look-up table for the control voltages associated with any desired phase for a fixed loss of 9 dB was accomplished by measurements and is depicted in Fig. 14.15c. We observe that the control linearity is not very good demanding for a DAC with high resolution. The input and output return losses over the whole phase range vary between 6 dB and 22 dB. Better impedance matching is not feasible due to the simple combiner/divider concept. An 1 dB input compression point within the whole phase range of higher than 16.5 dBm was measured.

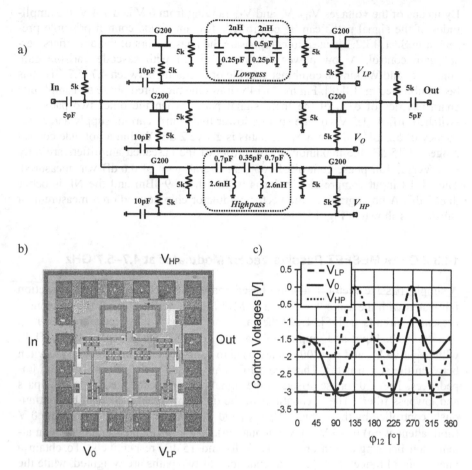

Fig. 14.15a,b. Simple, passive, vector modulator based phase shifter with 360° phase control: **a** schematic; **b** photo, overall chip size: 1×1 mm²; **c** look-up table for control voltages vs phase control range at 5.2 GHz and loss fixed to S_{21}=-9 dB, compare [Ell45] © IEEE 2000

14.3.3 GaAs MESFET Active Vector Modulator with Single Control Voltage

Vector modulator based phase shifters exhibit one drawback compared to other approaches – they require at least two control voltages. This increases the control complexity and the costs. Approaches with single control voltage are preferred. A smart method as demonstrated in [Ell20] at L-band and illustrated in Fig. 14.16a can be applied to circumvent this disadvantage. The idea is as follows: the amplitude tuning characteristics of series and shunt attenuators have opposite properties. If the control voltage is high, the series cold FET exhibits low attenuation, whereas the shunt attenuator has a high one. Opposite behaviour is observed for the control voltage being low. Consequently, the control voltages of both attenuators can be combined to achieve complementary and monotone amplitude weighting with one single control voltage. In Fig. 14.16b, the corresponding characteristics of the employed attenuators are plotted. The attenuations can be varied from 0.6 dB to 17 dB. Active FETs are used as divider and combiner and to isolate the impedance mismatch generated by the attenuators. The phase offsets of 130° are realised by lowpass and highpass structures. Within a phase control range beyond 120°, a gain of 1.6±0.6 dB is measured at a centre frequency of 825 MHz, current consumption of 5 mA and supply voltage of 1.8 V. The circuit covers a bandwidth of 100 MHz. One drawback of the concept is that no relative amplitude compensation can be accomplished. However, the presented circuit has demonstrated that the amplitude errors are small. The chip photo is plotted in Fig. 14.16c.

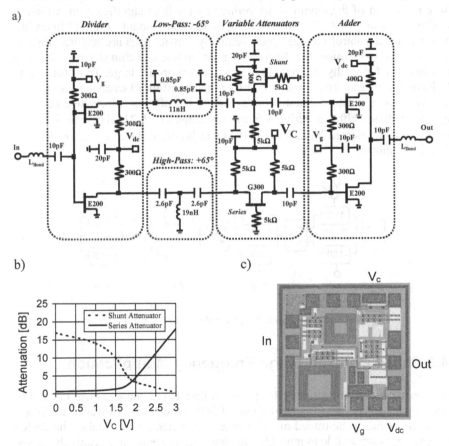

Fig. 14.16a–c. Vector modulator based phase shifter with single control voltage (V_C) and 120° phase control range: **a** schematic E200: enhancement FET, w_g=200 μm. G300: deep depletion FET, w_g=300 μm; **b** characteristics of attenuators; **c** IC photograph, overall chip size is 0.9×0.9 mm^2, compare [Ell20] © IEEE 2002

14.4 Digitally Adjustable Phase Shifters

Up to now we have focussed our discussions on phase shifters with continuously adjustable phase. Major advantage is the high phase resolution attributed to the fact that every individual phase can be set by means of analogue voltages given that the resolution of the DACs is high enough. However, this approach requires power consuming and costly DACs for the control. Similar to the digital attenuators treated in Sect. 13, the need for DACs can be circumvented by using circuits, which require only "digital" control voltages with only two states being high or low. The idea is to switch between different phase paths by means of SPDT switches. Typically exhibiting opposite signs, the path delays can be realised by means of delay lines, or better since more compact by lumped highpass/lowpass structures. An example of a 4 bit phase shifter is illustrated in Fig. 14.17. The phase resolution of the circuit is determined by the least significant bit, which in the shown example is 22.5° corresponding to 360°/bit-amount². Eight (2 times bit-amount) digital control voltages and eventually contact pads are required for this task. Needless to mention that the overall insertion loss and chip size increase with bit resolution. Generally, the required lines or inductors are large. Thus, these kind of phase shifters require a large circuit size thereby in turn increasing the costs as well. We may take into account a loss per bit of 1–2 dB. For information concerning a 5 bit phase shifter covering a frequency range of 17–21 GHz with 5±0.6 dB insertion loss fabricated in 0.25-µm PHEMT technology, the reader is encouraged to study [Cam00].

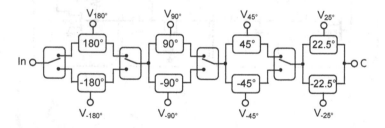

Fig. 14.17. Digital adjustable phase shifter with 4 bit resolution

14.5 Phase Control Range/Frequency Multiplication

One important design parameter for phase shifters is the phase control range. For many applications a phase control range of 360° is required. However, the phase control range has to be traded off with other parameters, which makes the design challenging. Is it possible to multiply the phase control range in a relatively simple way?

Consider that frequency multiplication divides the signal period and thus the equivalent phase duration of a signal. Consequently, the phase control range is multiplied by the factor of the frequency multiplication. We have shown in previous circuit discussions that phase control range, loss and loss variations are opposite goals. The method of frequency/phase multiplication relaxes the constraints concerning the phase control range allowing the optimisation for low amplitude variations. This principle has been successfully applied in an adaptive antenna receiver, where the phase is tuned in the LO section [Ell18]. Drawback is the additional power consumed by the frequency multipliers, which should have good linearity and low noise to avoid deteriorations of the VCO signal.

Table 14.1. Comparison of phase shifters

Principle	Phase control	Band-width	Gain/ ripple	Power supply	Control voltages	Circuit area/ technology	Ref.
TLPS	90°	4 GHz– 6 GHz	−1.2 dB ±0.5 dB	~0 mW	1	0.5-mm^2/0.6-μm GaAs MESFET	[Ell25]
RTPS with complementary bias	90°	40 GHz– 60 GHz	−4 dB ±0.4 dB	~0 mW	2	1.5-mm^2/0.3-μm GaAs PHEMT	[Nam01]
All pass network using two λ/4 lines	180°	12 GHz– 14 GHz	−3.6 dB ±1.1 dB	~0 mW	1	3-mm^2/0.3-μm GaAs MESFET	[Hay02]
RTPS	210°	6.1 GHz– 6.3 GHz	−5.3dB ±1.4 dB	~0 mW	1	0.9-mm^2/0.6-μm GaAs MESFET	[Ell26]
RTPS using active inductors	225°	4.7 GHz– 6.7 GHz	−0.6 dB ±0.2 dB	>100 mW	1	0.1-mm^2/0.1-μm InP HEMT	[Hay96]
Tuned CPW transmission line	360°	5 GHz– 20 GHz	−4.4 dB ±0.6 dB	~0 mW	1	Large/GaAs Schottky diode	[Nag99]
Active variable resonance	360°	2.38 GHz– 2.42 GHz	2 dB ±0.7 dB	>90 mW	2	2.3-mm^2/0.3-μm GaAs MESFET	[Hay99]
Active vector modulator	360°	4.8 GHz– 5.8 GHz	0 dB n.a.	9 mW	3	2.4-mm^2/0.6-μm GaAs MESFET	[Ell29]
Active vector modulator	360°	5.1 GHz– 5.3 GHz	0.6 dB n.a.	10 mW	3	1.3-mm^2/0.6-μm GaAs MESFET	[Ell27]
Passive vector modulator	360°	4.7 GHz– 5.7 GHz	−9 dB n.a.	~0 mW	3	1-mm^2/0.6-μm GaAs MESFET	[Ell45]
RTPS	360°	5.15 GHz– 5.7 GHz	−6.4 dB ±3 dB	~0 mW	1	0.9-mm^2/0.6-μm GaAs MESFET	[Ell23]
Distributed transmission line PS	360°	75 GHz– 110 GHz	−5 dB n.a.	~0 mW	1	n.a./MEMS	[Bar00]
TLPS	360°	5 GHz– 6 GHz	−4 dB ±1.7 dB	~0 mW	1	0.8-mm^2/0.6-μm GaAs MESFET	[Ell19]

TLPS: Tuned Transmission Line Phase Shifter, RTPS: Reflective Type Phase Shifter, CPW: Coplanar Wave

14.6 Comparison of Phase Shifters

In Table 14.1 the key characteristics of different phase shifters published in litera-ture are summarised. All of them have advantages and disadvantages. Hence, de-pendent on the application and the available technology, the designer has to care-fully choose the concept and optimise the circuit from system point of view.

14.7 Tutorials

1. What are the applications and requirements for phase shifters?
2. Why do we observe amplitude variations versus phase control?
3. Comment on stability issues when using phase shifters in systems with gain.
4. Explain the basic principles of the tuned transmission line and reflective type phase shifters, and vector modulators. What are their pros and cons in terms of phase coverage range, losses/gain and the associated variations, control com-plexity, power consumption, large signal performance, etc?
5. Why do we need a coupler in reflective type phase shifters? Consider the im-pedance matching. Discuss the impact of different reflective loads concerning the phase control range.
6. Can we use vector modulators for both phase and amplitude control? What are the requirements and what are the drawbacks when considering the required look-up table for the control voltages?
7. What is the advantage to employ lowpass sections for transmission line tuned phase shifters compared to highpass structures? Consider the biasing and bandwidth.
8. What are the pros and cons regarding analogue and "digital" approaches with respect to control complexity?
9. Draw the equivalent circuit of a 4 bit phase shifter with overall phase coverage range of 90°.

References

[Bar00]   N. S. Barker and G. M. Rebeiz, "Optimization of distributed MEMS transmission-line phase shifters-U-band and W-band designs", IEEE Transactions on Microwave Theory and Techniques, Vol. 48, No. 11, pp. 1957–1966, Nov. 2000.

[Cam00]   C. F. Campell, S. A. Brown, "A compact 5-bit phase-shifter MMIC for K-band satellite communication systems", IEEE Transactions on Microwave Theory and Techniques, Vol. 48, No. 12, pp. 2652–2656, Dec. 2000.

[Ell18]   F. Ellinger and W. Bächtold, "Compact and low power consuming frequency/phase multiplier MMICs for wireless LAN at S-band and C-band", IEE Proceedings on Devices, Circuits and Systems, Vol. 150, No. 3, June 2003.

[Ell19]   F. Ellinger, H. Jäckel and W. Bächtold, "Varactor loaded transmission line phase shifter at C-band using lumped elements", IEEE Transactions on Microwave Theory and Techniques, Vol. 51, No. 4, pp. 1135–1140, April 2003.

[Ell20]   F. Ellinger, W. Bächtold, "Novel principle for vector modulator based phase shifter operating with one control voltage", IEEE Journal of Solid-State Circuits, Vol. 37, No. 10, pp. 1256–1259, Oct. 2002.

[Ell21]   F. Ellinger, W. Bächtold, "Adaptive antenna receiver module for WLAN at C-Band with low power consumption", IEEE Microwave and Wireless Components Letters, Vol. 12, No. 9, pp. 348–350, Sept. 2002.

[Ell23]   F. Ellinger, R. Vogt and W. Bächtold, "Ultra compact reflective type phase shifter MMIC at C-band with 360° phase control range for smart antenna combining", IEEE Journal of Solid-State Circuits, Vol. 37, No. 4, pp. 481–486, April 2002.

[Ell25]   F. Ellinger, R. Vogt and W. Bächtold, "Ultra compact, low loss, varactor tuned phase shifter MMIC at C-band", IEEE Microwave and Wireless Components Letters, Vol. 11, No. 3, pp. 104–107, March 2001.

[Ell26]   Ellinger, R. Vogt and W. Bächtold, "Compact reflective type phase shifter MMIC for C-band using a lumped element coupler", IEEE Transactions on Microwave Theory and Techniques, Vol. 49, No. 5, pp. 913–917, May 2001.

[Ell27]   F. Ellinger, U. Lott and W. Bächtold, "An antenna diversity MMIC vector modulator for HIPERLAN with low power consumption and calibration capability", IEEE Transactions on Microwave Theory and Techniques, Vol. 49, No. 5, pp. 964–969, May 2001.

[Ell29]   F. Ellinger, R. Vogt and W. Bächtold, "Calibratable adaptive antenna combiner at 5.2 GHz with high yield for PCMCIA card integration", IEEE Transactions on Microwave Theory and Techniques, Special Issue, Vol. 48, No. 12, pp. 2714–2720, Dec. 2000.

[Ell32]   F. Ellinger, "A 15 GHz reflective type phase shifter MMIC fabricated on 0.25 µm SiGe BiCMOS technology", IEEE International Telecommunication Conference, CD-ROM, May 2005.

[Ell44]   F. Ellinger, R. Vogt, W. Bächtold, "Ultra compact reflective type phase shifter MMIC for L-band", Workshop on Compound Semiconductor Devices and Integrated Circuits Europe, Cagliari, pp. 5–7, Mai 2001.

[Ell45]   F. Ellinger, R. Vogt, W. Bächtold, "A high yield, ultra small, passive, vector modulator based phase shifter for smart antenna combining at C-band", IEEE/CSIRO Asia Pacific Microwave Conference, Sydney, pp. 794–797, Dec. 2000.

[Ell48]   F. Ellinger, U. Lott, W. Bächtold, "A calibratable 4.8 – 5.8 GHz MMIC vector modulator with low power consumption for smart antenna receivers", IEEE MTT-S International Microwave Symposium, Boston, pp. 1277–1280, June 2000.

[Gar74]   R. V. Garver, "Design equations and bandwidth of loaded line phase shifters", IEEE Transactions on Microwave Theory and Techniques, MTT-22, pp. 561–563, May 1974.

[Hay96]   H. Hayashi, M. Muraguchi, Y. Umeda, T. Enoki, A high-Q broadband active inductor and its application to a low-loss analog phase shifter, IEEE Transactions on Microwave Theory and Techniques, Vol. 44, No. 12, pp. 2369–2374, Dec. 1996.

[Hay99]   H. Hayashi and M. Muraguchi, "An MMIC active phase shifter using a variable resonant circuit", IEEE Transactions on Microwave Theory and Techniques, Vol. 47, No. 10, pp. 2021–2026, Oct. 1999.

[Hay02]   H. Hayashi, T. Nakagawa, K. Araki, "A miniaturized MMIC analog phase shifter using two quarter-wave-length transmission lines", IEEE Transactions on Microwave Theory and Techniques, Vol. 50, No. 1, pp. 150–154, Jan. 2002.

[Kou91]   S. K. Koul and B. Bhat, Microwave and Millimeter Wave Phase Shifters, Vol. II, Artech House, 1991

[Nam01]   S. Nam, A. W. Payne, I. D. Robertson, "RF and microwave phase shifter using complementary bias techniques", Electronic Letters, Vol. 37, No. 18, pp. 1124–1125, Aug. 2001.

[Nag99]   A. S. Nagra, R. A. York, "Distributed analog phase shifters with low insertion loss", IEEE Transactions on Microwave Theory and Techniques, Vol. 47, No. 9, pp. 1705–1711, Sept. 1999.

15 RF Measurement Basics

When a scientist states that something is possible,
he is almost certainly right.
When he states that something is impossible,
he is very probably wrong.
Derived from Arthur C. Clarke, science fiction author

Proper characterisation of RF circuits is an important step within the design process. The measurements verify if the simulations hold and the IC hardware has been fabricated without significant process variations. Equipment with performances beyond that of the DUT (Device Under Test) is required making the measurement of high-speed circuits challenging and expensive. Today's commercial equipment allows measurements at frequencies of up to 100 GHz. The costs for the basic equipment in an RF measurement laboratory may easily exceed 500 k€. In this section, we will review the key measurement techniques and components. Detailed discussions can be found in the specific literature [Bai85, Bry88, Sch99].

15.1 Overview

The DUT and the measurement equipment have to be connected with low losses. We can identify two major principles, which are the on-wafer and the SMA (Sub Miniature Adaptor) connections.

A typical on-wafer measurement setup is shown in Fig. 15.1. For characterisation, the IC is connected via measurement probes and cables to the measurement equipment. Advantages are the low losses of the connections and the simple assembling making the approach well suited for prototype testing. Moreover, the ground connection is very good.

Good mechanical stability is obtained by assembling the IC on an RF test substrate, which is subsequently connected to robust SMA adapters, which in turn can easily be connected to the measurement equipment via cables. An example of an SMA measurement setup is shown in Fig. 15.2. The impact of the inductive bond connections must be considered. If the measured chip is not intended for bond-wire connections, we have to de-embed the bond wires. However, for many chips, the bond-wires are required in the final applications. In this case, it is fruitful to employ similar bond connections. One example is, e.g. the connection between an antenna or an external filter, and an LNA IC. To optimise the grounding, as many ground connections as possible have to be implemented to reduce the parasitic inductance. On the other hand, the ground connections require pads, which consume circuit area. Thus, one must strive for a tradeoff.

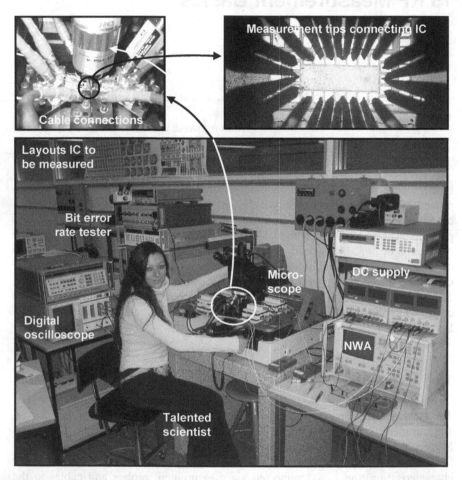

Fig. 15.1. Typical on-wafer measurement setup consisting of a probe station, small signal network analyser (NWA) and large signal bit error tester. Insert shows measurement tips connected by cables. Photographs made at the Electronics Laboratory of the ETH Zürich. Upper right photograph is kindly provided by L.C. Rodoni

Due to their individual requirements and specifications, various RF measurements as outlined in Table 15.1 have to be performed for full characterisation of the circuits. The corresponding measurements can be segmented into linear (small signal) and nonlinear (large signal) characterisations. S-parameter and linear noise characterisations belong to the first group, whereas the latter one mainly includes the measurement of the large signal RF and DC power, the transient properties, the intermodulation products, harmonics, and phase noise. Typically, the losses of RF cable and the connectors are in the range of 0.3–1.5 dB around 5 GHz. They rise towards higher frequencies. The losses, and eventually also the phase contributions and time delays of all interconnects have to be taken into account. A frequency, power and temperature dependent calibration has to be performed for all measurement units.

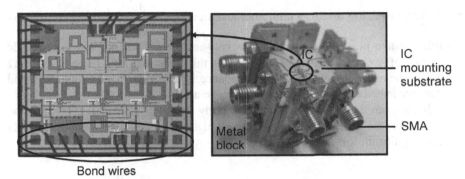

Fig. 15.2. Assembling and bonding of IC on substrate mounted on metal block, connection via bond wires and subsequently via transmission lines to SMA connectors

15.2 The 50 Ω Reference Impedance

Throughout this book, we have outlined that the reference impedance for measurements is 50 Ω, since it is a good compromise between a short and an open required for common measures such as Z-parameters. But why exactly 50 Ω? We can prove that coax cables exhibit a minimum insertion loss at this impedance. Especially at the highest frequencies, the losses of cables can be a limiting factor for measurements. At 100 GHz the losses of state-of-the-art cables are well above 3 dB. Thus, it is aimed at minimising the loss L being proportional to the total skin resistance divided by the characteristic impedance Z_0. The total skin resistance is the sum of the skin resistance of the inside and the shield conductor, which are both anti-proportional to their diameters d_1 and d_2, respectively. Hence, we deduce the following relation:

$$L \propto \frac{\dfrac{1}{d_1} + \dfrac{1}{d_2}}{Z_0}.$$

(15.1)

The characteristic impedance of coax structures is given by

$$Z_0 = \frac{60}{\sqrt{\varepsilon_r}} \ln\left(\frac{d_2}{d_1}\right).$$

(15.2)

Using the previous two equations yields

$$L \propto \frac{\sqrt{\varepsilon_r}}{60} \cdot \frac{1}{d_2} \cdot \frac{1 + \dfrac{d_2}{d_1}}{\ln\left(\dfrac{d_2}{d_1}\right)}.$$ (15.3)

The latter equation reveals a minimum insertion loss at $\dfrac{d_2}{d_1}$ =3.5911. Typical coax cables feature polyethylene insulations between the conductors with a dielectric constant of ε_r=2.25. Referring to Eq. (15.2), this leads to Z_0=51.1 Ω and Z_0=50 Ω after rounding. Consequently, this impedance is frequently used as interface and reference impedance, and for measurement setups.

It turns out that the loss minimum is fairly flat. Impedances of ±50% increase the loss by less than 20%. However, much higher deviations with respect to 50 Ω cause significant losses.

15.3 Basic Equipment

Among the key components for RF measurements are signal generators, vector network analysers, spectrum analysers and power meters, which will be treated now.

15.3.1 Signal Generator

The signal generator is based on a synthesiser comprising a VCO stabilised by a PLL and a crystal low frequency oscillator. In this regard, details can be found in Chaps. 11 and 12.

15.3.2 Vector Network Analyser

Vector network analysers (NWAs) are commonly used to measure the S-parameters of a DUT. S-parameter basics have been elaborated in Sect. 3. One of the working horses is the commercial HP 8510A device allowing precise measurements up to 40 GHz. The device has been on the market since 1984. Measurement frequencies above 100 GHz are possible with amendment units based on frequency converters and multipliers.

In Fig. 15.3, the functional principle is explained. The device employs an RF synthesiser, an LO synthesiser, two directional power splitters, two signal couplers, four IF channels including mixers and amplifiers, two mixer based comparators and switches to select between two measurement modes. Adjustable attenuators are applied in front of the DUT to optimise the power of the injected test signals. To minimise the impact of noise within the measurement setup, the power of the signals should be as high as possible. On the other hand, the test signals should not saturate the DUT or the measurement equipment. This would generate undesired nonlinearities. For simple signal processing in the digital domain, the signals are down-converted to a low IF of 100 KHz. For simplicity, the required sampling latches, analogue to digital converters and signal processing unit are not included in Fig. 15.3.

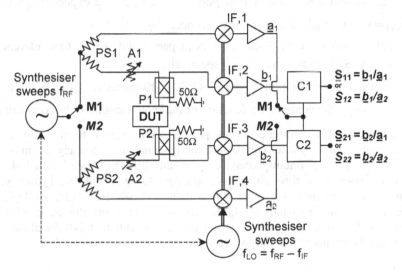

Fig. 15.3. Simplified schematics of network analyser, DUT: Device Under Test, n: number of channel, Pn: port, Cn: comparators, PSn: power splitters, An: adjustable attenuators, M1: mode to calculate $\underline{S}_{11}$ and $\underline{S}_{21}$, M2: mode to calculate $\underline{S}_{22}$ and $\underline{S}_{12}$

First, let us consider the upper position of the switches denoted by mode 1. An RF signal is fed into the DUT port P1, as well as into the IF channel 1 representing the known reference signal $\underline{a}_1$. The signal $\underline{b}_1$ reflected at P1 is transmitted via the power splitter and subsequently through IF channel 2, and compared with $\underline{a}_1$ to determine the magnitude and phase of $\underline{b}_1$. Thus, finally, $\underline{S}_{11} = \dfrac{\underline{b}_1}{\underline{a}_1}$ can be calculated. Furthermore, $\underline{b}_2$ is available at IF channel 3 and also compared with $\underline{a}_1$ providing $\underline{S}_{21} = \dfrac{\underline{b}_2}{\underline{a}_1}$. In similar fashion, $\underline{S}_{22}$ and $\underline{S}_{12}$ can be extracted by setting both switches in the lower position. It is worthwhile mentioning that the amplitude and phase relations of the a/b ratios are not changed by the mixing processes since all values exhibit the same amplitude and phase transformations.

To account for measurement errors associated with setup asymmetries, temperature changes, etc., a calibration is required. Usually, these errors increase with frequency making calibration routines mandatory for frequencies above

1 GHz. The SOLT (Short, Open, Load, Through) calibration procedure can be employed for this purpose. It is based on a comparison with exactly known reference measures. The following measurements are accomplished, which are partly redundant:

1. Short circuit by means of a ground connection at port 1 and port 2. In this case we expect $|\underline{S}_{11}| = 1$ and $\varphi_{11} = -\pi$, and $|\underline{S}_{22}| = 1$ and $\varphi_{22} = -\pi$, respectively.

2. Open without any connection at port 1 and port 2. We expect $|\underline{S}_{11}| = 1$ and $\varphi_{11} = 0$, and $|\underline{S}_{22}| = 1$ and $\varphi_{22} = 0$, respectively.

3. Load termination with 50-Ω resistors at port 1 and port 2. Our reference is given by $|\underline{S}_{11}| = 0$ and $|\underline{S}_{22}| = 0$, respectively.

4. Through connecting port 1 and port 2. Consequently, $|\underline{S}_{21}| = |\underline{S}_{12}| = 1$ and $|\varphi_{21}| = |\varphi_{12}|$, the phases are a function of the known length of the connection.

By readjusting the amplitudes and phases with respect to these reference measurements, the system can be calibrated and the accuracy is improved significantly. For maximum precision, the nonidealities associated with 1–4., e.g. the small losses of the through connections, have to be considered. Typically, the accuracy of calibrated NWAs is better than ±0.1 dB and ±0.5°. At high RF frequencies, such precision would hardly be feasible on the basis of other characterisation measures such as Z-parameters. Recall that NWAs allow only small signal measurements. Large signal measurements are not feasible.

15.3.3 Power Meter

Power meters are employed for precise measurement of the large signal RF power. Different elements and methods are used for the detector circuit, which is the core of the apparatus. In thermistors, the fed RF power raises the temperature of the device material leading to a decrease of the resistance. Suitable circuitries allow the monitoring of the associated power.

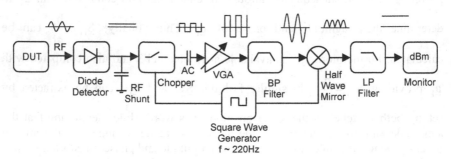

Fig. 15.4. Simplified schematics of a diode based power meter

Thermoelements are also sensitive regarding the heating associated with the applied RF power. A DC voltage being proportional to the applied RF power is generated across the junction of two materials. Optionally, a Schottky diode can be used as detector since the applied RF power generates a DC voltage as well. The advantage is the high sensitivity due to the quadratic characteristics at low input signals. The typical dynamic range of a single diode covers −70 to −20 dBm. Higher power handling is possible by cascading of diodes and by using attenuators. Further advantages of diodes are the ability for integration and the low costs. Thus, today, most power meters employ diodes for detection. The simplified schematic of a diode based power meter is illustrated in Fig. 15.4. DC offsets of the processing circuits caused by impact of temperature, process variations and aging can significantly worsen the measurement accuracy at input signals with weak amplitude. For that reason, the DC signal provided by the diode is translated to AC (Alternate Current) by means of a chopper synchronised with a square wave generator. After amplification up to an optimum signal level, filtering, mirroring to positive half waves and further filtering, the DC signal can be recovered and monitored representing the power level. No phase or frequency information is provided by power meters.

15.3.4 Spectrum Analyser

A spectrum analyser is capable of measuring signal amplitudes including harmonics and intermodulation products versus frequency. The principle is explained in Fig. 15.5. To avoid saturation of the processing stages, the attenuation of an incoming signal is adjusted, pre-selectively filtered and mixed down to IF. To allow a fixed IF simplifying further fine filtering, the LO signal is swept together with the RF signal. For this purpose, a sweep generator drives the VCO stabilised by a PLL and a crystal oscillator. After IF amplification and filtering, the signal amplitude is detected and monitored versus frequency.

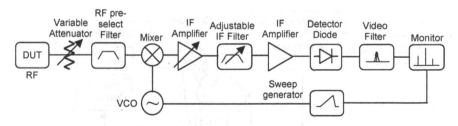

Fig. 15.5. Simplified schematic of diode based power meter

Today, commercial devices allow measurements up to around 70 GHz. The frequency resolution of spectrum analysers is very good due to the PLL stabilisation. However, because of the wideband operation, the accuracy of the amplitude detection is limited. Errors of up to ±1.5 dB may occur. For amplitude sensitive characterisations, calibrations based on signal generator and power meters are recommended. Typically, the latter two components have significant higher amplitude accuracy than spectrum analysers and are therefore well suited for calibration.

15.3.5 Oscilloscopes

The NWA and the spectrum analyser allow characterisations in the frequency domain. For measurements of signal waveforms vs time, an oscilloscope is employed. The oscilloscope allows the extraction of the signal frequency and DC components. Moreover, it can be used for the evaluation of the waveform, distortions and phase versus time. Helpful tutorials with further information can be found on the web [Tek05, Tcd05].

15.3.5.1 Analogue Oscilloscope

Figure 15.6 illustrates a simple analogue oscilloscope. The amplified voltage is applied to the vertical plates of a cathode ray tube causing a dot to glow on the monitor. A positive voltage causes the dot to move up while a negative voltage causes the dot to move down. The signal also travels to the trigger system to trigger a horizontal sweep. Triggering the horizontal system causes the horizontal time base to move the glowing dot across the screen from left to right within a specific time interval. Many sweeps in rapid sequence cause the movement of the glowing dot to blend into a solid line. Together, the horizontal sweep and the vertical deflection trace a graph of the signal on the screen. The trigger is necessary to stabilise the periods during signal repetition ensuring that the sweep always begins at the same point of a repeating signal resulting in a clear picture. The speed of an oscilloscope is determined by the writing properties of the cathode ray tube. At high speed the display becomes to dim to see the signal. Frequencies of up to 1 GHz are feasible.

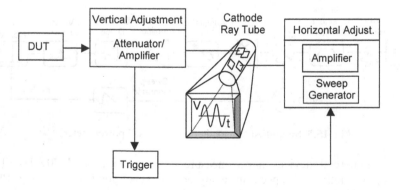

Fig. 15.6. Schematics of analogue oscilloscope

15.3.5.2 Digital Sampling Oscilloscope

In contrast to analogue oscilloscopes, the digital counterparts use an ADC to convert the measured analogue voltage into digital information. They acquire and store the waveform as series of samples until enough samples have accumulated. Then the digital oscilloscope re-assembles the waveform to be displayed on a compact screen. The speed of digital sampling oscilloscopes depends on the sampling clock. For slowly changing signals, enough sample points are available to construct the waveform. However, for signal frequencies not well below the sampling rate, the oscilloscope fails to collect enough sample points for the complete waveform. Up to a certain speed, smart techniques can be applied to recover the signal. Interpolation can be used to increase the number of points generating the waveform to be displayed. Moreover, since the digital oscilloscope has a memory, time multiplexing can be used to extract the full waveform. Prerequisite for the latter technique is that the signal is static meaning that it does not change its frequency or shape vs. time. Figure 15.7 depicts the typical architecture of a digital sampling oscilloscope.

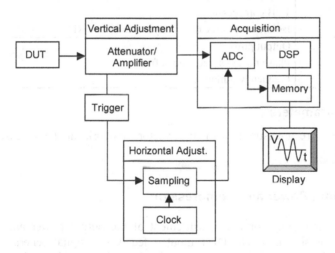

Fig. 15.7. Schematics of digital sampling oscilloscope

15.4 Key Measurements

Key RF measurements accomplished for devices, circuits and systems are listed in Table 15.1. All properties are a function of the applied DC bias. Further information can be found in the following sub-sections.

Table 15.1. Key RF measurements

DUT	Typical measurements
Transistors and diodes	S-parameter Intermodulation Harmonics
Passives	S-parameter Highly linear, usually no nonlinear measurements required
LNA	S-parameter determining gain, impedance matching and isolation Linear noise IIP3 1 dB compression
PA	S-parameter determining gain, impedance matching 1 dB output compression as measure for maximum output power OIP3, intermodulation, adjacent channel power
Mixer	Conversion gain Conversion noise 1 dB compression Isolation of ports, especially LO to IF and RF
VCO	Output power Phase noise vs frequency offset Frequency tuning

15.4.1 S-Parameters

S-Parameters are measured with the vector network analyser described in Sect. 15.3.2.

15.4.2 Output Power and Compression

In Fig. 15.8, the setup for the measurement of the output power and the 1 dB compression is sketched. The DUT input is fed with a signal generator and the DUT output is monitored by a power meter. To make sure that only the desired (e.g. the fundamental) power is measured, a bandpass filter is inserted at the output. Alternatively, a spectrum analyser can be employed. In this case, no filter is required. However, the power accuracy of spectrum analysers is limited. This makes the calibration on the basis of a signal generator and power meter necessary.

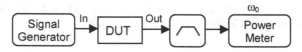

Fig. 15.8. Measurement setup for output power and compression measurements

15.4.3 Intermodulation and Conversion Gain

To measure the intermodulation properties, two nearby frequency components are combined and fed into the DUT input as sketched in Fig. 15.9a. The output spectrum is measured by the spectrum analyser. The third order intermodulation is extracted on the basis of Eq. (4.56). With a similar setup (see Fig. 15.9b), the conversion gain of a down mixer given by the relation between IF and RF power can be measured. The mixer combines the two input signals.

a)

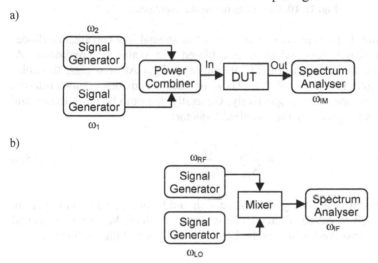

Fig. 15.9a,b. Measurement setups: **a** intermodulation; **b** mixer conversion gain

15.4.4 Noise Figure

Suppose a typical setup as shown in Fig. 15.10. According to the equation of Friis (refer to Sect. 4.3.5), the noise figure F_1 of a DUT can be determined by

$$F_1 = F_{21} - \frac{F_2}{G_1},$$ (15.4)

where F_{21} is the system noise figure, F_2 denotes the noise figure of the measurement unit and G_1 is the gain of the DUT. The noise figures are based on the relations of the noise powers, which are linearly dependent on T. Thus, we can simplify the representations of the latter equation using the equivalent noise temperatures:

$$T_1 = T_{21} - \frac{T_2}{G_1}.$$ (15.5)

The unknown T_1 can be determined by means of T_{21}, T_2 and G_1. To get these parameters, we perform measurements with two different levels of source input noise. These two power levels allow the operation with power ratios rather than with absolute values. This is an advantage since power ratios can be measured more accurately than the absolute power since any power offsets within the measurement equipment cancel out.

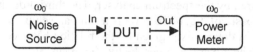

Fig. 15.10. Noise figure measurement setup

The most frequently used noise sources consist of special low capacitance diodes that generate resistive noise when reverse biased into avalanche breakdown. At reverse bias, the diode has a noise temperature of $T_{s,on}$. At zero bias, the noise temperature is given by $T_{s,off} \ll T_{s,on}$. These on- and off- modes are also referred to as cold and hot operation, respectively. Generally, the ratio between the on and off noise powers are given by the so-called Y-factor:

$$Y = \frac{N_{on}}{N_{off}} = \frac{T_{s,on}}{T_{s,off}} .$$

(15.6)

For high measurement accuracy, the Y-factor should be high. Due to its important role, the approach is called Y-factor measurement method. Moreover, a second key parameter is involved, which is the ENR (Excess Noise Ratio) defined as

$$ENR = \frac{T_{s,on} - T_{s,off}}{T_0} = \frac{T_{s,on} - T_0}{T_0}$$

(15.7)

with $T_{s,off} = T_0 = 290$ K after pre-calibration. Typical values for $T_{s,on}$ are around 10 000 K. We can also find the relation in logarithmic scale with $ENR_{dB} = 10 \log ENR$. After these definitions, let us calculate T_2, which is the noise temperature of the measurement unit. This measurement is done without the DUT by directly connecting the noise source to the input of the measurement unit yielding a Y-factor of

$$Y_c = \frac{N_{c,on}}{N_{c,off}} = \frac{T_{s,on} + T_2}{T_{s,off} + T_2} .$$

(15.8)

The calibration on and off noise powers $N_{c,on}$ and $N_{c,off}$ are displayed by the measurement unit, which can be a broadband power meter, a spectrum analyser or special detectors. From Eq. (15.8) we can calculate the wanted noise temperature of the measurement unit:

$$T_2 = \frac{T_{s,on} - Y_c T_{s,off}}{Y_c - 1} .$$

(15.9)

In a next step, the DUT is inserted and the associated Y_{21} factor given by

$$Y_{21} = \frac{N_{21,on}}{N_{21,off}} \qquad (15.10)$$

is measured. Similar to Eq. (15.8), we can determine the associated total noise temperature:

$$T_{21} = \frac{T_{s,on} - Y_{21}T_{s,off}}{Y_{21} - 1} . \qquad (15.11)$$

The gain G_1 of the DUT can be calculated by the ratio of noise powers including and excluding the DUT:

$$G_1 = \frac{N_{21,on} - N_{21,off}}{N_{c,on} - N_{c,off}} . \qquad (15.12)$$

Now, according to Eq. (15.5), T_1 can be calculated employing the known values of T_2, T_{12}, and G_1. The noise factor is given by $F_1 = 1 + \frac{T_1}{T_0}$. A fruitful tutorial dealing with noise measurements has been reported in [Ag105, Ag205].

15.4.5 Noise Parameters

In the preceding section we have presented a method to measure the NF at given source impedance, which in most cases is the 50 Ω reference impedance. However, for circuit simulations and optimisations, it is important to predict the noise characteristics versus any source impedance, which does not equal 50 Ω. According to Sect. 4.3.6, the impedance dependant noise factor of a device can be determined by the noise parameters comprising F_{min}, the complex optimum source impedance Y_{opt}, and R_n indicating the sensitivity on non-optimum source impedance. These parameters can be measured by systematically varying the source impedance. First, Y_{opt} and the associated F_{min} are determined by automatic gradient and random optimisation procedures. Then, R_n is calculated on basis of test measurements around F_{min}.

15.4.6 Conversion Noise Figure

The Y-factor method can also be applied for nonlinear DUT such as mixers and frequency multipliers, which have different input and output frequencies. A feasible measurement setup is shown in Fig. 15.11a. We have to distinguish between SSB (Single Sideband) and DSB (Double Side Band) measurements. In double sideband measurements, noise from both the USB (Upper Sideband) and the LSB (Lower Sideband) will be converted to the same IF frequency, as indicated in Fig. 15.11b. The NF instrument measures the average of the USB and LSB noise.

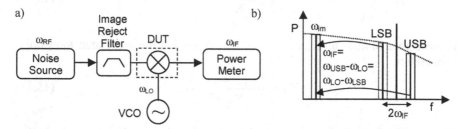

Fig. 15.11a,b. Conversion gain: **a** measurement setup for mixer as DUT; **b** sideband mixing

For many applications the important measure is the SSB NF determined by either the USB or the LSB noise. If the image rejection of the mixer by itself is not sufficient, we have to insert an image reject filter in the front of the DUT. The second option is to make a DSB measurement and to extrapolate the SSB NF. DSB measurements are often preferred because they do not require a specific image rejection filter, which may not be available in the laboratory. Assuming equal noise characteristics for both the USB and the LSB, the SSB noise figure will be a factor of 3 dB lower than the DSB measurement. The accuracy of this assumption depends on the flatness of the DUT and the setup vs frequency. Typically the error increases with increasing IF representing the distance between the USB and the LSB.

15.4.7 Phase Noise

As discussed in Sect. 11.4, phase noise in oscillators significantly worsens the performance of wireless transceiver and has to be carefully considered. According to Sect. 11.4.1.4, the phase noise performance is characterised by $L(\Delta\omega) = 10\log\left(\dfrac{S_n}{P_{LO}}\right)$. The two dominant techniques employed to measure the phase noise are the direct spectrum analyser approach and the phase detector approach [Pay05] The latter approach can be segmented into the delay line discriminator and the reference source method.

15.4.7.1 Direct Spectrum Analyser

The simplest method to measure phase noise is the direct spectrum analyser method. As shown in Fig. 15.12a, the signal from the DUT is fed into a spectrum analyser tuned to the DUT frequency. The sideband noise power can be directly measured and related to the carrier signal power to obtain $L(\Delta\omega)$. However, there is one inaccuracy associated with this approach. The total sideband noise power consisting of phase and amplitude noise is measured. Fortunately, the noise associated with amplitude fluctuations is much lower than the phase noise. Thus, with reasonable accuracy, phase noise is given by the total noise. Finally, from system point of view, we are anyway interested in the total noise independent of the origin. The sensitivity of this approach is limited by the internal noise of the analyser.

a)   b)

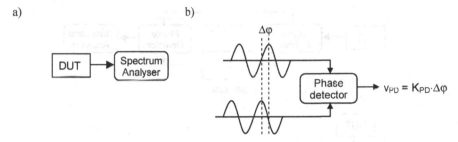

Fig. 15.12a,b. Phase noise measurements: **a** direct spectrum approach; **b** phase detector principle, $\Delta\varphi$ phase difference between two signals, v_{PD}: phase detector output voltage, K_{PD}: phase constant

15.4.7.2 Phase Detector Techniques

Phase detector techniques have the advantage that they are capable of extracting the phase noise out of the total noise. As illustrated in Fig. 15.12b, the phase detector converts the phase difference of the two input signals into an output voltage. Within a certain range, the voltage is proportional to the phase difference, which in turn describes the level of phase noise. In this context refer to Sect. 12.1.1. Hence, the associated phase noise power can be determined. When the nominal phase difference is set to 90°, which is also known as quadrature condition, the voltage output will be zero (see Fig. 12.2b). Any phase fluctuation from the quadrature condition will result in a voltage fluctuation at the output. Double balanced mixers can be employed to accomplish the detector function. Adding low noise amplifiers prior to the signal splitter or in front of the phase detector are advantageous if the signal power is weak. However, the additive noise of the amplifiers increases the noise floor. In the following, we elaborate two specific approaches.

A. Delay Line Discriminator
Figure 15.13a illustrates the principle of the delay line discriminator approach [Yuv81, Gol87]. Two major tasks have to be performed before signal detection. First, we have to split the signals into two branches. Then, we have to implement a time delay in one of the paths. Important point is that this time delay is frequency dependant. Moreover, it serves as memory. Frequency fluctuations, which are the source of phase noise are converted into phase changes. Mixing with the origin signal results in a voltage, which dependents on the phase difference between the two paths. Subsequently, the phase noise power can be determined. To optimise

the mixer sensitivity towards quadrature operation, a variable phase shifter can be inserted in the second branch. The contribution of the amplitude noise is suppressed by the phase detector, which ideally does not produce output signals variations with respect to amplitude fluctuations. In other words, amplitude-wise, it is operated in saturation.

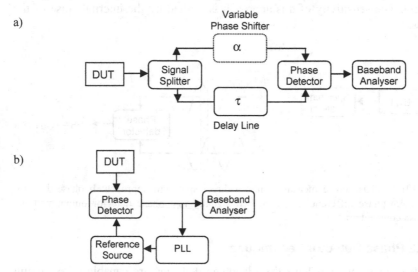

Fig. 15.13a,b. Phase detector based phase noise measurement approaches: **a** delay line discriminator method, τ: delay, α: phase shift; **b** PLL/reference source

B. Reference Source
With this technique, another source is used to provide the reference phase signal for the phase detector. Figure 15.13b shows an architecture of such a system. A PLL is used to establish stable phase quadrature at the phase detector. By means of the level of the detector output voltage, the phase noise can be quantified. The measurement accuracy is limited by the phase noise performance of the reference source.

15.5 Tutorials

1. Describe and compare on-wafer and mounted-IC measurement setups. What are the pros and cons? In a system used in practice, which setup is more realistic?
2. List key measurement components. Which one are small signal and which one large signal instruments? What is the difference between small signal and large signal measurements? Is there a difference between small signal and large signal DC operation points?
3. Which are the relevant measurements to be performed for LNAs, PAs, mixers and VCOs? Explain them and sketch the measurement setups.

4.   How can we measure the 1 dB compression point of a circuit? How about the IIP3 and OIP3?
5.   Explain the characterisation of the conversion gain of a mixer. Which externally fed parameters impact the performance of mixers significantly? What about a sweep?
6.   Explain the architecture and functional principle of a vector network analyser.
7.   How do analogue and digital oscilloscopes work? Which one can be applied for higher frequencies?
8.   Which effects are exploited for power meters? Are power meters capable of determining phases? How can we measure the large signal phase of a circuit?
9.   How does a spectrum analyser work? What about the frequency and amplitude resolution?
10. Why are the noise factor and the temperature equivalent parameters?
11. Describe the Y-factor method. What is the benefit to working with power ratios rather than with absolute values?
12. How can we measure the noise of mixers? What is the difference compared to the characterisation of LNA noise?
13. How can we make sure we are measuring DSB or SSB noise? What is the difference?
14. Explain both the discriminator and PLL-reference source based techniques for phase noise measurements. What are the limits?
15. Your measurement frequency is much higher than that of your measurement equipment. Which self-made components can you add to solve the problem, e.g. to enhance the frequency range of a spectrum analyser? What are the drawbacks?
16. You measure gain ($S_{21}>0$) and negative return losses ($S_{11,22}>0$) for a resistor used as DUT. What do you do?: a) go for a publication; b) go home; c) recalibrate your measurement setup; d) drink a beer. Multiple answers are possible.

References

[Ag105] Agilent tutorial notes, fundamentals of RF and microwave noise figure measurements, application note 57-1, 2005.
[Ag205] Agilent tutorial notes, Noise figure measurement accuracy – the Y-factor method, application note 57-2, 2005.
[Bai85] A. E. Bailey, Microwave Measurements, Peter Peregrinus Ltd., London, 1985.
[Bry88] G. H: Bryant, Principles of Microwave Measurements, Peter Peregrinus Ltd., London, 1988.
[Gol87] S. Goldman, "Analyze effects of system delay lines", Microwave & RF, pp. 149–156, June 1987.

[Pay05] G. Payne, Practical considerations for modern RF & microwave phase
         noise measurements, Agilent applications note, 2005.
[Sch99] B. Schiek, Grundlagen der Hochfrequenz Messtechnik, Springer, 1999.
[Tek05] www.tek.com
[Tcd05] www.cs.tcd.ie/courses/baict/bac/jf/labs/scope/oscilloscope.html
[Yuv81] I. Yuval, "Proper delay line discriminator design avoids common
         pitfalls", Microwave System News, pp. 109–114, Dec. 1981.

Abbreviations and Symbols[1]

A

ADC	Analogue digital converter
ADS	Advanced design system of Hewlett Packard
a_i	Current gain
Al	Aluminium
a_v	Voltage gain

B

BJT	Bipolar junction transistor
BP	Bandpass
BPSK	Binary PSK
BW	Bandwidth

C

C	Capacitor
CAD	Computer aided design
CASE	Center for Advanced Silicon Electronics
C_{bc}	Base collector capacitance
C_{be}	Base emitter capacitance
C_{ce}	Collector emitter capacitance
C_{ds}	Drain source capacitance
C_{gd}	Gate drain capacitance
C_{gs}	Gate source capacitance
C_{in}	Input capacitance
C_m	Miller capacitance

CMOS	Complementary metal oxide semiconductor
C_{ox}	Oxide capacitance

D

DAC	Digital analogue converter
DC	Direct current
D-FET	Depletion FET
D_n	Diffusion constant of electrons
D_p	Diffusion constant of holes
d_{sm}	Distance substrate to metal conductor
DSB	Double side band
DRC	Design rule check
DUT	Device under test

E

E	Electrical field
E-FET	Enhancement FET

F

F	Noise factor
f	Frequency
f_c	Corner/cutoff frequency
FET	Field effect transistor
f_{max}	Maximum frequency of oscillation
F_{min}	Minimum F

---

[1] General comments concerning the notation of parameters
- Large signal: capital letters, e.g. current " I "
- Small signal in time domain: small letters, e.g. current " i "
- Small signal in frequency domain: complex capital letters, e.g. complex current " $\underline{I}$ "

FMCW	Frequency modulated continuous wave
FOM	Figure of merit
FOX	Field oxide isolation layer
f_r	Resonance frequency
f_{SR}	Self resonance frequency
f_t	Transit frequency

G

G	Gain
GaAs	Gallium arsenide
g_d	Channel conductance
g_{d0}	Channel conductance at V_{ds}=0 V
GFSK	Gaussian frequency shift keying
g_m	Transconductance
GPS	Global positioning system
GSM	Global system for mobile communication

H

HBT	Heterojunction bipolar transistor
HEMT	High electron mobility transistor
HFSS	High frequency structure simulator
HIPER	High performance radio
LAN	LAN
HP	Hewlett Packard

I

I_b, i_b	Base current
IC	Integrated circuit
I_c, i_c	Collector current
I_d, i_d	Drain current
I_{dc}	DC current
I_e, i_e	Emitter current
I_{fl}	Flicker noise current
I_g, i_g	Gate current
I_s, i_s	Source current
I_{dc}	DC current
I_{dp}	Drain peak current

I_{dq}	Drain quiescence current
IIP3	Third order intercept point at input
IF	Intermediate frequency
IM	Intermodulation product
IM3	Third order intermodulation product
I_{max}	Maximum I_d
InP	Indium phosphide
I_{th}	Threshold current
IV	Current voltage

K

K	Rollet's factor for stability investigations
k	Boltzmann constant
k_{SL}	Substrate loss factor
k_{SR}	Self resonance loss factor

L

L	Inductor
LAN	Local area network
l_g	Gate length
LINC	Linear amplification with nonlinear components
LMDS	Local multipoint distribution service
L_n	Diffusion length of electrons
LNA	Low noise amplifier
LO	Local oscillator
L_p	Diffusion length of holes
LP	Lowpass
LPS	Local positioning system
LVS	Layout vs schematics

M

MAG	Maximum available gain
MBS	Mobile broadband services
MEMS	Micro-electrical mechanical system
MES	Metal semiconductor
MESFET	Metal semiconductor field effect transistor

MMAC	Multimedia mobile access communication
MMIC	Microwave monolithic integrated circuits
MOS	Metal oxide semiconductor
MOSFET	Metal oxide semiconductor field effect transistor
MSG	Maximum stable gain
MVDS	Mobile video distribution system

N

n	Number, e.g. of turns, or integer variable 1, 2, 3 ...
N_A	Acceptor doping concentration
N_D	Donator doping concentration
n_{oF}	Number of stages for minimum F
n_{oG}	Number of stages for maximum gain
NF	Noise figure, logarithmic scale
NF_{min}	Minimum NF
NiCr	Nickel chrome

O

| OFDM | Orthogonal frequency division multiplexing |
| OIP3 | Third order intercept point at output |

P

PAE	Power added efficiency
PD	Phase detector
P_{dc}	DC power
PHEMT	Pseudomorphic HEMT
PLL	Phase locked loop
P_{LO}	LO power
P_{L1}	Power at fundamental frequency

P_m	Maximum output power
p_n	Normalised power factor
P_n	Maximum available noise power
P_{out}	RF output power
PPM	Pulse position modulation
P_{RF}	RF power
PSK	Phase shift keying
P_{1dB}	1 dB compression point
PUF	Power utilisation factor

Q

Q	Quality factor
q	Elementary charge
Q_{max}	Maximum Q

R

R	Resistor
R_{bb}	Base series resistor
r_{ce}	Small signal output impedance
r_{ds}	Small signal output impedance not considering R_{ds}
R_{ds}	Parasitic ohmic channel resistance
RF	Radio frequency
RFIC	RF integrated circuit
RFID	RF identification
R_L	Load resistor
RMS	Root mean square
r_{NFmin}	Source reflection coefficient where NF_{min} is achieved
R_S	Source resistor
RX	Receive/receiver
r_o	$r_{ds} \| R_{ds}$

S

SFDR	Spurious free dynamic range
SPDT	Single pole double throw
SPST	Single pole single throw
Si	Silicon
SiGe	Silicon germanium

S_n	Spectral noise power density	V_g, v_g	Gate voltage
SNR	Signal to noise ratio	V_{gd}, v_{gd}	Gate drain voltage
SOI	Silicon on insulator	V_{gs}, v_{gs}	Gate source voltage
SSB	Single side band	VLSI	Very large scale integration
		V_{max}	Maximum V_{ds}
S_{11}	Input scattering parameter	vs	Versus
S_{22}	Output scattering parameter	V_{supply}	DC supply voltage
		V_T	Temperature voltage
S_{21}	Transmission scattering parameter	V_{th}	Threshold voltage
S_{12}	Reverse transmission scattering parameter		

W

w	Device width
w_b	Base width
w_g	Gate width

T

t	Time
t_{ox}	Thickness of gate oxide
TAS	Transadmittance stage
TIS	Transimpedance stage
TUD	Dresden University of Technology
TWA	Travelling wave amplifier
TX	Transmit/transmitter
t_v	Capacitance control ratio

X

x	Unknown parameter, length, x-axis

Y

Y	Admittance

U

UMTS	Universal mobile telecommunication standard
UWB	Ultra wideband

Z

Z	Impedance
Z_0	Reference impedance, usually 50 Ω
Z_w	Characteristic impedance of transmission line

V

V_{bc}, v_{bc}	Base collector voltage
V_{be}, v_{be}	Base emitter voltage
V_{ce}, v_{ce}	Collector emitter voltage
V_{dc}	DC voltage
VCO	Voltage controlled oscillator
V_{dp}	Drain source peak voltage
V_{dq}	Drain source quiescence voltage
V_{ds}, v_{ds}	Drain source voltage

Special characters

α	Conduction angle
φ	Phase
θ	ωt
$\Delta\omega$	Frequency control range
λ	Wave length
η	Efficiency
ω	Angular frequency
ω_{skin}	Skin effect corner frequency
ω_t	Angular transit frequency
δ	Gate noise coefficient
γ	Drain noise coefficient
ρ	Resistivity
Γ	Reflection coefficient

Index

About the Author

Frank Ellinger was born in Friedrichshafen, Germany, in April 1972. He graduated in electrical engineering (EE) in 1996 from the University of Ulm, Germany. He received the Master in Business and Administration (MBA), and the PhD degree in EE from the ETH Zürich, Switzerland, in 2001. For his habilitation thesis Mr. Ellinger obtained the Venia Legendi (university teaching degree) in high frequency circuit design from the ETH in 2004.

Since August 2006 he has been full professor and head of the Chair for Circuit Design and Network Theory at the Dresden University of Technology, Germany. Mr. Ellinger has been the project manager and initiator of the EU funded projects RESOLUTION (Reconfigurable Systems for Mobile Local Communication and Positioning) and MIMAX (Advanced MIMO Systems for MAXimum Reliability and Performance) started in Feb. 2006 and Jan. 2008, respectively.

From 2001 to 2006 he has been head of the RFIC design group of the Electronics Laboratory at the ETH, and a project leader of the IBM/ETH Competence Center for Advanced Silicon Electronics at IBM Research in Rüschlikon. During his MBA in 2001, he was with the wireless marketing division of Infineon, Germany.

He is mainly interested in the design of ICs for high-speed wireless and optical communication. In this area Mr. Ellinger has been lecturer at the ETH between 2002 and 2006.

In the time frame between 2005-2006, he has been engaged as associated editor for the IEEE Microwave and Wireless Component Letters. Prof. Ellinger acts as reviewer for several IEEE Journals and the European Community, and has served as a technical consultant for Sony-Ericsson regarding law suits. He has contributed to the organization of several conferences, e.g. as program chair for the WOCSDICE (Workshop on Circuits and Semiconductor Devices Europe) 2004, and as co-chair for the SCD (Semiconductor Conference Dresden) 2008.

He has published more than 100 refereed scientific papers, most of them IEEE journal contributions, and 3 patents. One of his publications has been among the three most-read (downloaded) papers in the IEEE Journal on Solid-State Circuits 2004.

For his work he has received several awards including the ETH Medal, the Denzler Award of the Swiss Federal Association of Electrical Engineers, the Rohde&Schwarz/Agilent/Gerotron EEEfCOM Innovation Award, and the Young

PhD Award of the ETH (Bonus 29). His students at the chair have obtained several awards, e.g. the Best Student Paper Award at the IEEE IMOC (International Microwave and Optoelectronic Conference).